T0139965

Communications
in Computer and Information Science   1880

Rationale

The CCIS series is devoted to the publication of proceedings of computer science conferences. Its aim is to efficiently disseminate original research results in informatics in printed and electronic form. While the focus is on publication of peer-reviewed full papers presenting mature work, inclusion of reviewed short papers reporting on work in progress is welcome, too. Besides globally relevant meetings with internationally representative program committees guaranteeing a strict peer-reviewing and paper selection process, conferences run by societies or of high regional or national relevance are also considered for publication.

Topics

The topical scope of CCIS spans the entire spectrum of informatics ranging from foundational topics in the theory of computing to information and communications science and technology and a broad variety of interdisciplinary application fields.

Information for Volume Editors and Authors

Publication in CCIS is free of charge. No royalties are paid, however, we offer registered conference participants temporary free access to the online version of the conference proceedings on SpringerLink (http://link.springer.com) by means of an http referrer from the conference website and/or a number of complimentary printed copies, as specified in the official acceptance email of the event.

CCIS proceedings can be published in time for distribution at conferences or as post-proceedings, and delivered in the form of printed books and/or electronically as USBs and/or e-content licenses for accessing proceedings at SpringerLink. Furthermore, CCIS proceedings are included in the CCIS electronic book series hosted in the SpringerLink digital library at http://link.springer.com/bookseries/7899. Conferences publishing in CCIS are allowed to use Online Conference Service (OCS) for managing the whole proceedings lifecycle (from submission and reviewing to preparing for publication) free of charge.

Publication process

The language of publication is exclusively English. Authors publishing in CCIS have to sign the Springer CCIS copyright transfer form, however, they are free to use their material published in CCIS for substantially changed, more elaborate subsequent publications elsewhere. For the preparation of the camera-ready papers/files, authors have to strictly adhere to the Springer CCIS Authors' Instructions and are strongly encouraged to use the CCIS LaTeX style files or templates.

Abstracting/Indexing

CCIS is abstracted/indexed in DBLP, Google Scholar, EI-Compendex, Mathematical Reviews, SCImago, Scopus. CCIS volumes are also submitted for the inclusion in ISI Proceedings.

How to start

To start the evaluation of your proposal for inclusion in the CCIS series, please send an e-mail to ccis@springer.com.

Zhiwen Yu · Qilong Han · Hongzhi Wang ·
Bin Guo · Xiaokang Zhou · Xianhua Song ·
Zeguang Lu

Editors

Data Science

9th International Conference
of Pioneering Computer Scientists, Engineers
and Educators, ICPCSEE 2023
Harbin, China, September 22–24, 2023
Proceedings, Part II

 Springer

Editors
Zhiwen Yu
Harbin Engineering University
Harbin, China

Hongzhi Wang
Harbin Institute of Technology
Harbin, Heilongjiang, China

Xiaokang Zhou
Shiga University
Shiga, Japan

Zeguang Lu
National Academy of Guo Ding Institute
of Data Science
Beijing, China

Qilong Han
Harbin Engineering University
Harbin, China

Bin Guo
Northwestern Polytechnical University
Xi'an, China

Xianhua Song
Harbin University of Science and Technology
Harbin, China

ISSN 1865-0929   ISSN 1865-0937 (electronic)
Communications in Computer and Information Science
ISBN 978-981-99-5970-9   ISBN 978-981-99-5971-6 (eBook)
https://doi.org/10.1007/978-981-99-5971-6

This Springer imprint is published by the registered company Springer Nature Singapore Pte Ltd.
The registered company address is: 152 Beach Road, #21-01/04 Gateway East, Singapore 189721, Singapore

Paper in this product is recyclable.

Preface

As the chairs of the 8th International Conference of Pioneering Computer Scientists, Engineers and Educators 2023 (ICPCSEE 2023, originally ICYCSEE), it is our great pleasure to welcome you to the conference proceedings. ICPCSEE 2023 was held in Harbin, China, during September 22–24, 2023, and hosted by the Harbin Engineering University, the Harbin Institute of Technology, the Northeast Forestry University, the Harbin University of Science and Technology, the Heilongjiang Computer Federation, and the National Academy of Guo Ding Institute of Data Sciences. The goal of this conference series is to provide a forum for computer scientists, engineers, and educators.

This year's conference attracted 244 paper submissions. After the hard work of the Program Committee, 66 papers were accepted to appear in the conference proceedings, with an acceptance rate of 27%. The major topic of this conference is data science. The accepted papers cover a wide range of areas related to basic theory and techniques of data science including mathematical issues in data science, computational theory for data science, big data management and applications, data quality and data preparation, evaluation and measurement in data science, data visualization, big data mining and knowledge management, infrastructure for data science, machine learning for data science, data security and privacy, applications of data science, case studies of data science, multimedia data management and analysis, data-driven scientific research, data-driven bioinformatics, data-driven healthcare, data-driven management, data-driven e-government, data-driven smart city/planet, data marketing and economics, social media and recommendation systems, data-driven security, data-driven business model innovation, and social and/or organizational impacts of data science.

We would like to thank all the Program Committee members, a total of 261 people from 142 different institutes or companies, for their hard work in completing the review tasks. Their collective efforts made it possible to attain quality reviews for all the submissions within a few weeks. Their diverse expertise in each research area helped us to create an exciting program for the conference. Their comments and advice helped the authors to improve the quality of their papers and gain deeper insights.

We thank the team at Springer, whose professional assistance was invaluable in the production of the proceedings. A big thanks also to the authors and participants for their tremendous support in making the conference a success.

Besides the technical program, this year ICPCSEE offered different experiences to the participants. We hope you enjoyed the conference.

July 2023

Rajkumar Buyya
Hai Jin
Zhiwen Yu
Qilong Han
Hongzhi Wang
Bin Guo
Xiaokang Zhou

Organization

General Chairs

Rajkumar Buyya   University of Melbourne, Australia
Hai Jin   Huazhong University of Science and Technology,
     China
Zhiwen Yu   Harbin Engineering University, China

Program Chairs

Qilong Han   Harbin Engineering University, China
Hongzhi Wang   Harbin Institute of Technology, China
Bin Guo   Northwestern Polytechnical University, China
Xiaokang Zhou   Shiga University, Japan

Program Co-chairs

Rui Mao   Shenzhen University, China
Min Li   Central South University, China
Wenguang Chen   Tsinghua University, China
Wei Wang   Harbin Engineering University, China

Organization Chair

Haiwei Pan   Harbin Engineering University, China

Publications Chairs

Xianhua Song   Harbin University of Science and Technology,
     China
Zeguang Lu   National Academy of Guo Ding Institute of Data
     Science, China

Secretary General

Zeguang Lu   National Academy of Guo Ding Institute of Data
                               Science, China

Under Secretary General

Xiaoou Ding   Harbin Institute of Technology, China

Secretary

Zhongchan Sun   National Academy of Guo Ding Institute of Data
                               Science, China
Dan Lu   Harbin Engineering University, China

Executive Members

Cham Tat Huei   UCSI University, Malaysia
Xiaoju Dong   Shanghai Jiao Tong University, China
Lan Huang   Jilin University, China
Ying Jiang   Kunming University of Science and Technology,
                               China
Weipeng Jiang   Northeast Forestry University, China
Min Li   Central South University, China
Junyu Lin   Institute of Information Engineering, CAS, China
Xia Liu   Hainan Province Computer Federation, China
Rui Mao   Shenzhen University, China
Qiguang Miao   Xidian University, China
Haiwei Pan   Harbin Engineering University, China
Pinle Qin   North University of China, China
Xianhua Song   Harbin University of Science and Technology,
                               China
Guanglu Sun   Harbin University of Science and Technology,
                               China
Jin Tang   Anhui University, China
Ning Wang   Xiamen Huaxia University, China
Xin Wang   Tianjin University, China
Yan Wang   Zhengzhou University of Technology, China
Yang Wang   Southwest Petroleum University, China
Shengke Wang   Ocean University of China, China

Yun Wu	Guizhou University, China
Liang Xiao	Nanjing University of Science and Technology, China
Junchang Xin	Northeastern University, China
Zichen Xu	Nanchang University, China
Xiaohui Yang	Hebei University, China
Chen Ye	Hangzhou Dianzi University, China
Canlong Zhang	Guangxi Normal University, China
Zhichang Zhang	Northwest Normal University, China
Yuanyuan Zhu	Wuhan University, China

Steering Committee

Jiajun Bu	Zhejiang University, China
Wanxiang Che	Harbin Institute of Technology, China
Jian Chen	ParaTera, China
Wenguang Chen	Tsinghua University, China
Xuebin Chen	North China University of Science and Technology, China
Xiaoju Dong	Shanghai Jiao Tong University, China
Qilong Han	Harbin Engineering University, China
Yiliang Han	Engineering University of CAPF, China
Yinhe Han	Chinese Academy of Sciences, China
Hai Jin	Huazhong University of Science and Technology, China
Weipeng Jing	Northeast Forestry University, China
Wei Li	Central Queensland University, China
Min Li	Central South University, China
Junyu Lin	Chinese Academy of Sciences, China
Yunhao Liu	Michigan State University, USA
Zeguang Lu	National Academy of Guo Ding Institute of Data Sciences, China
Rui Mao	Shenzhen University, China
Qiguang Miao	Xidian University, China
Haiwei Pan	Harbin Engineering University, China
Pinle Qin	North University of China, China
Zheng Shan	PLA Information Engineering University, China
Guanglu Sun	Harbin University of Science and Technology, China
Jie Tang	Tsinghua University, China
Tian Feng	Chinese Academy of Sciences, China
Tao Wang	Peking University, China

Hongzhi Wang	Harbin Institute of Technology, China
Xiaohui Wei	Jilin University, China
Lifang Wen	Beijing Huazhang Graphics & Information Co., Ltd., China
Liang Xiao	Nanjing University of Science and Technology, China
Yu Yao	Northeastern University, China
Xiaoru Yuan	Peking University, China
Yingtao Zhang	Harbin Institute of Technology, China
Yunquan Zhang	Chinese Academy of Sciences, China
Baokang Zhao	National University of Defense Technology, China
Min Zhu	Sichuan University, China
Liehuang Zhu	Beijing Institute of Technology, China

Program Committee

Witold Abramowicz	Poznan University of Economics, Poland
Chunyu Ai	University of South Carolina Upstate, USA
Jiyao An	Hunan University, China
Ran Bi	Dalian University of Technology, China
Zhipeng Cai	Georgia State University, USA
Yi Cai	South China University of Technology, China
Zhao Cao	Beijing Institute of Technology, China
Wanxiang Che	Harbin Institute of Technology, China
Wei Chen	Beijing Jiaotong University, China
Hao Chen	Hunan University, China
Xuebin Chen	North China University of Science and Technology, China
Chunyi Chen	Changchun University of Science and Technology, China
Yueguo Chen	Renmin University, China
Siyao Cheng	Harbin Institute of Technology, China
Byron Choi	Hong Kong Baptist University, China
Vincenzo Deufemia	University of Salerno, Italy
Gong Dianxuan	North China University of Science and Technology, China
Xiaofeng Ding	Huazhong University of Science and Technology, China
Jianrui Ding	Harbin Institute of Technology, China
Hongbin Dong	Harbin Engineering University, China
Lei Duan	Sichuan University, China

Xiping Duan	Harbin Normal University, China
Xiaolin Fang	Southeast University, China
Ming Fang	Changchun University of Science and Technology, China
Jianlin Feng	Sun Yat-sen University, China
Jing Gao	Dalian University of Technology, China
Yu Gu	Northeastern University, China
Qi Han	Harbin Institute of Technology, China
Meng Han	Georgia State University, USA
Qinglai He	Arizona State University, USA
Wei Hu	Nanjing University, China
Lan Huang	Jilin University, China
Hao Huang	Wuhan University, China
Feng Jiang	Harbin Institute of Technology, China
Bin Jiang	Hunan University, China
Cheqing Jin	East China Normal University, China
Hanjiang Lai	Sun Yat-Sen University, China
Shiyong Lan	Sichuan University, China
Hui Li	Xidian University, China
Zhixu Li	Soochow University, China
Mingzhao Li	RMIT University, China
Peng Li	Shaanxi Normal University, China
Jianjun Li	Huazhong University of Science and Technology, China
Xiaofeng Li	Sichuan University, China
Zheng Li	Sichuan University, China
Min Li	Central South University, China
Zhixun Li	Nanchang University, China
Hua Li	Changchun University of Science and Technology, China
Rong-Hua Li	Shenzhen University, China
Cuiping Li	Renmin University of China, China
Qiong Li	Harbin Institute of Technology, China
Yanli Liu	Sichuan University, China
Hailong Liu	Northwestern Polytechnical University, China
Guanfeng Liu	Macquarie University, Australia
Yan Liu	Harbin Institute of Technology, China
Zeguang Lu	National Academy of Guo Ding Institute of Data Sciences, China
Binbin Lu	Sichuan University, China
Junling Lu	Shaanxi Normal University, China
Jizhou Luo	Harbin Institute of Technology, China

Li Mohan	Jinan University, China
Tiezheng Nie	Northeastern University, China
Haiwei Pan	Harbin Engineering University, China
Jialiang Peng	Norwegian University of Science and Technology, Norway
Fei Peng	Hunan University, China
Jianzhong Qi	University of Melbourne, Australia
Shaojie Qiao	Southwest Jiaotong University, China
Qingliang Li	Changchun University of Science and Technology, China
Zhe Quan	Hunan University, China
Yingxia Shao	Peking University, China
Wei Song	North China University of Technology, China
Yanan Sun	Oklahoma State University, USA
Minghui Sun	Jilin University, China
Guanghua Tan	Hunan University, China
Yongxin Tong	Beihang University, China
Xifeng Tong	Northeast Petroleum University, China
Vicenç Torra	Umeå University, Sweden
Leong Hou	University of Macau, China
Hongzhi Wang	Harbin Institute of Technology, China
Yingjie Wang	Yantai University, China
Dong Wang	Hunan University, China
Yongheng Wang	Hunan University, China
Chunnan Wang	Harbin Institute of Technology, China
Jinbao Wang	Harbin Institute of Technology, China
Xin Wang	Tianjin University, China
Peng Wang	Fudan University, China
Chaokun Wang	Tsinghua University, China
Xiaoling Wang	East China Normal University, China
Jiapeng Wang	Harbin Huade University, China
Huayu Wu	Institute for Infocomm Research, China
Yan Wu	Changchun University of Science and Technology, China
Sheng Xiao	Hunan University, China
Ying Xu	Hunan University, China
Jing Xu	Changchun University of Science and Technology, China
Jianqiu Xu	Nanjing University of Aeronautics and Astronautics, China
Yaohong Xue	Changchun University of Science and Technology, China

Li Xuwei	Sichuan University, China
Mingyuan Yan	University of North Georgia, USA
Yajun Yang	Tianjin University, China
Gaobo Yang	Hunan University, China
Lei Yang	Heilongjiang University, China
Ning Yang	Sichuan University, China
Xiaochun Yang	Northeastern University, China
Bin Yao	Shanghai Jiao Tong University, China
Yuxin Ye	Jilin University, China
Xiufen Ye	Harbin Engineering University, China
Minghao Yin	Northeast Normal University, China
Dan Yin	Harbin Engineering University, China
Zhou Yong	China University of Mining and Technology, China
Lei Yu	Georgia Institute of Technology, USA
Ye Yuan	Northeastern University, China
Kun Yue	Yunnan University, China
Peng Yuwei	Wuhan University, China
Xiaowang Zhang	Tianjin University, China
Lichen Zhang	Shaanxi Normal University, China
Yingtao Zhang	Harbin Institute of Technology, China
Yu Zhang	Harbin Institute of Technology, China
Wenjie Zhang	University of New South Wales, Australia
Dongxiang Zhang	University of Electronic Science and Technology of China, China
Xiao Zhang	Renmin University of China, China
Kejia Zhang	Harbin Engineering University, China
Yonggang Zhang	Jilin University, China
Huijie Zhang	Northeast Normal University, China
Boyu Zhang	Utah State University, USA
Jian Zhao	Changchun University, China
Qijun Zhao	Sichuan University, China
Bihai Zhao	Changsha University, China
Xiaohui Zhao	University of Canberra, Australia
Jiancheng Zhong	Hunan Normal University, China
Fucai Zhou	Northeastern University, China
Changjian Zhou	Northeast Agricultural University, China
Min Zhu	Sichuan University, China
Yuanyuan Zhu	Wuhan University, China
Wangmeng Zuo	Harbin Institute of Technology, China

Contents – Part II

Social Media and Recommendation Systems

Education Using Big Data, Intelligent Computing or Data Mining, etc.

Contents – Part I

Big Data Mining and Knowledge Management

Data Visualization

Data-Driven Security

Infrastructure for Data Science

Machine Learning for Data Science

Multimedia Data Management and Analysis

Data-Driven Healthcare

Better Fibre Orientation Estimation with Single-Shell Diffusion MRI Using Spherical U-Net

Hang Zhao, Chengdong Deng, Yu Wang, and and Jiquan Ma[✉]

Department of Computer Science and Technology, Heilongjiang University, Harbin, China
majiquan@hlju.edu.cn

Abstract. Diffusion MRI is an important technology for detecting human brain nerve pathways, aiding in neuroscience and clinical diagnosis. However, the Multi-Shell Multi-Tissue Constrained Spherical Deconvolution (M-CSD) method, which is a significant technique for reconstructing the fibre orientation distribution function (fODF), requires multishell data with a considerable number of gradient directions to achieve high accuracy. As multishell data are not easy to acquire, the Single-Shell Single-Tissue CSD (S-CSD) suffers from the Partial Volume Effect (PVE). It would be more convenient if we could use single-shell data to reconstruct better fODFs.

We propose a novel method that utilizes the spatial structure and anisotropy of dMRI data through a spherical convolution network. We reduce the need for high-quality data by utilizing $b = 1000$ s/mm^2 with 60 gradient directions or even less. Our results show that our method outperforms the traditional S-CSD when compared to the M-CSD results as our gold standard.

Keywords: Deep learning · Fibre orientation estimation · Diffusion MRI · Spherical U-Net

1 Introduction

Diffusion nuclear magnetic resonance imaging (dMRI) is a technique that maps the direction of water molecule movement or diffusion in living tissue. This imaging technology, which mainly includes DWI and DTI [29], provides a unique research platform for neuroanatomy and is the only noninvasive method that can probe brain microstructure in vivo [1].

Fibre tracking technology utilizes continuous streamlines to display the direction and distribution of fibres [2], enabling the acquisition of white matter tissue structure and details of fibre bundles. This technology is increasingly important in clinical and neuroscience research. The reconstruction of fODF from DWIs is a critical step of probabilistic fibre tracking [25], as fODFs contain meaningful information regarding connections, such as orientations and volume fractions.

Since DTI can only detect primary orientation [3], various DWI methods for multifiber configuration and fODF reconstruction have been proposed. However, the high

Z. Yu et al. (Eds.): ICPCSEE 2023, CCIS 1880, pp. 3–12, 2023.
https://doi.org/10.1007/978-981-99-5971-6_1

hardware requirements and long acquisition time hinder clinical application. High Angular Resolution Diffusion Imaging (HARDI) [17, 24] requires high b-values and numerous gradient directions to probe complex structures; otherwise, the results will be unreliable. Multi-Shell Multi-Tissue Constrained Spherical Deconvolution [4], the state-of-the-art nonlearning method for fODF reconstruction, requires multiple b-values and sufficient gradient directions. Its previous version, Single-Shell Single-Tissue Constrained Spherical Deconvolution [5], could only use single-shell data. However, S-CSD is vulnerable to partial volume effect (PVE), as the signal obtained by each voxel represents the average value of signals from different types of tissues within the voxel. Therefore, when there are multiple fibres within the voxel or when there are more nonwhite matter tissues within the voxel, the accuracy of the CSD method decreases. Although multishell data are more compatible with high-definition fibre tractography than single-shell data, dMRI is often acquired in a single-shell due to clinical limitations [23].

Convolutional neural networks (CNNs) have made significant progress and shown advantages in medical image processing [6, 26, 27]. Drawing inspiration from the works of Zhao et al. [7, 8], which provided the neighborhood positions of various vertices of the icosahedron, we attempted to let the deep learning network leverage the spatial anisotropy information of dMRI data at each voxel position. We constructed a spherical convolution network that better suited our task, as detailed in Sect. 4.

2   Related Work

Initially, a few researchers conducted an initial investigation using a straightforward network. Koppers et al. [9] substituted the traditional model fitting procedure with a CNN approach that first determines the number of fibre directions based on Schultz et al. [10], and then directly estimates fibre orientations from the raw data (signal). Prior to entering the network, the data were converted into a 2D form with cyclic signal shifting. While the network structure may vary, the primary objective of almost all deep learning methods for fibre orientation estimation is to reduce data requirements. This is evident in many aspects, such as utilizing fewer gradient directions. Koppers et al. [11] proposed a novel CNN approach that leverages adjacent spatial information of voxels with fewer gradient directions, outperforming the CSD method. Lin et al. [12] also proposed a method to reconstruct the fODF from downsampled DWIs, utilizing a 3D CNN and achieving appreciable results. Similarly, Zeng et al. [13] presented FOD-Net, which utilizes a larger spatial patch to obtain high-quality tractography from single-shell data with few gradient directions. These studies demonstrate the significant contribution of spatial information in reconstructing fODF.

Alternatively, DL methods aim to achieve satisfactory results even when lower b-value data are available, and have demonstrated notable success. Patel et al. [14] combined convex optimization and unsupervised learning to achieve improved fODF results from low b-value data. Nath et al. [15] introduced a residual DNN and residual CNN to estimate 8-order 3-shell M-CSD data from single-shell data and recover high-definition fibre tractography. This suggests that single-shell dMRI data may contain more information than previously anticipated. Lucena et al. [16, 20] proposed a HighResNet architecture to enhance single-shell data tractography, establishing regression from the S-CSD to the M-CSD spherical harmonic coefficients. Perhaps multishell data can be recovered from single-shell data, similar to how Jha et al. [18] used low b-value data to reconstruct high b-value data through MSR-Net, which is another remarkable means of improving data quality.

In dMRI data, different measurement directions produce varying values, and q-space information is also crucial [28]. Therefore, we can design the model from this perspective. Karimi et al. [21, 22] utilized a multilayer perceptron to directly reconstruct the fODF from DWI measurements in q-space and achieved significant results. Sedlar et al.'s [19] work also projects dMRI signals onto a sphere, but our network is more lightweight and straightforward.

3 Data and Implementation

We utilized the publicly available Human Connectome Project (HCP) dataset [31] for our study. The provided data include three b-values, namely 0 s/mm^2, 1000 s/mm^2, 2000 s/mm^2 and 3000 s/mm^2, with 18, 90, 90 and 90 gradient directions respectively. We used the M-CSD method in MRtrix3 [32] to calculate our learning target, utilizing all b-values and gradient directions mentioned above. The output consists of a set of 8-order spherical harmonic coefficients, with each voxel requiring 45 coefficients to represent its fODF [30]. For SH coefficients, a higher order is associated with a greater expressible accuracy. In this paper, it is sufficient to set order $L_{max} = 8$.

Our dMRI raw data can be considered a discrete function defined on a sphere, with each gradient direction corresponding to a sampling point on the spherical function. To express the dMRI data with spherical harmonic coefficients and continue the discrete spherical function, we utilized MRtrix3's *amp2sh*. We then resampled the spherical function using MRtrix3's *sh2amp* based on these SH coefficients. The positions of the sampling points were determined by the vertices of the regular icosahedron's subdivisions (see Fig. 1). Projecting the dMRI data on the icosahedron's subdivisions, which gradually approximates spherical shape, only required the 2nd subdivision due to the spatial simplicity of the spherical function. As a result, 162 values were generated for each voxel of the input data.

We used data from the entire brain without distinguishing between its various tissues. Therefore, our model is almost end-to-end and does not require additional operations to discriminate between different tissues. However, we remove the outer skull using an eroded mask.

We utilized 15 objects from the HCP dataset, with 8 objects allocated for training and 7 for testing. The model was trained for 200 epochs using an AdamW optimizer with an initial step of 0.001, and the mean squared error was used as the loss function. All experimental results presented in this paper were obtained using Python-3.9 and Pythoch-1.11.0, and were executed on an NVIDIA GeForce RTX 2080 SUPER graphics card.

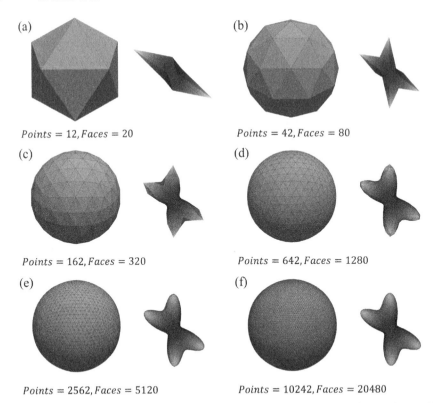

Fig. 1. The icosahedron and its subdivisions. (a) depicts an undivided regular icosahedron, while (b) through (f) display the icosahedron divided sequentially once, twice, thrice, four times, and five times using the *loop-subdivide* method. The number of vertices and faces of each solid structure is marked below. On the right, a rough representation obtained by sampling fODF based on the corresponding vertices is also shown.

4   Methods

In our spherical U-Net, our input data require two downsampling blocks and two upsampling blocks as the encoder and decoder, respectively, while the predicted fODF coefficients can be obtained by connecting two fully connected layers. The purpose of these FC layers is to reduce the dimensionality of the output from the U-shaped part, while the input data undergo a convolutional layer to expand its dimensionality. For the pipeline and spherical U-Net structure of our work, please refer to Fig. 2.

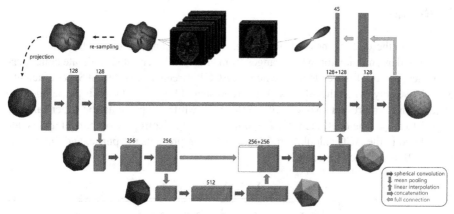

Fig. 2. The network structure of this article. The input data consist of two dimensions, namely (batch-size, points), and after the first convolution operation, it transforms into a three-dimensional format, i.e., (batch-size, points, channel). The last 3D data will go through the fully connected layer, resulting in dimension reduction to (batch-size, points). Alternatively, when the data are two-dimensional, the points dimension can be utilized as the channel dimension as well.

Each downsampling block contains two spherical convolution layers and one mean-pooling layer, while each upsampling block contains two spherical convolution layers and one-time linear interpolation. Initially, the input data consist of 162 points on each sphere (i.e., each voxel), representing 162 dimensions. After the first downsampling block, the number of points is reduced to 42. Subsequently, after the second downsampling block, it is further reduced to 12. The upsampling block will generate more vertices. The change in point dimension corresponds to the number of vertices of the icosahedron's subdivisions. In the convolution layer, the convolution core takes the shape of a ring, functioning as a sliding window on a sphere surface that scans each vertex and its adjacent vertices. The structure of the ring-shape convolutional kernel is based on the work of Zhao et al. [7]. We have made some adjustments and improvements to better align with our research objectives.

5 Results

To evaluate the effectiveness of our model, we use the angular correlation coefficient (ACC) [33] as an indicator, which ranges from -1 to 1. Higher values indicate better performance. The fODFs are identical when the ACC between the two sets of SH coefficients is 1. Conversely, when the ACC is -1, the represented fibre orientation is perpendicular. Here, u and v represent two sets of coefficients, while l represents the order ranging from 1 to L_{max}. The coefficient for $l = 0$ represents the isotropic component and is not considered in the ACC calculation. Meanwhile, m represents the phase, and there are a total of $2l + 1$ phases for each l.

$$ACC = \frac{\sum_{l=1}^{L_{max}}\sum_{m=-l}^{l}u_{lm}v_{lm}^{*}}{\sqrt{\sum_{l'=1}^{L_{max}}\sum_{m'=-l'}^{l'}|u_{l'm'}|^{2}} \cdot \sqrt{\sum_{l''=1}^{L_{max}}\sum_{m''=-l''}^{l''}|v_{l''m''}|^{2}}} \tag{1}$$

When S-CSD uses single-shell data, a higher b-value does not necessarily mean better results. In fact, the selection of the b-value should be optimized based on the type of tissue being studied and the resolution needed. While a higher b-value can provide greater angular resolution and improved fibre bundle separation, it will increase noise in the data and reduce the signal-to-noise ratio (SNR). When the SNR is very low, using a higher b-value may result in errors in the estimated orientation of fibres. It is generally believed that the higher the number of gradient directions, the better the data quality is, and the easier it is to obtain a more accurate fODF.

The final results demonstrate that our method outperforms the S-CSD method under various gradient direction numbers when using only single-shell data. Both the S-CSD method and our method achieve optimal results when the b-value is 2000 s/mm^2 and the number of gradient directions is 90 (see Table 1). We plotted the ACC map under the two data conditions of the worst and best results. The results were selected from three objects and presented from different perspectives, as shown in Fig. 3.

Table 1. Averaged angular correlation coefficient of the whole brain.

b-value	1000 s/mm2			2000 s/mm2			3000 s/mm2		
The number of gradient directions	30	60	90	30	60	90	30	60	90
S-CSD	0.457	0.521	0.564	0.502	0.556	0.587	0.478	0.528	0.559
Spherical-UNet	0.713	0.776	0.820	0.720	0.796	0.836	0.668	0.754	0.792

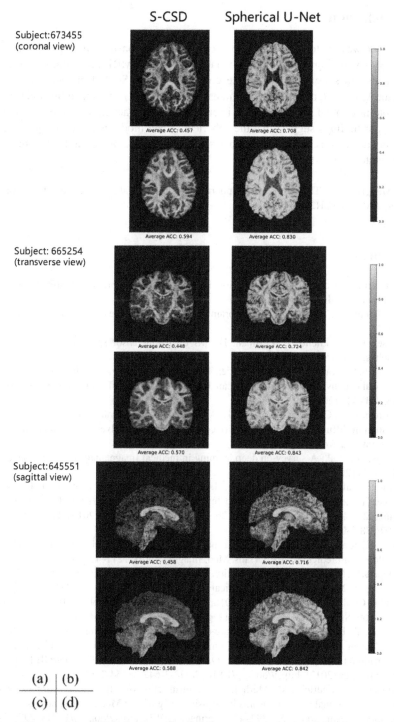

Fig. 3. ACC maps from different views. The four images for each view show the results of two different sets of parameters. Specifically, (a) and (b) show the results obtained with a b value of 1000 s/mm² and 30 gradient directions, while (c) and (d) show the results obtained with a b value of 2000 s/mm² and 90 gradient directions. To enhance visualization, we plotted values less than zero as zero.

6 Conclulsion

In this work, we developed a novel approach for fODF reconstruction using the spherical U-Net. By leveraging the spherical spatial structure of dMRI data within a voxel, our model achieved superior performance compared to the S-CSD method with single-shell data. Moreover, our approach demonstrated excellent generalizability and is highly applicable in clinical settings, which is crucial for the practical utility of the method. Overall, our findings suggest that the spherical U-Net model is a promising model for improving the accuracy of fODF reconstruction and advancing our understanding of the brain's white matter structure.

Acknowledgements. This work was supported by the Natural Science Foundation of Heilongjiang Province (LH2021F046).

References

1. Le Bihan, D.: Looking into the functional architecture of the brain with diffusion mri. Nat. Rev. Neurosci. **4**(6), 469–480 (2003)
2. Garyfallidis, E.: Towards an accurate brain tractography. PhD thesis, University of Cambridge (2013)
3. Basser, P.J., Mattiello, J., LeBihan, D.: Mr diffusion tensor spectroscopy and imaging. Biophys. J . **66**(1), 259–267 (1994)
4. Jeurissen, B., Tournier, J.-D., Dhollander, T., Connelly, A., Sijbers, J.: Multitissue constrained spherical deconvolution for improved analysis of multishell diffusion mri data. Neuroimage **103**, 411–426 (2014)
5. Tournier, J.-D., Calamante, F., Connelly, A.: Robust determination of the fibre orientation distribution in diffusion mri: nonnegativity constrained superresolved spherical deconvolution. Neuroimage **35**(4), 1459–1472 (2007)
6. Litjens, G., et al.: A survey on deep learning in medical image analysis. Med. Image Anal. **42**, 60–88 (2017)
7. Zhao, F., et al.: Spherical u-net on cortical surfaces: methods and applications. In: Information Processing in Medical Imaging: 26th International Conference, IPMI 2019, Hong Kong, China, 2–7 June 2019, Proceedings 26, pp. 855–866. Springer (2019). https://doi.org/10.1007/978-3-030-20351-1_67
8. Zhao, F., et al.: Spherical deformable u-net: Application to cortical surface parcellation and development prediction. IEEE Trans. Med. Imaging **40**(4), 1217–1228 (2021)
9. Koppers, S., Merhof, D.: Direct estimation of fibre orientations using deep learning in diffusion imaging. In: Machine Learning in Medical Imaging: 7th International Workshop, MLMI 2016, Held in Conjunction with MICCAI 2016, Athens, Greece, 17 October 2016, Proceedings 7, pp. 53–60. Springer (2016). https://doi.org/10.1007/978-3-319-47157-0_7
10. Schultz, T.: Learning a reliable estimate of the number of fibre directions in diffusion mri. In: Medical Image Computing and Computer-Assisted Intervention–MICCAI 2012: 15th International Conference, Nice, France, October 1–5, 2012, Proceedings, Part III 15, pp. 493–500. Springer (2012). https://doi.org/10.1007/978-3-642-33454-2_61
11. Koppers, S., Friedrichs, M., Merhof, D.: Reconstruction of diffusion anisotropies using 3d deep convolutional neural networks in diffusion imaging. In: Modelling, Analysis, and Visualization of Anisotropy, pp. 393–404. Springer (2017). https://doi.org/10.1007/978-3-319-61358-1_17

12. Lin, Z., et al.: Fast learning of fibre orientation distribution function for mr tractography using convolutional neural network. Med. Phys. **46**(7), 3101–3116 (2019)
13. Zeng, R., et al.: Fod-net: A deep learning method for fibre orientation distribution angular super resolution. Med. Image Anal. **79**, 102431 (2022)
14. Patel, K., Groeschel, S., Schultz, T.: Better Fiber ODFs from suboptimal data with autoencoder based regularization. In: Frangi, A.F., Schnabel, J.A., Davatzikos, C., Alberola-López, C., Fichtinger, G. (eds.) MICCAI 2018. LNCS, vol. 11072, pp. 55–62. Springer, Cham (2018). https://doi.org/10.1007/978-3-030-00931-1_7
15. Nath, V., Pathak, S.K., Schilling, K.G., Schneider, W., Landman, B.A.: Deep learning estimation of multitissue constrained spherical deconvolution with limited single shell dw-mri. In: Medical Imaging 2020: Image Processing, vol. 11313, pp. 162–171. SPIE (2020)
16. Lucena, O., Vos, S.B., Vakharia, V., Duncan, J., Ourselin, S., Sparks, R.: Convolutional neural networks for fibre orientation distribution enhancement to improve single-shell diffusion mri tractography. In: Computational Diffusion MRI: MICCAI Workshop, Shenzhen, China, October 2019, pp. 101–112. Springer (2020). https://doi.org/10.1007/978-3-030-52893-5_9
17. Tuch, D.S., Reese, T.G., Wiegell, M.R., Makris, N., Belliveau, J.W., Wedeen, V.J.: High angular resolution diffusion imaging reveals intravoxel white matter fibre heterogeneity. Magn. Reson. Med. Official J. Internat. Soc. Magn. Reson. Med. **48**(4), 577–582 (2002)
18. Jha, R.R., Nigam, A., Bhavsar, A., Pathak, S.K., Schneider, W., Rathish, K.: Multishell d-mri reconstruction via residual learning utilizing encoder-decoder network with attention (msr-net). In: 2020 42nd Annual International Conference of the IEEE Engineering in Medicine & Biology Society (EMBC), pp. 1709–1713 (2020). IEEE
19. Sedlar, S., Papadopoulo, T., Deriche, R., Deslauriers-Gauthier, S.: Diffusion mri fibre orientation distribution function estimation using voxelwise spherical u-net. In: Computational Diffusion MRI: International MICCAI Workshop, Lima, Peru, October 2020, pp. 95–106. Springer (2021). https://doi.org/10.1007/978-3-030-73018-5_8
20. Lucena, O., et al.: Enhancing the estimation of fibre orientation distributions using convolutional neural networks. Comput. Biol. Med. **135**, 104643 (2021)
21. Karimi, D., Vasung, L., Jaimes, C., Machado-Rivas, F., Warfield, S.K., Gholipour, A.: Learning to estimate the fibre orientation distribution function from diffusion-weighted mri. Neuroimage **239**, 118316 (2021)
22. Karimi, D., et al.: A machine learning-based method for estimating the number and orientations of major fascicles in diffusion-weighted magnetic resonance imaging. Med. Image Anal. **72**, 102129 (2021)
23. Canales-Rodríguez, E.J., et al.: Sparse wars: a survey and comparative study of spherical deconvolution algorithms for diffusion mri. Neuroimage **184**, 140–160 (2019)
24. Descoteaux, M., Angelino, E., Fitzgibbons, S., Deriche, R.: Apparent diffusion coefficients from high angular resolution diffusion imaging: estimation and applications. Magn. Reson. Med. Official J. Inter. Soc. Magn. Reson. Med. **56**(2), 395–410 (2006)
25. Wilkins, B., Lee, N., Gajawelli, N., Law, M., Leporé, N.: Fibre estimation and tractography in diffusion mri: development of simulated brain images and comparison of multifibre analysis methods at clinical b-values. Neuroimage **109**, 341–356 (2015)
26. Puttashamachar, N., Bagci, U.: End to end brain fibre orientation estimation using deep learning. arXiv preprint arXiv:1806.03969 (2018)
27. Ye, C., Prince, J.L.: Fibre orientation estimation guided by a deep network. In: Medical Image Computing and Computer Assisted Intervention-MICCAI 2017: 20th International Conference, Quebec City, QC, Canada, 11–13 September 2017, Proceedings, Part I, pp. 575–583. Springer (2017). https://doi.org/10.1007/978-3-319-66182-7_66
28. Golkov, V., et al.: Q-space deep learning: twelve-fold shorter and model-free diffusion mri scans. IEEE Trans. Med. Imaging **35**(5), 1344–1351 (2016)

29. Hagmann, P., Jonasson, L., Maeder, P., Thiran, J.-P., Wedeen, V.J., Meuli, R.: Understanding diffusion mr imaging techniques: from scalar diffusion-weighted imaging to diffusion tensor imaging and beyond. Radiographics **26**(suppl_1), 205–223 (2006)
30. Dell'Acqua, F., Tournier, J.-D.: Modelling white matter with spherical deconvolution: How and why? NMR Biomed. **32**(4), 3945 (2019)
31. Glasser, M.F., et al.: The minimal preprocessing pipelines for the human connectome project. Neuroimage **80**, 105–124 (2013)
32. Tournier, J.-D., et al.: Mrtrix3: A fast, flexible and open software framework for medical image processing and visualization. Neuroimage **202**, 116137 (2019)
33. Anderson, A.W.: Measurement of fibre orientation distributions using high angular resolution diffusion imaging. Magn. Reson. Med. Official J. Inter. Soc. Magn. Reson Med. **54**(5), 1194–1206 (2005)

A Method for Extracting Electronic Medical Record Entities by Fusing Multichannel Self-Attention Mechanism with Location Relationship Features

Hongyan Xu[1], Hong Wang[1], Yong Feng[1(✉)], Rongbing Wang[1], and Yonggang Zhang[2]

[1] College of Information, Liaoning University, Shenyang 110036, China
fengyong@lnu.edu.cn
[2] Key Laboratory of Symbolic Computation and Knowledge Engineering of Ministry of Education, Jilin University, Changchun 130012, China

Abstract. With the implementation of the "Internet+" strategy, electronic medical records are generally applied in the medical field. Deep mining of electronic medical record content data is an effective means to obtain medical knowledge and analyse patients' states, but the existing methods for extracting entities from electronic medical records have problems of redundant information, overlapping entities, and low accuracy rates. Therefore, this paper proposes an entity extraction method for electronic medical records based on the network framework of BERT-BiLSTM, which incorporates a multichannel self-attention mechanism and location relationship features. First, the text input sequence was encoded using the BERT-BiLSTM network framework, and the global semantic information of the sentence was mined more deeply using the multichannel self-attention mechanism. Then, the position relation characteristic was used to extract the local semantic message of the text, and the position relation characteristic of the word and the position embedding matrix of the whole sentence were obtained. Next, the extracted global semantic information was stitched with the positional embedding matrix of the sentence to obtain the current entity classification matrix. Finally, the proposed method was validated on the dataset of Chinese medical text entity relationship extraction and the 2010i2b2/VA relationship corpus, and the experimental results indicate that the proposed method surpasses existing methods in terms of precision, recall, F1 value and training time.

Keywords: entity extraction · location relationship feature · electronic medical record · self-attention

1 Introduction

With the continuous promotion of the "Internet +" strategy in China, electronic medical records [1] are extensively used in major hospitals, such as clinical diagnosis, and online consultation. The mining and analysis of electronic medical records has become one of the main ways to obtain medical knowledge. Since manual extraction of entity messages can consume considerable manpower, natural language processing technologies are introduced to entity extraction to solve the above problems.

Over the years, the prosperity of deep learning neural network-based entity extraction methods for electronic medical records has become mainstream [2]. Compared with machine learning methods, deep learning methods can reduce the selection of manual features and improve the efficiency of entity extraction. The more mainstream ones are BiLSTM-CRF methods based on the BiLSTM network [3], which have achieved better results in entity extraction of electronic medical record datasets. Lu, Y [4] proposed a method to extract deeper information by fusing the BiLSTM network and self-attention mechanism to improve the extraction accuracy of unstructured text. Tang, B [5] proposed a self-attention CNN-LSTM-CRF method to identify entities in electronic medical records by introducing an attention layer to extract contextual text in sentence vectors for word vectors and location information. However, the above methods have shortcomings in the extraction of textual entities from electronic medical records, such as overlapping entity relationships, cumulative propagation of errors and missed extraction of unstructured text.

In summary, this paper proposed an electronic medical record entity extraction method (BBMSP) that fused a multichannel self-attention mechanism with location relationship features. First, the electronic medical record text was transformed to quantization by BERT-BiLSTM network. Second, the input vector was extracted by multichannel self-attention to obtain the global feature vector matrix. Third, the input vectors were extracted utilizing pooling and position relation characteristics to gain the term position features and sentence location embedding matrix. Finally, the global feature vector matrix was pooled with the sentence location embedding matrix to obtain the current entity classification matrix. Comparative experiments were conducted on the Chinese medical text entity relationship extraction dataset and the 2010i2b2/VA relationship corpus, and the experimental results indicate that the proposed method in this paper possesses preferable functionality compared with the present conventional methods.

2   Related Studies

2.1   Self-attention Mechanism

Currently, the self-attention mechanism can better simplify and abstract the sentential semantic message [6]. When the addressing process is carried out in the semantic decoding process, the self-attention mechanism first transforms the input text into the form of an embedding vector when the electronic medical record text is vectorized data input, and then obtains the query vector (Q), the key vector (K) and the value vector (V) based on the embedding vector. When Q is determined, the V corresponding to it can be found based on the correlation between Q and K, , and the value vector (V) is calculated by Eq. (1).

$$V = sim(K_i, Q) = \frac{K_i^T Q}{\sqrt{d}} \tag{1}$$

Weighting normalization may enable the weights of significant components to be better. The weight factor value vector α is calculated by Eq. (2), and the last attention value is calculated by Eq. (3).

$$\alpha_i = Soft\max(sim_i) = \frac{e^{sim_i}}{\sum_{i=1}^{L_x} e^{sim_i}} \tag{2}$$

$$A = \sum_{i=1}^{L_x} \alpha_i V_i \tag{3}$$

The self-attention network assigns higher weights to the keywords identified by the entities, thus noting the differences among words and obtaining attention-weighted new vectors [7]. In electronic medical record entity extraction, the self-attention mechanism obtains the global context of the sentence by extracting the word vector of the electronic medical record entity with the sentential semantic message.

2.2 Pooling Operation

Current research is investigating how to better simplify sentence semantic information through a new pooling layer. For example, downsampling is performed using a combined form of maximum pooling and average pooling, and this approach can further improve network performance but is not universal. He, S [8] used a self-attention mechanism to focus on the importance between words on the entity extraction task, and averaged pooling of the important word to gain the sentential entity message. The sentential entity message is calculated by Eq. (4).

$$Pel = \text{Sigmoid}(W_r[Avgpool(Y(s)); Avgpool(Soft\max(\frac{QK^T}{\sqrt{dk}})V)] \in R^{n_r \times 1} \tag{4}$$

In this paper, the above methods were improved, and the effective methods were retained (using pooling and self-attention mechanisms to extract the entity information of sentences, etc.). An entity extraction method for extracting local and global information was established. This method not only retained the advantages of the above methods, but also effectively solved the problems of information redundancy, entity overlap, and low accuracy rate in classical entity extraction methods.

3 Method Description

To solve the problems of information redundancy, entity overlap and low accuracy of existing entity extraction methods in electronic medical records, this paper proposed the BBMSP method, and the overall framework of the method is shown in Fig. 1. The method consisted of five parts: BERT layer, BiLSTM layer, multichannel self-attention layer, pooling-based location relationship feature layer and fully connected layer. The idea of the method was to obtain the input word vector and location feature information from the text through the BERT layer, and to extract the contextual information of the text from the word vector in the BiLSTM layer to obtain the sentence vector. The sentence vector obtained from the BiLSTM layer was used as the input of the multichannel self-attention layer, and the global feature vector matrix was obtained after several cycles of self-attention operations. In addition, the sentential local features were gained after maximum pooling and similar pooling operations. The BERT layer location feature information was subjected to a word attention operation to obtain the feature representation of the same entity in different relationships and stitched with the local features of the sentence to obtain the positional embedding matrix of the sentence. The global feature vector matrix was stitched with the positional embedding matrix of the sentence in the fully connected layer to obtain the current entity classification matrix. Finally, entity classification prediction was completed by the entity classification matrix.

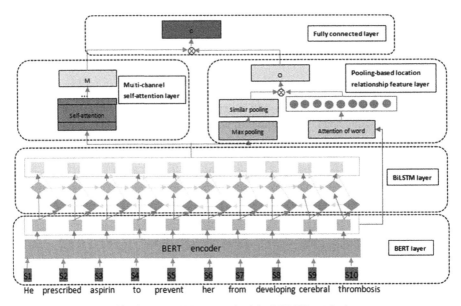

Fig. 1. Overall framework of the BBMSP method

3.1 BERT Layer

Due to the complexity of medical texts. Therefore, this layer uses the BERT architecture as the encoder of the input text to obtain the input text vector, and BERT obtains the hidden layer state vector calculated by Eq. (5).

$$x_t = Bert_{base}(w_t) \qquad (5)$$

Denote it as sequence $X = (x_1, x_2, ..., x_n)$.

3.2 BiLSTM Layer

In this paper, BiLSTM is used to encode from back to front to capture bidirectional semantic dependencies. The sequence X obtained by the BERT network is used as the input of each time step of BiLSTM to obtain the forward hidden state sequence $\overrightarrow{h}_t$ and backwards hidden state sequence $\overleftarrow{h_t}$ of the BiLSTM layer. The corresponding memory unit h_t is brought into the hidden layer and calculated by Eq. (6). The hidden layer vector output result h_{t_1} of BiLSTM is calculated by Eq. (7).

$$h_t = \sigma(w_0 \cdot [h_{t-1}, x_t] + b_0) \otimes \tanh(f_t \otimes C_{t-1} + i_t \otimes \tilde{C}_t) \tag{6}$$

$$h_{t_1} = [\overrightarrow{h}_t, \overleftarrow{h_t}] \tag{7}$$

h_{t-1} denotes the output of the hidden layer of the prior LSTM unit, C_{t-1} denotes the state result of the prior LSTM unit, x_t denotes the result of character vector input, f_t denotes the output of the forget gate, σ denotes sigmoid activation function, i_t and $\tilde{C}_t$ denotes the output of the input gate and C_t denotes the state value of the current cell.

3.3 Multichannel Self-Attention Layers

In the electronic medical record entity extraction region, neural networks based on self-attention mechanisms have gradually become mainstream [10]. The traditional self-attention mechanism calculates a set of weights for sentences, which lacks the capture of the overall characteristics of sentences. Because medical text is complicated, multichannel self-attention is introduced to capture sentence-level semantics [11].

The execution process of the multichannel self-attention mechanism is as follows:

(1) BiLSTM outputs hidden layer vector $Z_t = (h_1, h_2, ..., h_n)$, and the Z_t dimension is $n \times q$.

(2) The output Z_t of the BiLSTM is used as the input of multichannel self-attention. The orthodox self-attention with attention weight α is calculated by Eq. (8).

$$\alpha = soft\text{max}(w_j \cdot \tan(W_z \cdot Z_t^T)) \tag{8}$$

W_z and w_j are learnable parameter matrices with dimensions $a \times 2q$ and a, respectively.

(3) Current research on how to expand the learning ability of each component information in sentences. This paper adopts the method of multichannel self-attention mechanism method. Given the number of channels Q, the output of Bi-LSTM is calculated by the self-attention weight α_{q_i} in each channel $\mathcal{Q}_i$. The self-attention weight α_{q_i} is calculated by Eq. (9).

$$\alpha_{Q_t} = soft\text{max}(w_j^{Q_i} \cdot \tanh(W_Z \cdot Z_t^T)) \tag{9}$$

$w_j^{Q_i}$ denotes a set of learnable weight matrices in channel $\mathcal{Q}_i$.

α_{Q_i} can be seen as the importance of the i component in the target sentence. Therefore, if there is less key information in the sentence, the increase in the number of channels will lead to too much redundant information in the training process of the network, thus affecting the effect of downstream tasks. The experimental part can intuitively illustrate the ability of the multichannel attention mechanism to trap critical messages in sentences.

The calculation of the self-attention of each channel is independent of each other, and multichannel parallelization can be realized by matrix operation. The multichannel self-attention weight B is calculated by Eq. (10).

$$B = \alpha_Q = softmax(W_j \cdot \tanh(W_z \cdot Z_t^T)) \tag{10}$$

Q denotes the number of channels given, $Q \in [1, r]$, W_j denotes a 2-dimensional matrix with parameter $w_j^{Q_i}$ expanded to $r \times a$ dimensional scale.

Finally, weighted summation is carried out to obtain the global feature vector M, which is calculated by Eq. (11).

$$M = (B^T \cdot Z)^T = ((softmax(W_j \cdot \tanh(W_Z \cdot Z_t^T)))^T \cdot (h_1, h_2, ..., h_n))^T \tag{11}$$

M denotes the global feature vector with dimension $r \times 2a$.

3.4 Pooling-Based Location-Related Feature Layer

Extracting the most principal characteristics from multiple phrases in a sentence using the maximum pooling technique after the BiLSTM layer, the maximum pooling p_i is calculated by Eqs. (12) and (13).

$$p_i = \max\{h_{i+1}, h_{i+2}, ..., h_{i+f_1}\} \tag{12}$$

$$p = (p_1, p_2, ..., p_{n-f_1+1}) \tag{13}$$

f_1 denotes the filter length,n denotes the enter sentence size,p_i denotes the maximum value in the sentence vector for the *ith* phrase of length f_1, and p denotes the output of the entire sentence after maximum pooling.

Since the maximum pool will lose the location information of the feature, this may reduce the learnability. In this paper, a new method of location relation feature extraction based on similarity pooling is proposed. First, the input is the output h_t of the BiLSTM.h_t can solve the gradient vanishing problem by replacing RNN with GRU. We compare it with the local representation generated by the pooling layer. The higher the similarity, the greater the similarity weight assigned to the local representation. The similarity weight α is calculated by Eqs. (14) and (15). The final sentence matrix is calculated by Eq. (16).

$$s_i = \cos(p_i, h_{t_i}) = \frac{\sum_{i=1}^{T}(p_i \times h_{t_i})}{\sqrt{\sum_{i=1}^{T}(p_i)^2} \times \sqrt{\sum_{i=1}^{T}(h_{t_i})^2}} \tag{14}$$

$$\alpha_i = \frac{\exp(s_i)}{\sum_{i=1}^{T}\exp(s_i)} \tag{15}$$

$$e = \sum_{i=1}^{T}\alpha_i \cdot p_i = \frac{\exp(\frac{\sum_{i=1}^{T}(p_i \times h_{t_i})}{\sqrt{\sum_{i=1}^{T}(p_i)^2} \times \sqrt{\sum_{i=1}^{T}(h_{t_i})^2}})}{\sum_{i=1}^{T}\exp(\frac{\sum_{i=1}^{T}(p_i \times h_{t_i})}{\sqrt{\sum_{i=1}^{T}(p_i)^2} \times \sqrt{\sum_{i=1}^{T}(h_{t_i})^2}})} \cdot \max\{h_{t+1}, h_{t+2}, ..., h_{t+f_1}\} \tag{16}$$

s_i denotes the similarity.

The word attention operation at this layer is based on the 'QKV' method [12], and the method framework is shown in Fig. 2. The query matrix is a vector matrix $Q_{r\times1}$ randomly sampled by uniform distribution, r is the output vector dimension of the BERT layer, and the key matrix is a feature matrix generated by the word vector of the subwords in the sentence. The weight value of the attention mechanism is obtained by the query matrix and the key matrix. The attention output vector matrix is calculated by Eq. (17).

$$A_{r_{r\times1}} = (softmax(k_{w_{n\times r}} \times q_{w_{r\times1}}))^T \times v_{w_{n\times1}})^T \tag{17}$$

softmax denotes the normalization; $k_{w_{n\times r}}$ denotes the K-vector matrix; $q_{w_{r\times1}}$ denotes the Q-vector matrix; $v_{w_{n\times1}}$ denotes the V-vector matrix-the output vector matrix of the BERT layer.

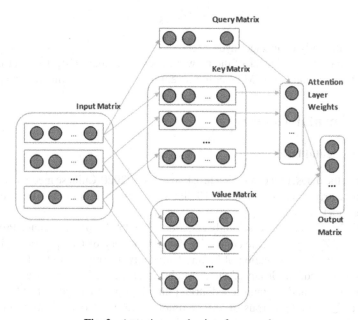

Fig. 2. Attention mechanism framework

In addition, the relationship between the pooling output and BERT output is established [9, 13] to distinguish the feature representation of the same entities among different relationships and make the final relationship classification prediction.

For each word in the output sequence of BERT, calculate its distance from the trigger word of the current pooled output sequence, and then randomly initialize the location embedding matrix F according to the maximum sentence length m and location feature size n. Obtain the relational location feature of each word by querying the location embedding matrix F. The relational position feature of each word is calculated by Eq. (18).

$$F_{f_t} = F_{f_r} - F_{f_w} \tag{18}$$

f_r denotes the position of the relational trigger word; f_w denotes the position of the BERT output sequence word.

The position embedding matrix O is calculated by Eq. (19).

$$O = (A_{r_{r \times 1}} \oplus F) \otimes e \tag{19}$$

$\oplus$ denotes a combination operation, and $\otimes$ denotes a splicing operation.

3.5 Fully-Connected Layer

Finally, the multichannel self-attention mechanism output matrix M is stitched with the location embedding matrix O in the fully connected layer to obtain the entity classification matrix C calculated by Eq. (20).

$$C = f(M; O) \tag{20}$$

$f(\cdot)$ denotes the fully connected layer.

In this paper, the optimizer of the network is chosen Adam [14]. Dropout is added to the BERT layer and BiLSTM layer to prevent overfitting during the training process.

4   Experimental Analysis

4.1   Experimental Datasets

In this paper, the dataset for evaluation task two of CHIP2020-Chinese medical text entity relationship extraction [15], and the 2010i2b2/VA relational corpus [16] are selected for method verification.

CHIP2020: The dataset was jointly constructed by the Natural Language Processing Laboratory of Zhengzhou University, the Key Laboratory of Computational Linguistics of the Ministry of Education of Peking University, Harbin Institute of Technology, and the Smart Medical Group of the Artificial Intelligence Research Center of Peng Cheng Laboratory, and contains a pediatric training corpus (derived from 518 pediatric diseases) and a training corpus of 100 common diseases (derived from 109 common diseases). After preprocessing, approximately 40,000 data points were gained to form 44 relationship classi-fications, and the training set and test set were divided according to a ratio of 75:25. Table 1 describes the relationship categories and related statistical information in detail.

2010i2b2/VA: This corpus is one of the recognized datasets for evaluating entity relationships in electronic medical records, derived from discharge summaries of three hospitals, and contains eight types of entity relationships. The training set and test set are divided according to the ratio of 80:20. Table 2 describes the meaning of the relationship categories and the related statistical information in detail.

Table 1. Information related to the CHIP2020 dataset

Class	Train size	Test size	Class	Train size	Test size
Prevention	91	30	Stage	88	29
Causes	1874	625	Synonyms	2268	756
Screening	68	23	Endoscopy	129	43
Transfer site	155	52	Risk factors	807	269
Chemotherapy	91	30	Radiotherapy	88	29
Drug therapy	3070	1023	Mortality rate	59	20
Complications	1423	474	Medical history	95	32
Genetic factors	89	30	Pathogenesis	44	15
Adjuvant treatment	1050	350	Image analysis	964	321
Related (caused)	1049	350	Related (symptoms)	261	87
Pathophysiology	21	7	Prognostic status	145	48
Age of onset	193	64	Multiple areas	148	49
Multiple groups	380	127	Incidence rate	271	90
Multiple seasons	53	18	Surgical treatment	587	196
Differential diagnosis	881	294	Auxiliary examination	377	126
Transmission routes	39	13	Clinical manifestations	8067	2689
Prognostic survival rate	52	17	Laboratory examination	1264	421
Correlation (transformation)	486	162	Symptoms after treatment	97	32
Department of treatment	29	10	The location of the disease	848	283
External invasion site	85	28	Histological examinations	217	72
Pathological classification	1280	427	Symptoms of surrounding tissue	35	12
Gender tendency of onset	92	31			
Risk assessment factors	340	113	Total	29750	9917

4.2 Evaluation Indexes and Comparison Methods

(1) Evaluation indexes

Table 2. Information related to the i2b2 corpus

Class	Define	Train size	Test size
TrIP	Treatment to improve medical problems	165	41
TrWP	Treatment of worsening medical problems	109	26
TrCP	Treatment leads to medical problems	436	108
TrAP	Treatment for medical problems	2131	532
TrNAP	No treatment due to medical problems	140	34
TeRP	Tests show medical problems	2457	614
TeCP	Conducting tests to verify medical problems	409	101
PIP	Medical issues indicate medical issues	1776	443
None	No relationship	52211	13045
Total	/	59831	11944

In this experiment, precision, recall, F1 and training time are used as evaluation metrics for entity extraction. The precision and recall are calculated by Eqs. (21) and (22). When comparing the precision and recall of different models, it is not good to directly judge the performance when each of the two metrics is high or low, so in the field of entity extraction, the F1 value is generally compared, which is calculated by Eq. (23).

$$precision = \frac{TP}{TP + FP} \tag{21}$$

$$recall = \frac{TP}{TP + FN} \tag{22}$$

$$F1 = \frac{2 \times precision \times recall}{precision + recall} \tag{23}$$

TP denotes the number of actual and predicted true,FP denotes the number of actual false but predicted true,FN denotes the number of actual true but predicted false and TN denotes the number of actual and predicted false.

(2) Comparison method

A more updated and efficient method was applied to consider whether to contain a multichannel self-attention mechanism, location-relational feature embedding and pooling operation. The following methods are selected as comparison methods.

BERT-BiLSTM-CRF method [17]: The semantic representation is enhanced using a BERT pretrained network, and then the BiLSTM network is combined with a CRF layer to use it based on word vectors.

BLSTM and ABLSTM methods [18]: The BiLSTM network is used to obtain the contextual information of the utterance, and then combined with maximum pooling and attention pooling for relation extraction.

BERT-BiLSTM-ATT method [19]: using the BERT encoding vector of the head entity as a condition, the BiLSTM-ATT network is constructed to finish the detection of sentential relations and tail entities, and the relations are modelled as a mapping function of the sentential head-entity to tail-entity for relation extraction.

The CRNN-max and CRNN-att methods [20]: combine BLSTM and CNN to learn contextual and local information of text utterances, respectively, and use max-pooling and attention pooling for relation extraction at the end, respectively.

BLSTM-Mcatt method [21]: a "recurrent + transformer" network on account of a multichannel self-attention mechanism with two weight-based training aids: cross-entropy loss function with weight learning and weight-based location embedding for relation extraction.

(3) Experimental parameter settings

The parameter settings of the BBMSP method network gained by parameter tuning experiments on the validation set are shown in Table 3.

Table 3. Parameter settings of the BBMSP method

parameter	Parameter setting value
Batch _ Size	64
Blstm _ Size	128
Position vector size	30
Dropout _ Rate	0.5
Leaning _ Rate	0.001
Epoch	150
Optimizer	Adam
Activation function	tanh
Number of self-attention channels	30

When training, the minimum number of samples for word training (batch _ size) was set to 64, the output dimension of the Bilstm hidden layer (blstm _ size) was set to 128, the position vector size was set to 30, the dropout rate was set to 0.5, the incipient learning rate was set to 0.001, the number of iterations per epoch was 150, the optimizer used Adam [24], the activation function used was tanh and the number of self-attention channels was set to 30.When testing, the test set is used for testing, and the precision, recall and F1 are calculated.

4.3 Experimental Results and Analysis

(1) Selection of the number of self-attention channels

Current research on how to expand the ability of various components to learn and store messages. This paper adopts the multichannel self-attention mechanism method. If the number of channels is too small, we cannot learn the sentential semantic message completely, while too many channels overload the competency of the BBMSP method to store messages and enhance the complication of the BBMSP. Therefore, the number of channels of the multichannel self-attention mechanism needs to be determined experimentally, and the experimental results are shown in Table 4.

Table 4. Experimental results for different numbers of self-attention channels

r	Precision/%	Recall/%	F1-Score/%
1	86.61	84.02	85.30
10	**86.97**	84.47	85.61
20	86.54	84.98	85.75
30	86.96	**85.12**	**86.03**
40	86.61	85.04	85.82

As shown in Table 4, when the number of self-attention channels is increased from 1 to 10, the precision, recall and F1 values of the BBMSP method are improved to a certain extent, indicating that increasing the number of channels helps the BBMSP method capture multifaceted semantic information in the sentence compared with traditional self-attention, so that the global semantic information of the sentence can be mined more completely, which verifies the effectiveness of the multichannel self-attention mechanism. Nevertheless, when the number of channels is increased from 20 to 40, the F1 value shows an upwards trend followed by a downwards trend, which indicates that too many channels will store too much information and lead to overfitting of the network, thus degrading the performance of the network. Therefore, the number of self-attention channels of the BBMSP method is determined to be 30 by experiment, and its F1 value is 86.03% at maximum.

(2) Experimental results

This section evaluates the effect of entity extraction by applying precision, recall, F1 value and training time for the CHIP2020 dataset and the 2010i2b2/VA relational corpus. The different methods are compared as detailed in Table 5.

From Table 5, the performances of the BBMSP method in entity extraction are better than the methods in the comparison tests, indicating that the method in this paper captures global features using multichannel self-attention and local features of words using pooled position-relational features, respectively, and the two aspects are fused. It can efficiently promote the effect of entity extraction and proves the validity of the method in this assignment. The training time column counts the training time required for each method to achieve the best classification effect, and it can be seen that the BBMSP (0.49 h, 1.51 h) method realizes efficiency promotion compared with BLSTM-Mcatt (0.57 h, 1.57 h) on two datasets. This experimental result shows that compared with the mainstream BLSTM in the field of electronic medical record entities, the experimental results show that compared with the mainstream BLSTM-Mcatt method in the field of electronic medical record entity extraction, the BBMSP method can further increase effectiveness and decrease the integral complexity on the basis of ensuring the performance of downstream relation extraction.

Table 5. Experimental results

Methods	CHIP2020				2010i2b2/VA			
	P/%	R/%	F1 /%	T/h	P/%	R/%	F1 /%	T/h
BERT-BiLSTM-CRF	82.11	82.45	82.28	0.71	69.06	59.04	63.66	1.59
BLSTM	82.91	82.91	82.69	0.50	69.74	59.87	64.43	**1.37**
ABLSTM	83.07	83.17	82.92	0.51	70.13	60.11	64.73	1.44
BERT-BILSTM-ATT	83.57	83.69	83.63	0.64	70.80	60.47	65.23	2.15
CRNN + att	83.54	83.34	83.21	0.75	70.76	57.08	63.19	1.97
CRNN + max	84.19	84.24	83.94	0.69	71.61	62.09	66.64	1.86
BLSTM-MCatt	86.51	**86.46**	**86.48**	0.57	71.12	67.47	69.72	1.57
BBMSP	**86.96**	85.12	86.03	**0.49**	**71.74**	**68.96**	**70.32**	1.51

On the 2010i2b2/VA dataset, the BBMSP method acquires the best results in precision, recall and F1 value, and the F1 value is 0.60% higher than the optimal baseline method. It exceeds the current best method. The effectiveness of the BBMSP method in entity extraction is demonstrated. The better performance of the method in this paper is because the multichannel attention mechanism is better in feature extraction compared to other networks.

On the CHIP2020 dataset, based on the Bi-LSTM network framework, BBMSP has higher F1 values than the BERT- Bi-LSTM-ATT, BLSTM, and ABLSTM methods, indicating that combining Bi-LSTM with pooling operations in this dataset enhances the advantages of text semantic extraction and improves the overall performance of the method. BLSTM- MCatt has a slightly higher F1 value than the BBMSP method tested, indicating that the weight-based auxiliary training method is preferable in the BLSTM-MCatt network in the CHIP2020 dataset, and BBMSP decreases network performance due to its more complicated structure. At the same time, due to the size of the data volume, BBMSP cannot bring out its performance advantage to a certain extent, so BLSTM-MCatt can achieve better results under this dataset. The specific reasons are as follows:(1) The number of texts in the CHIP2020 dataset is nearly halved compared to the 2010i2b2/VA corpus. Therefore, in the case of small data size, the network structure of the BLSTM-MCatt method is more concise and efficient, and is more suitable for small-scale dataset experiments. The BBMSP method is more complex and cannot play a greater role in small-scale datasets, so the effect is not as good as that of the BLSTM-MCatt method. (2) In the small-scale dataset, the combination of the location-based similarity pooling operation mechanism of the BBMSP extraction method and the multichannel self-attention cannot play a greater role. Therefore, the BLSTM-MCatt extraction method has better results on R and F1 values. (3) Compared with the BBMSP method, the BLSTM-MCatt method uses two weight-based auxiliary training methods based on the multichannel self-attention mechanism: the weighted learning cross-entropy loss function and the weight-based position embedding for extraction. The method has significantly improved the precision, recall rate and F1 value. Because it has completed coding and decoding, it can obtain more comprehensive text features and improve the breadth and accuracy of recognition. The BBMSP method combines pooling operations on the basis of multichannel attention mechanism, extracts more text features through a variety of different pooling operations and pays attention to hidden features at the sentence level, enhances the comprehensiveness of extracted features, provides multilevel corpus features, and improves the accuracy and practicability of the method. Therefore, the BBMSP method has obvious advantages in performance improvement on the 2010i2b2/VA corpus (large number of texts). On the CHIP2020 dataset (small number of texts), the performance advantage is weak.

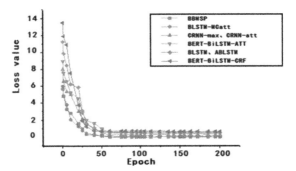

Fig. 3. Training loss function of different methods

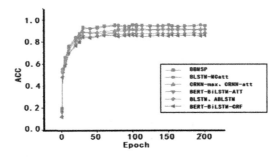

Fig. 4. ACC values of different methods under the test dataset

Figure 3 shows the loss function plots during training on the CHIP2020 dataset, and each method constriction in the final training loss. Comparatively, the BERT-BiLSTM-CRF method converges the fastest, but the final loss value is also the largest; in contrast, the training loss value after convergence of the BBMSP and BLSTM-MCatt methods in this paper is smaller.

Figure 4 shows the changes in the ACC value training procedure of various methods under the test set, and constriction to a stabilized value, and the BBMSP and BLSTM-MCatt methods achieve better ACC values. Overall, the BBMSP method works slightly better.

(3) BBMSP method parameter analysis experiment

The values of blstm _ size and batch_ size of the BBMSP method were taken. In this paper, the analysis experiments of these two parameters are carried out on the CHIP2020 dataset, and the experimental results are shown in Fig. 5 and Fig. 6. The values of blstm_ size are 16, 32, 64, 128, and 256, and the values of batch _ size are 16, 32, 64, 80, and 96.

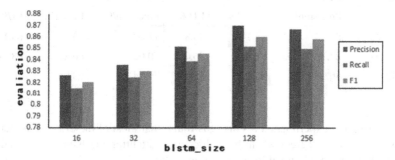

Fig. 5. Influence of parameter blstm _ size

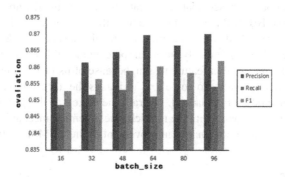

Fig. 6. Influence of parameter batch_ size

Figure 5 shows that the best results are obtained when blstm _ size is 128 from the three metrics of accuracy, recall and F1-Score value. Meanwhile, the test process of blstm _ size from 16 to 128 shows that the overall effectiveness of the method increases at the beginning when blstm _ size increases, but the effect peaks when blstm _ size is 128, and the effectiveness of the BBMSP method decreases as blstm _ size continues to grow.

Figure 6 shows that compared with the value of blstm _ size, the influence of batch _ size on the validity of BBMSP method in entity extraction is inconspicuous, and the improvement of batch_ size from 16 to 96 is less than 1% in terms of the F1 value.

(4) Ablation experiments

To verify the effects of multichannel attention mechanism and pooling-based relational location feature operations on the performance of the BBMSP method, ablation experiments are conducted on both datasets, and the comparison results are shown in Table 6, where BBMSP-ATT indicates that the multichannel attention mechanism network architecture is retained, and BBMSP-AP indicates that the pooling-based relational location feature network is retained.

Table 6. Results of ablation experiments on the two datasets

Methods	CHIP2020			2010i2b2/VA		
	Precision/%	Recall/%	F1/%	Precision/%	Recall/%	F1/%
BBMSP-ATT	87.04	84.74	85.87	71.25	65.23	68.11
BBMSP-AP	87.18	84.85	86.00	70.02	66.87	68.41
BBMSP	86.96	85.12	86.03	71.74	68.96	70.32

Comparing the experimental results, it is found that most of the evaluation metrics decrease to different degrees when the multichannel attention mechanism is not added or the pooling-based relational location feature network is not used. Among them, the pooling-based relational location feature network has a greater impact on the accuracy, indicating that the pooling-based relational location feature network improves the extraction accuracy of the BBMSP method by enriching the relationships of feature vectors; observing the data reveals that the accuracy and recall rates of the BBMSP-ATT network architecture differ significantly, so it can be concluded that after performing the pooling operation with location feature fusion, because of its increase in the individual dependencies between vectors, to some extent, makes the accuracy and recall results of the BBMSP method more balanced. Overall, the fusion of the pooling-based relational location feature network with the multichannel attention mechanism network has a significant effect on the performance improvement of the BBMSP method.

5 Conclusion

In this paper, aiming at the shortcomings of redundancy, overlap and low accuracy of traditional entity extraction methods, this paper proposed the BBMSP method. The global feature vector matrix was obtained by extracting the information. Next, the pooling-based location relationship feature approach was applied to extract the local semantic information from the electronic medical record text, and the position relation characteristics of words and the position embedding matrix of the whole sentence were obtained. Eventually, the extracted global feature vector matrix was combined with the position embedding matrix of the sentence to obtain the current entity categorization matrix. Through comparative tests, the BBMSP method realized excellent comparative results on both public datasets, and the main performance was improved compared with the current mainstream methods, which can availably address the difficulty of lost semantic message of word-level and phrase-level methods and better extract context information. Since medical terms have complicated, similar and deeper semantic information, the next work in this paper can consider combining medical knowledge terms and so on into the method to better express the semantic features of medical texts.

Acknowledgement. This work is partly supported by the General Project of Scientific Research Funds of Liaoning Provincial Department of Education under Grant Nos. LJKZ0085, and LJKMZ20220447; the Project of Public Welfare Research Fund for Science (Soft Science Research Program) of Liaoning Province under Grant No.2023JH4/10700056; and the Key Laboratory of Symbolic Computation and Knowledge Engineering of Ministry of Education, Jilin University under Grant No.93K172018K01.

References

1. Zhang, F., Qin, Q., Jiang, Y., et al.: Research on named entity recognition of Chinese electronic medical records based on RoBERTa-WWM-BiLSTM-CRF. Data Anal. Knowl. Dis. **6**(Z1), 251–262 (2022)
2. Cui, B., Jin, T., Wang, J., et al.: Overview of information extraction of free-text electronic medical records. J. Comput. Appli. **41**(4), 1055–1063 (2021)
3. Hu, Q., Liu, N., Wang, J., et al.: An overlapping sequence tagging mechanism for symptoms and details extraction on Chinese medical records. Comput. Electr. Eng. **91**, 107019 (2021)
4. Lu, Y., Yang, R., Yin, C., et al.: A military entity relation extraction method combining pre-training model and attention mechanism. J. Inform. Eng. Univ. **23**(01), 108–114 (2022)
5. Tang, B., Wang, X., Yan, J., et al.: Entity recognition in Chinese clinical text using attention-based CNN-LSTM-CRF. BMC Med. Inform. Dec. Making **19**(S3) (2019)
6. Yuan, Y., Zhou, X., Pan, S., et al.: A relation specific attention network for joint entity and relation extraction. Proceedings of the 29th International Joint Conference on Artificial Intelligence. Yokohama: IJCAI.org, 561, (2021)
7. Niu, Z., Zhong, G., Yu, H.: A review on the attention mechanism of deep learning. Neuro Comput. **452**, 48–62 (2021)
8. He, S., Wang, T., Liang, J., et al.: Entity relation extraction based on self-attention mechanism to simulate entity information. Comput. Syst. Appli. **32**(2), 364–370 (2023)
9. Zheng, L., Hong, Y., Zheng, G., et al.: Medical named entity recognition based on multi feature fusion of BERT.SU S B. In: Proceedings of 2021 4th International Conference on Big Data Technologies, pp. 86–91. Association for Computing Machinery, New York (2021)
10. Ren, H., Wang, X.: Review of attention mechanism. Comput. Appli. **6**(20), 1–7 (2021)
11. Lou, X., Xia, X., et al.: Chinese clinical entity recognition combined with multi-head self-attention mechanism and BiLSTM-CRF. J. Hunan Univ. (Nat. Sci. Ed.) **48**(04), 45–55 (2021)
12. Qi, Pei, Z., Lou, Z.: LBERT Chinese named entity recognition method with self-attention. Comput. Eng. Design **44**(02), 605–611 (2023)
13. Chen, W., Zhang, R., Yin, Z.: BERT model combined with entity vector knowledge graph entity extraction method.Microcomput. Syst. 43 (08), 1577–1582,(2022)
14. Geng, R., Wu, Y., Xiao, Q., et al.: Research on resource prediction of space-based information network based on improved GRU algorithm. J. Northeastern Univ. (Nat. Sci. Edn.) **44**(03), 305–314 (2023)
15. Gan, Z., Guan, T., Li, W., et al.: CHIP2020 Assessment task 2 overview: chinese medical text entity relationship extraction. Chin. J. Inform.. **36**(06), 101–108 (2022)

16. Veysel, K., David, T.: Accurate clinical and biomedical named entity recognition at scale. Softw. Impacts. **13**, 100373 (2022)
17. Zhang, W., Jiang, S., Zhao, S., et al.: A BERT-BiLSTM-CRF model for Chinese electronic medical records named entity recognition. In: 2019 12th International Conference on Intelligent Computation Technology and Automation (ICICTA), pp. 166–169. IEEE, Piscataway (2019)
18. Sahu, S., Anand, A.: Drug-drug interaction extraction from biomedical texts using long short-term memory network. J. Biomed. Inform. **86**, 15–24 (2018)
19. Li, D., Li, Z., Yan, L.: Research on a joint entity relation extraction method for Chinese. Microcomput. Syst. **43**(12), 2479–2486 (2022)
20. Rajd, S.S., Anand A.: Learning local and global contexts using convolutional recurrent network model for relation classification in biomedical text. In: Proceedings of the 21st Conference on Computational Natural Language Learning, pp. 311–321 Vancouver, CoNLL (2017)
21. Ning, S., Teng, F., Li, T.: Entity relation extraction of electronic medical records based on multi-channel self-attention mechanism. J. Comput. Sci. **43**(05), 916–929 (2020)

Application of Neural Networks in Early Warning Systems for Coronary Heart Disease

Yanhui Fang[1], Wei Fang[2(✉)], and Weizhen Yang[3]

[1] Baoding University of Technology, Baoding 071000, China
[2] Cardio-vascular 3rd Department, Baoding No. 1 Central Hospital, Baoding 071000, China
33956283@qq.com
[3] Department of General Surgery, Baoding No. 1 Central Hospital, Baoding 071000, China

Abstract. This paper presents a BP neural network-based algorithm for the identification of coronary heart disease through the clinical data of cardiology for many years and the personal physiological attributes easily obtained in daily life. The goal of this paper is to judge whether it may have coronary heart disease by testing the attribute values of the tester. First, through the training of samples, the network model structure is designed, and a relatively good neural network model is obtained. Second, according to the model, the possibility of coronary heart disease was calculated.

Keywords: Data mining · Neural network · Early warning system · Coronary heart disease

1 Introduction

BP is an artificial neural network with supervisory feedforward operation. It is composed of the learning process of the error algorithm, which includes the connection between the input layer, hidden layer, output layer and nodes of each layer, forward propagation of information and backwards propagation. In the process of forward propagation, input information is processed from the input layer to the output layer through the implicit layer, and each layer of neurons only affects the output of the next layer of neurons. If the expected output cannot be obtained in the output layer, backwards propagation and chain number derivation are used to return the left and right connection error functions to the original derivative along the connection path. The error function is reduced by modifying the weight of each layer.

In China, the application of neural networks in coronary heart disease warning systems is still in its early stages. Researchers collected a large amount of coronary heart disease-related data and established a coronary heart disease warning model based on neural networks by analysing data characteristics and patterns. For example, in a research article published by Xiao Yueqiong in the Journal of Electrocardiogram in 2020, the artificial neural network technology used in the study was divided into different types of coronary heart disease according to the different characteristics of the data, achieving the goal of utilizing sample data. The application of artificial neural networks in the early

Z. Yu et al. (Eds.): ICPCSEE 2023, CCIS 1880, pp. 31–39, 2023.
https://doi.org/10.1007/978-981-99-5971-6_3

screening of coronary heart disease in grassroots medical care has significant clinical significance [1].

Compared to China, there has been relatively mature research on the application of neural networks in coronary heart disease warning systems abroad. Foreign researchers have applied the established neural network models to actual medical scenarios to verify their predictive accuracy and practicality. For example, in 2019, researchers in the United States published an article in the Journal of Cardiology introducing a deep neural network-based coronary heart disease prediction model and tested it on real clinical data, demonstrating its high accuracy and practicality [2].

In summary, the application of neural networks in coronary heart disease warning systems has made certain progress, and future research will focus on improving the accuracy and efficiency of models, developing new neural network models, and combining them with other technologies to ultimately achieve the prevention and treatment of coronary heart disease.

2   Contribution

The BP neural network is a widely used artificial neural network that establishes multiple hidden layers between input and output and adjusts the weights between neurons through an error backpropagation algorithm to achieve the optimization of output results. The BP neural network has the characteristics of adaptability, nonlinearity, and parallel processing, which make it the best choice for predicting coronary heart disease due to its excellent performance in data processing [3, 4].

(1) Ability to handle nonlinear problems: Coronary heart disease is a complex, multi-factor collaborative nonlinear disease. To accurately predict the risk of this disease, it is necessary to establish a model that can handle nonlinear problems.
(2) Adaptive learning ability: BP neural networks can continuously adjust and optimize weight values through backpropagation algorithms, thereby achieving adaptive learning of different data features and improving the predictive ability of the model.
(3) High precision prediction ability: Through multiple iterations and optimizations, the BP neural network can improve accuracy in numerical calculations, enabling us to obtain more accurate coronary heart disease prediction results.
(4) Parallel processing capability: BP neural networks can use parallel processing to accelerate training and prediction speed, making them more efficient when processing massive amounts of data [4].

In summary, the BP neural network has greatly promoted its application in the field of coronary heart disease risk prediction due to its strong universality, nonlinear modelling, and accurate prediction ability, making it the best choice for coronary heart disease prediction [6].

Regarding the application advantages of BP neural networks in predicting coronary heart disease, many studies have been conducted in the academic community at home and abroad and compared with other models or methods.

For example, in a study published in Computer Engineering and Design in 2018, researchers compared BP neural networks with algorithms such as support vector machines (SVMs) and decision trees, confirming the advantages of BP neural networks in predicting the risk of coronary heart disease in patients.

In addition, in a study published in Computer Engineering and Applications in 2020, researchers used an improved BP neural network (REBP) and traditional BP neural network and compared them with multilayer perceptron (MLP) and C4.5 decision tree. The results indicate that the REBP neural network has higher accuracy and predictive ability, demonstrating its advantages in predicting coronary heart disease [7].

In summary, the BP neural network has good adaptability, nonlinear fitting, and high-precision prediction ability in coronary heart disease prediction models and has been widely applied in the medical field, and its advantages have been confirmed by comparison with other models or methods.

There are changeable risk factors and immutable risk factors for coronary heart disease. Mastering and intervening in these risk factors is helpful to prevent coronary heart disease.

The changing risk factors are high blood pressure, overweight/obesity, dyslipidemia (high triglycerides, high total cholesterol, low density lipoprotein cholesterol, high density lipoprotein cholesterol), hyperglycemia/diabetes, and unhealthy lifestyle, including irrational diet (high fat, high cholesterol, high calorie, etc.), smoking, heavy drinking, and lack of physical activity. The risk factors that cannot be changed are age, sex, and family history [8].

Through the inquiry and case record of patients, we can have a full understanding of their living habits and medical history, which can be the basis of early warning of coronary heart disease.

This paper presents a BP neural network-based algorithm for coronary heart disease identification. The goal is to judge whether coronary heart disease may be affected by the attributes of the tester. $x = \{x1, x2..., y\}$ is used to represent a training sample, where Xi represents the input attribute of the sample, for example, $x1$ for gender, $x2$ for age, etc., and y for the output attribute of the sample, i.e., whether there is coronary heart disease. The workflow is shown in Fig. 1. First, the structure of the network model is designed and trained by samples. This is an iterative process. If the effect is not good, then the network structure should be modified and retrained.(" Bad effect means taking part of data as training sample set and using part of data as test sample set. Additionally, $\{x1, x2, x3...y\}$ is used to represent a test sample, where y is called the expected output. A model is trained by the training sample set, and then the model is tested by the test sample set. The $x1$, $x2$, and $x3$ in each test sample are given to the model as input, and $y0$ and $y0$ are output as the actual output. If the number of samples in the test sample set that does not match the actual output and the expected output reaches the threshold.) The second step is to calculate whether the tester may have coronary heart disease according to the model.

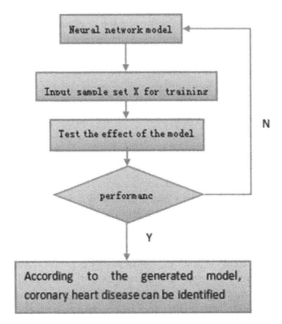

Fig. 1. Workflow of coronary heart disease identification based on a BP neural network

3 Experiments and Results

To train and test the model, 439 samples were taken from the hospital case and health files. A large amount of information was recorded in the file, some of which was related to the development of coronary heart disease. With the help of experts, this paper screened the attributes, including 13 attributes, 12 of which were independent attributes, including sex, age, smoking history, alcohol drinking history, hypertension history, height, weight, blood pressure, heart rate, hyperlipidemia history, diabetes history and family history. There is also a dependent variable attribute indicating whether the instance has coronary heart disease. A detailed description of each attribute is shown in Table 1.

The physical condition of the tested person can be described in detail by the data example in Table 1. To facilitate data mining, some data conversion work is needed to make it more applicable. Data conversion work is divided into two parts: the sample conversion of the neural network algorithm and the sample conversion of the Bayesian analysis and early warning algorithm.

For the samples of the neural network algorithm, the following conversion work is performed and numbered.

(1) For all Boolean attributes that can be expressed as "yes" or "no", including smoking history, drinking history, high blood pressure history, hyperlipidemia history, diabetes history, family history and coronary heart disease, 1 is used to indicate "yes" and 0 to indicate "no". In addition, men in gender are also represented by 1, and women are represented by 0.

Table 1. Detailed description of each attribute

Property name	Describe	Data examples
Gender	The test was conducted by gender	male
Age	he age of the subjects	56
Smoking history	Indicates whether the person has a history of smoking	yes
Drinking history	Indicates whether the person has a history of drinking	no
History of hypertension	Indicates whether the subject has or is suffering from high blood pressure	yes
height	The height of the subject	176
weight	The weight of the subjects	180
blood pressure	The test subjects blood pressure, including a low pressure and a high pressure	120/80
heart rate	The subject's heart rate per minute	66
Hyperlipidemia history	Indicates whether the subject has or is suffering from hyperlipidemia	yes
Diabetes history	Indicates whether the tester had or had diabetes	yes
Family history	It indicates whether there is coronary heart disease in the family of the tester	yes
Do you have coronary heart disease	Indicates whether the subject has coronary heart disease	no

(2) In this paper, the age of the oldest was 94, and the youngest was 22. The min max standardization method is used to normalize the test. The treatment method is shown in formula 3–1, where age_ Standard represents the result of normalization and age_ Old represents the original age.

$$age\_ standard = (age\_ old - 22)/(94 - 22) \tag{1}$$

(3) From the professional perspective, height and weight alone cannot reflect any problems, but it is of great significance to obtain BMI in combination with height and weight. The BMI calculation method is shown in formula 2.

$$BMI = weight\ (kg)/(height\ (m))2 \tag{2}$$

The body shape of the tester can be calculated by formula 3–2. In this paper, the height and weight in the original data are removed, and the body size items include too light, moderate, overweight, obesity and very fat. The data representation is represented by the value of the [0,1] interval. As shown in Table 2.

Table 2. Transformation table of the body shape of the neural network algorithm

shape	Male	female	Data representation
Too light	< 20	< 19	0.2
moderate	20–25	19–24	0.4
overweight	26–30	25–29	0.6
Obesity	30–35	29–34	0.8
Very fat	> 35	> 34	1

(4) Human blood pressure is composed of two parts: high pressure and low pressure. In the clinic, according to the value range of high pressure and low pressure, blood pressure can be divided into four levels: normal, grade 1 hypertension, grade 2 hypertension and grade 3 hypertension. In this paper, the values of high and low pressure are converted into blood pressure levels, and the data are expressed by the values of the [0,1] interval. The conversion is shown in Table 3.

Table 3. Neural network algorithm blood pressure conversion table

Blood pressure level	low pressure	relation	high pressure	Data representation
normal	< 90	AND	< 140	0.25
High voltage level	90–100	OR	140–160	0.5
High voltage secondary	100–110	OR	160–180	0.75
High voltage three stage	> 110	OR	> 180	1

(5) The normal range of heart rate is between 60 and 100, over 100 is too fast, and below 60 is too slow. In this paper, the heart rate is divided into three types: too slow, normal and too fast. The data are expressed by the value of the [0,1] interval. As shown in Table 4.

Table 4. Heart rate conversion table of the neural network algorithm

Heart rate	heart rate	Data representation
Too slow	< 60	0.33
normal	60–100	0.67
Too fast	> 110	1

Thus far, the work of data collation and conversion has been completed, and the original data have been converted into a data format acceptable to the neural network algorithm.

The input layer of the neural network consists of 11 nodes, covering 11 input attributes (the original height and weight attributes have been combined into one attribute, so the input attributes have changed from 12 to 11), which is represented by N1. The output layer has a node representing the output category, the output "1" represents illness, and "0" represents health. Generally, a single-layer neural network is sufficient to solve most of the problems, so this paper sets the number of hidden layers as one layer and sets the number of hidden layer nodes as N2. First, according to experience, set N2 as shown in formula 3.

$$N2 = K * N1, (K = 1, 2, 3, 4, 5 \ldots 9) \tag{3}$$

In this paper, cross validation is used as the test method of the model, which is widely used and generally believed to be a reliable test method. The principle of cross validation is that the training sample set is divided into m groups on average, and the M sample groups are represented by $\{s1, S2, S3\ldots\ldots sm\}$. One of the groups is taken as the test set at a time, and the other M-1 group is used as the training set. This process circulates m times and selects a different Si as the test set at a time. After M cycles, the average of the verification accuracy is taken as the final accuracy. In this experiment, the m value is set to 10.

In this paper, the mean square error (MSE) is used as the evaluation standard of the parameters. When k is different, the MSE is shown in Fig. 2.

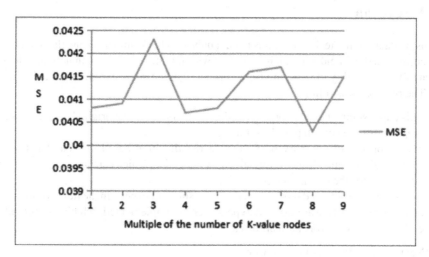

Fig. 2. Root mean square error

It can be seen from Fig. 2 that the effect is better when k is equal to 8, so at last, we choose to set the number of nodes in the hidden layer to $N2 = k * N1 = 8 * 11 = 88$.

The training result is 88 columns. Table 3–5 shows the partial parameters of the five nodes. One column represents a hidden layer node in the model, the row represents the input layer node, and the parameter is the weight of the input layer to hidden layer connection.

Table 5. Partial parameters of five nodes

	Node1	Node2	Node3	Node4	Node5
Gender	-1.338	-0.157	-0.057	2.179	-0.101
Age	-0.141	0.925	0.884	1.307	0.9015
Smoking history	1.170	0.164	0.355	-0.132	0.289
Drinking history	0.494	0.211	0.066	1.426	0.124
Hyperlipidemia history	0.45	0.266	-0.088	-2.136	0.003
height	0.509	0.021	0.282	-2.006	0.3003
weight	-0.351	0.346	0.543	2.95	0.352
blood pressure	0.348	0.125	0.031	1.7401	-0.0405
heart rate	0.162	0.622	0.640	1.326	0.581
History of hyperlipidemia	0.670	0.334	0.413	-2.830	0.366
Diabetes history	0.925	0.788	0.606	1.355	0.708

4 Discussion

By using data mining technology, this paper analyses a large amount of available data in the current medical field and obtains an early warning model, which can help people to warn of the possibility of suffering from coronary heart disease early in their daily life.

The results are as follows:

(1) This paper summarizes the process of data conversion for the original data table and prepares for the next step of data mining.
(2) The algorithm of coronary heart disease identification based on neural network technology is realized, and its effectiveness is verified, which can qualitatively judge whether it has the possibility of disease.
(3) Combining the obtained model with the suggestions of experts in the field, a coronary heart disease early warning prototype system is implemented, which can provide early warning of coronary heart disease.

Next, we can do the following work:

(1) The factors affecting the judgment of coronary heart disease are not comprehensive enough, and more attributes need to be introduced to better improve the accuracy of the judgment of coronary heart disease.
(2) The amount of data needs to be further increased, and the accuracy is higher with the training model.

Author Contributions. YanHui Fang performed the computer simulations. Wei Fang, YanHui Fang and WeiZhen Yang analysed the data. Wei Fang wrote the original draft. YanHui Fang and WeiZhen Yang revised and edited the manuscript. YanHui Fang edited the manuscript. All authors confirmed the submitted version.

Conflict of Interest. The authors declare that the research was conducted in the absence of any commercial or financial relationships that could be construed as a potential conflict of interest.

References

1. Xiao, Y.: Application of artificial neural network technology in early screening of coronary heart disease in grassroots medical care. J. Electrocardiogram **29**(6), 568–570 (2020)
2. Dey, D., Wang, A., Hung, O.Y., Slominski, M.: Development and validation of a deep neural network-based coronary heart disease prediction model. J. Cardiol. **74**(6), 537–542 (2019)
3. Wei, Y., Chen, C., Liang, M.: Using BP neural network model to predict the prevalence of coronary heart disease in China. BMC Public Health **21**(1), 1–8 (2021)
4. . Liu, Z., Li, Z. Y., Zhang, X.Q.: An improved BP neural network model for predicting coronary heart disease based on clinical risk factors. J. Healthcare Eng. **2019** (2019)
5. Chen, W., Gao, R., Liu, L., Zhu, M., Wang, W.: Prevention and treatment of coronary heart disease in China. Chin. J. Cardiovascular Dis. **47**(5), 392–399 (2019)
6. Wang, L., Guo, X., Chen, Y., Guo, Y.: Research on the application of BP neural network in coronary heart disease prediction. J. Healthcare Eng. **2020**, 1–8 (2020)
7. Yang, X., Cao, X., Chen, Y., Zhang, J.: A REBP neural network based framework for coronary heart diseases prediction. Comput. Eng. Appl. **56**(1), 162–169 (2020)
8. Hosseini, K., Mortazavi, S.H., Sadeghian, S., Ayati, A., Nalini, M., Aminorroaya, A., et al.: Prevalence and trends of coronary artery disease risk factors and their effect on age of diagnosis in patients with established coronary artery disease: Tehran Heart Center (2005–2015). BMC Cardiovasc. Disord. **21**(1), 477 (2021)

Research on Delivery Order Scheduling and Delivery Algorithms

Qian Hong and Yue Wang[✉]

School of Information, Central University of Finance and Economics, Beijing, China
yue1wang@163.com

Abstract. Online takeout has become the dining style of most urban residents. With the development of takeout industry, takeout delivery efficiency and customer satisfaction have attracted more and more attention from the industry and academia. The essence of takeout delivery problem is a vehicle routing problem under various constraints. Designing reasonable order dispatching and routing algorithms will help to improve the delivery efficiency and service quality for takeout platform. Firstly, by analyzing the order dispatching and delivery problem itself, we summarize the characteristics of the problem, compare the theoretical research of related problems, and describe it as a problem with the characteristics of soft time window, vehicle capacity limitation, dynamic order and 1–1 pick-up and delivery. Secondly, the problem scenario is described in detail, and the main elements and constraints of the problem are explained. Based on this, aiming at minimizing the overdue time, we divide the problem into two subtasks: order dispatching and routing, and propose an order dispatching strategy considering couriers' route and a two-stage routing algorithm. Specifically, this paper improves the calculation the cost in the previous heuristic algorithm based on the cost, that is, when dispatching an order, we consider whether there are orders close enough to the source location or destination location of the new order in every courier's route, and if so, he can pickup and delivery the new order incidentally. In addition, it is also necessary to punish the courier who have long routes and whose orders have large overdue time; When planning the path for each order, we firstly apply the nearest neighbor greedy algorithm for initial routing, which means iteratively find the time-consuming nearest feasible node in the remaining unplanned nodes, and then apply the tabu search algorithm to optimize the initial path. Finally, the static scenario and dynamic scenario are designed to verify the above algorithm. The static problem scenario solution verifies the effectiveness of the two-stage routing algorithm. The dynamic problem scenario proves the effectiveness of the order dispatching strategy considering couriers' route. We also analyze the ratio of the number of orders and couriers' influence on the performance of the algorithm, and recommend the quantity of couriers under the demand of orders with specific quantity. The experimental results can provide suggestions for the management of courier under the background of epidemic prevention measures such as closing-down due to the COVID-19. In addition, we propose that in a specific scenario, the order re-dispatching can make great improvement at small cost of computing resources.

This paper is supported by Engineering Research Center of State Financial Security, Ministry of Education, Central University of Finance and Economics, Beijing, 102206, China.

Keywords: takeaway delivery · order dispatching · route optimization · Greedy algorithm · Tabu search

1 Introduction

At present, the development of China's Internet shows a leap forward trend. In August 2021, China Internet Network Information Center released the 48th Statistical Report on China's Internet Development. According to the report, the scale of Internet users in China continues to expand. Online and offline consumption, physical and virtual consumption, etc. have achieved organic integration through digital consumption. Online takeout in urban life has gradually matured. The report data shows that by June 2021, the number of online takeout users in China has reached 469 million, accounting for 46.4% of the total Internet users. The business revenue and the number of members in the takeaway industry have increased significantly. There are large takeaway platforms such as Meituan and Lulema in China. The financial report of Meituan in 2021 shows that the annual revenue of Meituan is 179.128 billion yuan, compared with 114.795 billion yuan in the same period of 2020, a year-on-year increase of 56.0%. At the same time, the efficiency of the takeaway industry, the protection of the rights and interests of takeaway workers, and user experience have also received widespread attention.

In 2019, the COVID-19 epidemic swept across the country, affecting the production and life of the country, especially the catering industry. As an online sales channel for catering, online delivery services played a huge role in the national fight against the epidemic, ensuring the normal life of people. In April 2022, Shanghai will become another hardest hit area of the COVID-19 epidemic. Under strict epidemic prevention and control measures, people's livelihood will be more strained, and medical treatment, food supply and other issues need to be solved urgently. From the 15th, JD started "suicide one-way logistics" to help prevent and control the epidemic in Shanghai. It divided the couriers into several batches and followed JD Logistics into Shanghai every day in batches for material distribution services. Each batch of couriers will be isolated locally and will not return. Therefore, in today's realistic background, when similar major infectious disease events occur in cities, the problem of urban real-time distribution no longer only needs to consider economic benefits and user experience, but also needs to consider infectious disease prevention and control, such as how many distribution personnel can best meet the needs of urban users.

Based on the characteristics of the take away business, this paper studies the take away scheduling and distribution algorithm, and studies the distribution scheme under the real take away scenario, so as to provide suggestions on distribution efficiency, emergency management and other aspects for the take away industry.

The theoretical significance is as follows. This paper simulates a real take out service scenario, which is actually a variant of the Pickup and Delivery Problem (PDP) with multiple restrictions. PDP is essentially a variant of the Vehicle Routing Problem (VRP). The VRP problem is a classic NP problem in the academic world. The classical VRP problem can be solved with an accurate algorithm. With the emergence of problem variants and the increase of the problem scale, the search for an approximate solution algorithm has become a hot topic for scholars. The research content of this paper is the research of delivery scheduling and distribution algorithm, which is the algorithm design and analysis of VRP variant problem considering the actual delivery scenario, and has certain theoretical significance.

The practical significance is the following. In the current post epidemic era, when the Internet is developing rapidly and the takeout business is growing significantly, the study of the takeout order scheduling and the path planning of the takeout staff is of great practical significance to the efficiency of the takeout platform, user experience, the allocation of logistics resources, and the prevention and control of the epidemic.

2 Related Work

Since the development of the takeout industry, there are two main order distribution modes: the order grabbing mode and the order dispatching mode. Na Deng et al. [1] analyzed and summarized the characteristics of these two order distribution modes: the order grabbing mode is that after the user places an order, the takeout platform pushes the task of new orders to all the takeout workers through the distribution APP installed by the takeout workers, who decide whether to grab orders according to their own conditions; Dispatch mode means that the platform uses the order scheduler to allocate orders and notify the corresponding delivery clerk according to the mobile APP. Under the order grabbing mode, because the individual delivery staff cannot grasp the overall information, there will be too many orders for some delivery staff, too few orders for others, and orders will be snatched by the delivery staff farther away, resulting in low overall delivery efficiency; With the continuous development of the takeout industry, most platforms have turned to the order distribution mode, which can consider the overall factors such as the distance of the takeout staff, whether they are on the way, and the order holding situation, so as to improve the overall effect. Delivery staff need to pick up meals from businesses scattered in various regions of the city and deliver them to customers, with a relatively scattered starting point and end point; The maximum number of takeouts delivered by each delivery clerk at the same time shall not exceed about 20; Every morning, lunch and dinner is the peak time for takeout, and the delivery time of takeout is minutes.

1) **Order scheduling algorithms**

The large-scale order scheduling problem itself is a NP hard problem, and it is difficult to find the optimal solution with an accurate algorithm in a limited time. Most scholars choose graph theory, heuristic algorithm, neural network and other methods to help find an approximate solution. In the study of urban carpooling business, Ma et al. [2] proposed a large-scale urban taxi sharing system that accepts real-time taxi requests and matches appropriate taxis through time, load and other restrictions. The system divides roads into grids, and maintains three lists for each grid: the grid list with the least time to the grid, the grid list with the shortest distance, and a sequential list of taxis that arrive at the grid in time. When searching for a taxi for a real-time taxi request, search for idle taxis that meet the time constraints around the grid of the boarding and alighting points at the same time, and select the taxi distribution order with the least distance increase. Asghari et al. [3] used the auction based car sharing system.

Because there is a detour for some orders during car sharing, which has an impact on the interests of individual passengers and drivers, in this paper, the system considers the bidding preference of passengers, the matching degree of routes between passengers and drivers, etc. to allocate orders. Lin et al. [4] revised the current route planning by analyzing the historical passenger demand and estimating the possible future demand and profitability, and the results showed that it could speed up the response speed of the system. Jiang et al. [5] analyzed the background of China's Spring Festival travel, and some self driving people returned home with empty vehicles. Therefore, they proposed a national Spring Festival travel car pooling system based on online greedy algorithm, and completed the interface design of the web side and mobile side. The algorithm considers the starting point deviation, end point deviation, detour distance and departure date deviation between passengers and self driving people as matching weights, Greedily select the passenger with the highest matching degree and self driver for final matching each time. Wang et al. [6] proposed a scheduling method based on multiple prediction scenarios to optimize online real-time distribution order scheduling for the problem of one to many pickup and delivery between merchants and customers. Compared with forecasting a single scenario, this scheme predicts multiple distribution scenarios and integrates them. During the implementation process, it continues to optimize the tasks that have not yet been executed by using large neighborhood search. Many scholars divide the order distribution problem of large scale into small scale problems according to geographical location for solution. For example, Chen et al. [7] use clustering method to divide orders according to distribution destination, configure corresponding number of vehicles, and then carry out specific route planning. In the study of delivery business, Zhang et al. [8] used machine learning algorithms to mine historical order delivery data. This paper believes that the delivery staff can independently plan the path. This paper reveals the impact of perceived distance (different from the actual distance) on the specific decisions of the delivery staff, and conducts order scheduling based on the predicted path. In addition, many scholars [9–13] have established a variety of reinforcement learning models to solve the order scheduling problem of takeout or carpooling business.

In this paper, the above order scheduling algorithms are divided into cost based heuristic algorithm and learning based scheduling algorithm. The cost of the former refers to the cost related to time, path and cost generated after the match between the distributor and the order. When scheduling, select the matching corresponding to the minimum cost, and set the corresponding cost calculation method for different problems. This paper will improve the cost calculation method to adapt to this problem; The latter is a method to optimize the scheduling strategy by using data mining, machine learning, reinforcement learning, etc. It depends on the learning of a large number of historical data, and is difficult to apply directly.

2) Path planning algorithms

As early as 1959, Dantzig et al. [14] proposed the VRP (Vehicle Routing Problem), which has been widely studied and concerned by scholars at home and abroad for a long time. Scholars have conducted in-depth research on this problem and a series of variants with different methods. The simple VRP problem includes a certain number of customer nodes with different needs. The warehouse, as the distribution center, has goods that can meet customer needs. In order to meet the needs of customer nodes and other constraints, the fleet loading goods needs to plan the optimal driving route to minimize the cost of the planned route of the fleet, that is, the total mileage. With the development of the actual needs of society and the deepening of research, the VRP problem gradually evolves and extends to the actual social life, with many branch variations.

An important variant of VRP is the PDP (Pick and Delivery Problem). The classic PDP problem includes a certain number of orders. When a team completes an order, it needs to pick up the goods from the picking point, and then deliver them to the delivery point. The goal of the problem is to minimize the total cost, such as the total distance and total cost. Battarra et al. [15] summarized the classic PDPs into three main categories according to the type of order demand, namely, many to many (M-M), one to many to one (1-M-1), and one to one (1–1). In the M-M type PDP problem, the goods of each order may have multiple sources and destinations, and any location may be the source or destination of multiple goods. The corresponding application scenario is the transportation management of manufacturing parts; In the 1-M-1 type of PDP problem, the 1-M process corresponds to that some goods need to be transported from the distribution center to multiple customer nodes, and the M-1 process corresponds to that some goods need to be transported from multiple customer nodes to a destination. The corresponding application scenario is the distribution and recycling management of reusable goods (such as reusable glass drink bottles); Finally, in the 1–1 type PDP problem, each order has a specific starting point and destination, and the corresponding application scenario is urban real-time delivery services. By adding constraints to the PDP problem, multiple variant problems are also generated, such as pickup and delivery problem with time window [16], pickup and delivery problem with capacity limit [17], multi warehouse pickup and delivery problem [18], random demand pickup and delivery problem [19], etc.

PDP is also a NP hard problem, which can not give the optimal solution within the time range of polynomial. There are three kinds of algorithms to solve PDP problems: precision, traditional heuristic and meta heuristic algorithms.

Exact algorithm refers to the algorithm that can solve the optimal solution, including cutting plane method, branch and bound method, integer programming algorithm and dynamic programming algorithm [20]. The exact algorithm needs to find the optimal solution in limited time and steps, so it is generally used to solve small and medium-sized problems. The development of accurate algorithms for PDP problem largely benefits from the study of the Traveling Salesman Problem (TSP) with priority constraints. One node may have multiple predecessor nodes. Kalantari et al.[21] used the branch and bound method to solve the traveling salesman problem with pickup and delivery customers, which requires the pickup customer to visit before the relevant delivery customer. Desrosiers et al. [22] solved the single vehicle call problem with time window constraints by using the dynamic programming algorithm. This problem includes not only the boarding point, alighting point, capacity constraints, but also the predecessor node constraints of nodes. However, as the problem scale increases, the above algorithms cannot be solved in the effective time.

Unlike the exact algorithm, the traditional heuristic algorithm does not pursue the optimal solution, but constructs the algorithm based on experience, optimizes the goal as far as possible, reduces the cost, and gives a feasible solution to the problem to be solved, that is, to obtain an approximate optimal solution. The actual problem often does not need to obtain an optimal solution, but to obtain an approximate optimal solution under acceptable conditions. Many traditional heuristic algorithms have emerged in previous studies to solve classic PDP problems. Gajpal et al. [23] proposed a parallel saving heuristic algorithm. When merging two paths with the saving algorithm, it is necessary to check whether the vehicles will be overloaded after merging the paths. This paper proposed a cumulative load increase method to quickly judge the path feasibility. Jun et al. [24] defined a heuristic algorithm, which uses scanning based method to build a better initial path, uses node exchange within and between routes to optimize the path, and finally uses perturbation method to jump out of the local optimal solution, that is, delete a node and then find a new location to insert to obtain a better solution. Zheng et al. [25] proposed a two-stage algorithm based on clustering to solve the optimization problem of delivery path for takeout. The algorithm first sorts the order pickup points according to the remaining delivery time, and then speeds up the initial path planning according to the geographical location clustering. Finally, it makes fine adjustments according to the overdue time. The initial cluster accelerated planning of the algorithm can obtain a solution similar to the brute force algorithm.

Meta heuristic algorithm, also known as intelligent optimization algorithm, can optimize the feasible solution of a problem by changing a feasible solution locally or globally at an acceptable cost of time and space. Meta heuristic algorithm can not optimize the goal every iteration, but can accept bad solutions with a certain probability, avoid falling into local optimal problems, and explore the global optimal solution. Meta heuristic algorithms include genetic algorithm, simulated annealing algorithm, tabu search algorithm, ant colony algorithm, particle swarm optimization algorithm, etc. [26]. Meta heuristic algorithm is widely used in classical DPD problems. Chen et al. [27] proposed a record to record algorithm to optimize the path, analogy to the idea of tabu search. The algorithm relaxes the optimal solution, and if the candidate solution is the optimal solution after relaxation, it will be accepted. Hosny et al. [28] studied the PDP problem with hard time windows by using simulation data. When generating simulation data, they first generated a problem with a feasible solution to ensure that there is at least one feasible solution. Later, they added constraints such as time windows, and solved the problem by using genetic algorithm, simulated annealing algorithm, and mountain climbing heuristic algorithm. Tasan et al. [29] used genetic algorithms to solve the problem of simultaneous pickup and delivery requirements at the same node, and designed a numerical example to evaluate the performance. Demir et al. [30] used adaptive neighborhood search algorithms to solve the vehicle minimum pollution problem. This problem is a combination of vehicle routing problems with time windows and low-carbon goals, requiring the completion of distribution tasks while reducing carbon emissions and other costs. Su et al. [31] used the greedy algorithm to generate the initial solution and then improved the tabu search algorithm in the multi vehicle routing problem with a single starting point and multiple demand points with a time window. The solution of an example verifies that the proposed scheme can effectively solve the problem.

The delivery scheduling and distribution problem in this paper is a 1–1 type PDP with the characteristics of soft time windows, capacity constraints, and dynamic order generation. There are few studies with high degree of similarity to the problem constraints in this paper. In the work with roughly similar problem scenarios, most studies separate the research on order scheduling and path optimization. In this paper, the problem is divided into two tasks: order scheduling and path planning. Drawing on the heuristic scheduling algorithm of previous scholars to maintain the cost matrix, we improve the cost calculation method, and propose a combined order scheduling strategy that considers the route of the delivery clerk, while taking the route length and overdue situation as punishment; Combined with previous greedy and meta heuristic algorithms, this paper proposes a two-stage path planning algorithm; The two can effectively integrate and promote, and improve the distribution efficiency as a whole.

3 Problem Statement

The data used in this paper is the simulated data after desensitization of the real business scenario data of the *ele* delivery platform. In the Elime delivery scenario, the order scheduling platform will allocate the user's delivery orders to the delivery clerk every moment. The delivery clerk may also be distributing the historical orders allocated before the current scheduling time when receiving the new orders allocated by the scheduling platform. To complete a delivery order, the delivery clerk needs to arrive at the store, obtain food, and deliver the order to the user. Each delivery order has a promised delivery time estimated by the platform to the user. The delivery clerk should try to complete the order before the promised delivery time. Since a delivery clerk will receive new orders at any time, he/she also needs to adjust his/her delivery route in real time at each scheduling moment.

The process of delivery order scheduling and distribution problem can be described as follows: within each business district, the scheduling time is t_k, and the issuer pushes the newly launched orders and delivery staff to the order pool and delivery staff pool within the $[t_{k-1}, t_k]$ time period. At time t_k, there are O_k orders to be allocated, and there are C_k delivery staff who can provide services. The scheduler needs to allocate the O_k new orders to C_k delivery staff under the condition that the problem constraints are met, At the same time, the delivery routes of new orders and historical orders are re planned for the delivery staff, with the goal of minimizing the average overdue time of orders. The same process is carried out at the next scheduling time t_{k+1} until all orders in the business district are delivered.

The delivery order scheduling and distribution problem studied in this paper consists of several elements, including the issuer, scheduler, delivery order information, delivery clerk information, current planning information, etc.:

- *Order Issuer*: The order issuer pushes [tk-1, tk] newly launched orders and delivery staff into the order pool and delivery staff pool at each scheduling time.
- *Dispatcher*: According to the current order pool and the updated information of the vendor pool, the dispatcher allocates the new order to a vendor according to a certain allocation strategy, re-plans and his path.
- *Order*: Each take away order includes business circle ID, order ID, status, price, merchant location, user location, order placement time, estimated meal delivery time,

and promised delivery time. In the following algorithm description, the starting point of the order is the merchant location, and the end point of the order is the user location.

- *Delivery staff:* Each deliveryman includes business circle ID, deliveryman ID, current location, driving speed, maximum quantity of back orders, and currently planned deliveryman path.
- *Action node:* Each action node describes the action information to execute an order, including order ID, action type, whether it has been submitted, action start time, action completion time, and action location. Among them, there are three types of behaviors: arriving at the store, picking up meals, and delivering meals.
- *Plan routes:* The path of each delivery clerk consists of several orderly action nodes. If each delivery clerk has been assigned a new order, the delivery clerk path needs to be updated.
- *Current planning information solution:* The current planning information is in the path of all the takeout workers, and tk needs to submit the list of behavior nodes completed between $[t_{k-1}, t_k]$ at any time. Planning information cannot be modified after submission, but actions completed after tk do not need to be submitted.

4 Algorithms

4.1 Design of Delivery Scheduling Algorithms

Based on the idea of maintenance cost matrix, this paper uses the heuristic algorithm of previous maintenance cost matrix for reference, proposes a single order allocation strategy considering the route of the delivery clerk, and improves the cost calculation method. The following compares the algorithm design of the two order scheduling strategies.

A. *New order starting time consuming greedy allocation strategy*

At the scheduling time t_k, there are currently O_k new orders that need to be allocated to C_k delivery clerks. The simplest algorithm is to allocate the orders to the nearest delivery clerks at the current time. The specific process is: t_k, calculate the time required from the starting point of O_k new orders to the current position of C_k delivery clerks, maintain a new order starting time consuming matrix, select the order that takes the least time in the current matrix - delivery clerks matching each time, check whether the delivery clerks will exceed the maximum quantity of back orders after the new order is allocated, if not, perform the allocation, set the value related to the allocated new orders in the matrix to infinity, then continue to search for the order that takes the least time in the current matrix - the delivery clerk to match until all new orders are allocated.

B. *Considering the order routes combining strategy of the delivery staff*

Because at the time of dispatch t_k, each delivery clerk may be delivering other orders. With the execution of the delivery task, the delivery clerk closest to the starting point of the new order at the time of dispatch t_k may be far away from the new order over time; In addition, the end point of the new order is far away from all the route nodes of the delivery clerk, so the order combining strategy based on the original route of the delivery clerk is proposed. This algorithm takes into account the situation that the delivery clerk is delivering orders that are close to the start point or end point of the new order. If there is an order in the original route of a delivery clerk whose start point or end point is close enough to the new order, then "behavior combination" can be achieved. The delivery staff can achieve "incidental" delivery and pick up. Compared with other delivery staff, they can pay less to complete this new order.

The pseudocode of this algorithm is shown in Fig. 1. The first line of pseudocode indicates that the strategy maintains a cost matrix CostTable, and the fifth to eighth lines indicate that the cost cost of matching each delivery clerk with each new order in the current delivery clerk pool is calculated on the basis of considering the existing route of the delivery clerk, where minTimetoSrc calculates the shortest time between the starting point of the existing order of the delivery clerk and the starting point of the new order MinTimetoDst calculates the shortest time between the end point of the existing order and the end point of the new order for the delivery clerk. If the maximum quantity of back orders will be exceeded after the delivery clerk is assigned the current new order, set the cost to None. Lines 11–13 indicate that when making a final decision, always select the least cost best in the matrix_ The order corresponding to cost is matched with the delivery clerk to perform the actual allocation. After each allocation, set the cost related to the allocated order to infinity, update the corresponding delivery clerk path, and update the relevant elements of the delivery clerk in the cost matrix 错误!未找到引用源。.

In the above algorithm, if we only consider whether there are orders with the same starting point or the same destination in the original route of the delivery clerk, then there will be such a situation: there are many orders with the same starting point or the same destination that meet the requirements in the original route of the delivery clerk, then we may fall into the cycle of always distributing new orders to the delivery clerk, resulting in the route being too long, and eventually leading to the overdue of most orders, Therefore, it is necessary to prevent new orders from being assigned to delivery clerks whose routes are too long or whose current overdue time is too long.

4.2   Design of Route Planning Algorithms

A. *The principle of the nearest neighbor greedy algorithm*

The nearest neighbor greedy algorithm is a basic greedy heuristic algorithm to solve the shortest path problem. The running process of the algorithm is: the shortest path is an ordered list of nodes connected sequentially; First, add the starting node as the current node to the shortest path list, and then select the node with the shortest distance from the current node from the nodes that have not passed through each iteration, and add it to the end of the shortest path until all nodes are added to the shortest path list.

B. *The principle of Tabu search*

Tabu Search (TS) is a meta heuristic algorithm with good results, and its essence is a random search algorithm. The tabu search algorithm maintains a tabu table that is initially empty. The corresponding solution of each object in the tabu table will be tabbed for a period of time, making the search direction jump out of the local optimal solution, avoiding repeated searches for the local optimal solution, and making it more likely to search for the global optimal solution. However, for some specific cases, such

---

Algorithm 1

Input: new order O_k, delivery staff pool C_k at dispatch time t_k

1   CostTable←[]

2   for o in O_k:

3   costRow←[]

4   for c in C_k:

5   minTimetoSrc ←time from the closest pickup node in c.planRoutes to o.srcLoc

6   minTimetoDst ←time from the closest delivery node in c.planRoutes to o.dstLoc

7   cost←calculatePunishment(minTimetoSrc, minTimetoDst, c.planRoutes)

8   cost←None if exceed the maxLoad

9   costRow+=[cost]

10   CostTable+=[costRow]

11 while CostTable has minimum best_cost < ∞:

12   dealBestCost(best_cost)

13   update CostTable

Fig. 1. The algorithm considering the order routes combining strategy of the delivery staff

as when it is better than any historical optimal solution, the original taboo object can be unblocked if it meets the amnesty criteria. The algorithm first generates an initial feasible solution as the current solution, generates a series of neighborhood feasible solutions transformed from the current solution through the element exchange operator, selects the best solution from the neighborhood feasible solutions to update the current solution, and adds the object corresponding to the best solution to the end of the tabu list. If the tabu list exceeds the predefined length, the object that entered the tabu list earliest will be unblocked. At present, tabu search algorithm is widely used in path planning.

C. *The design of a two stage path planning algorithm*

The path planning in this paper uses a two-stage path planning algorithm, that is, the nearest neighbor greedy algorithm is used for initial path planning, and the tabu search algorithm is used for optimization based on the initial feasible solution.

The concrete application of the initial planning of the nearest neighbor greedy algorithm is as follows. Because this problem combines with specific business scenarios, there are many constraints, so these constraints need to be considered when planning the path. Before a delivery clerk accepts a new order, his delivery path is PlanRoutes. After accepting a new order, the dispatcher initializes three ActionNodes for the new order. The properties of the three new ActionNodes are respectively set as: new order ID, action type (arrival, picking, delivery), non delivery, 0, 0, and the starting point of the new order. The action node of the new order is inserted into PlanRoutes for the next greedy path planning of the nearest neighbor.

The specific application of tabu search algorithm optimization is as follows. When optimizing the path based on the tabu search algorithm, the constraints brought by specific business scenarios should also be considered. Some details of the algorithm need to be modified in application to adapt to this problem. Next, the key elements of the tabu search method are set in this paper, including the initial solution, neighborhood solution, tabu table, evaluation function of the solution, convergence criteria, amnesty criteria, etc.:

- *Initial solution:* When using the tabu search algorithm to optimize, use a feasible solution generated by the greedy algorithm of the nearest neighbor as the initial solution.
- *Neighborhood solution:* the neighborhood solution in this paper is generated according to the local optimal solution of each iteration. The generation of neighborhood solution is usually generated through the exchange of some action nodes of the local optimal solution by the search operator, which usually includes relocation operator, exchange operator, etc. This paper uses the exchange operator to generate the neighborhood solution of the local optimal solution, as shown in Fig. 5. Given the local optimal solution, randomly select two action nodes to exchange, and check whether the path is legal. If it is legal, it can be added to the neighborhood solution set, and the two nodes exchanged are the corresponding prohibited objects.
- *Evaluation function of a solution:* The objective of this paper is to minimize the overdue situation of all orders. The objective function is to minimize the average overdue time of orders. The path planning of each vendor is a solution. The average overdue time of orders involved in this path is calculated as the evaluation function of the solution. The overdue situation of the overall order is optimized by optimizing the overdue situation of orders involved in each path 错误!未找到引用源。
- *Tabu list:* Tabu list records the objects corresponding to the local optimal solution that is currently tabooed. Tabu list is an ordered list with a fixed length. The objects corresponding to the local optimal solution generated by each iteration will enter the tabu list to avoid repeated searching for the current local optimal solution. The length of the tabu list increases continuously during iteration. When the tabu list exceeds the fixed length, the objects that are tabooed first will be unblocked first, It can be searched later. The objects in the tabu list can be all the information of the solution, or the specific exchange method generated by the search operator, or the value of the solution evaluation function. In this paper, the exchange method is regarded as the taboo object, and the exchange method of the optimal neighborhood solution generated by each iteration is put into the taboo table.
- *Convergence criteria:* If no convergence criteria are set for tabu search, the search will be updated all the time. The convergence criteria in this paper is to limit the maximum number of iterations of the algorithm. Because this paper simulates a dynamic process, with more order accumulation in the later stage, the delivery clerk's path is longer, but there are fewer online orders in the earlier stage, and the delivery clerk's path is shorter, so when the tabu search algorithm is called to update the path in the earlier stage, if it is found that multiple searches cannot produce a feasible neighborhood solution,

for example, when using a random exchange operator to generate a neighborhood solution, if it cannot produce a feasible neighborhood solution after more than 10000 attempts, it will exit the search in a timely manner.

- *Amnesty criteria:* There are two cases of amnesty in this paper: the first case is that if the local optimal solution generated in an iteration is superior to the historical optimal solution and the object is in the tabu list, the object is amnesty, the historical optimal solution is updated, and then the object is added to the tail of the tabu list again; The second case is that if all feasible neighborhood solutions are taboo due to the lack of feasible neighborhood solutions in an iteration, the local optimal solution will be pardoned and its object will be added to the tail of the tabu list.

The pseudo code applied by the tabu search algorithm is shown in Fig. 2. The first line of pseudo code indicates that the algorithm maintains a tabu list; Lines 2–3 regard

---

Algorithm 2

Input: the delivery clerk initially plans the route planRoutes; Tabu table length tabu_ Len; Neighborhood solution length near_ Len; Neighborhood solution generating function $N(x)=\{xk, k=1,..., near\_len\}$; The evaluation function of solution x is f (x); The taboo object of solving x is obj (x); The maximum number of iterations is iteration_ num.

```
1  tabulist←[]
2  x_best_local←planRoutes
3  x_best_global←x_best_local
4  while i<iteration_num:
5      nearx←N(x_best_local)
6      neary←[f(x) for x in nearx]
7      xk←find_best(nearx,neary)
8      if f(xk)<f(x_best_global):
9          x_best_local←xk
10         x_best_global←xk
11         updatetabulist(tabu_len,tabulist,obj(xk))
12     else:
13         while obj(xk) not in tabulist:
14             remove(nearx,neary,xk)
15             xk←find_best(nearx,neary)
16         x_best_local←xk
17         updatetabulist(tabu_len,tabulist,obj(xk))
18 return x_best_global
```

Fig. 2. The application of Tabu search in the routing planning algorithm

the initial feasible solution planRoutes as the current local optimal solution x_ best_local and global optimal solution x_best_global, line 4 represents the convergence criteria; Lines 5–7 represent the neighborhood solution that produces the current local optimal solution, and use the evaluation function of the solution to find the optimal neighborhood solution xk; Line 8 judges whether the optimal neighborhood solution meets the amnesty criteria, that is, whether it is superior to the historical global optimal solution; Lines 9–11 indicate that if the amnesty criteria are met, the optimal neighborhood solution will be taken as the current local and global optimal solution, and the tabu list will be updated; Lines 13–17 indicate that if the amnesty criterion is not met, the solution that is not tabooed in the neighborhood solution will be found as the local optimal solution, and the taboo table will be updated; Line 18 returns the global optimal solution as the optimized delivery path 错误!未找到引用源。

5   Experiments

As the largest city in China, Shanghai has a high level of economy and Internet popularity, a high level of residents' consumption, a high consumption concept, a mature takeout industry, and a large demand for takeout. Therefore, for the universality of the algorithm, the data used in this paper is selected from the business data of Shanghai ele delivery platform in December 2019. The order time span is about 4 h, with more than 1000 orders and more than 50 delivery staff every day 错误!未找到引用源。

Select the order of a business district for a day for preliminary data description: the distribution of the delivery clerk and the order location is shown in Fig. 3. The blue and green dots represent the starting and ending points of the order respectively, and the red dot represents the position of the delivery clerk at the initial time. It can be seen from

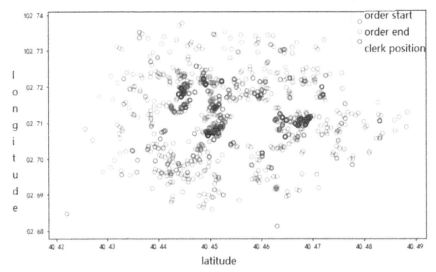

错误!未找到引用源。 错误!未找到引用源。

Fig. 3. Distribution of sellers and order positions inside and outside a business district in a day

the figure that in reality, the distribution of restaurants and beverage stores in a business district is relatively centralized, the starting points of the order are relatively centralized, and the end points of the order are relatively scattered. The order generation frequency is shown in Fig. 4. The data records the orders generated in about 4 h from 8:00 a.m. to 12:00 p.m., which shows that there are obvious peaks and peaks.

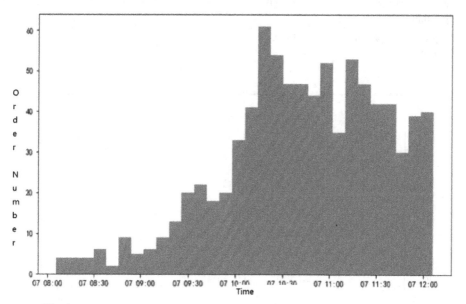

Fig. 4. Time distribution of order generation in a business district within one day

5.1 Static Problem Solving and Analysis

To explore the effectiveness of the above two-stage path planning algorithm design, this paper constructs a static problem by generating simulation data.

In this static problem experiment, the following path planning strategies are to be compared: first come first service, nearest neighbor greedy algorithm, two-stage path planning algorithm. The unit of scheduling is 60 s of fixed time step.

The results of the static experiment path planning algorithm are shown in Table 1 and Fig. 5. The average overdue time of the first come first serve method is 3136.6 s, and the average service time is 5003.4 s. The average overdue time of the nearest neighbor greedy algorithm is 2105.0 s, and the average service time is 3834.6 s. The iterative process of tabu search algorithm in two-stage path planning is shown in Fig. 10. After 5000 iterations, the average overdue time is reduced to 1829.8 s, which is 42% higher than the first come first service method, and 13% higher than the greedy initial planning route of the nearest neighbor; The average service time is 3682.8 s, which is 26% higher than the first come first serve, and 4% higher than the greedy initial planned route of the nearest neighbor. It can be seen that the performance of two-stage path planning algorithm has been greatly improved 错误!未找到引用源。.

Table 1. Comparison of running results of various algorithm settings in static experiment 1

Algorithm settings	Average overdue time (seconds)	Average service time (seconds)
First come, first served	3136.6	5003.4
Nearest neighbor greedy algorithm	2105.0	3834.6
Two-stage path planning algorithm	1829.8	3682.8

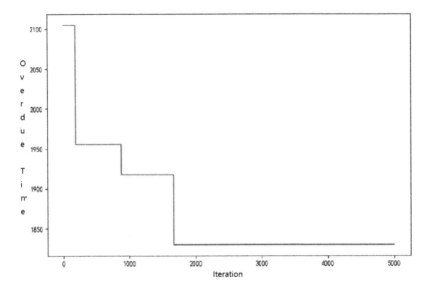

Fig. 5. Iteration process of static experiment tabu search method

5.2 Dynamic Problem Solving and Analysis

In order to explore the dynamic delivery order scheduling and distribution problem, this paper designs a dynamic problem scenario to verify the performance of this order scheduling algorithm.

In the dynamic problem scenario data, all the delivery staff are online at the initial time, while the order is dynamically online at the actual order placement time. The data used in the experiment samples 100 days of orders and delivery staff from the real data of a business district. Each day includes 20 delivery staff and 100 dynamic orders. The experimental results take the average of 100 days.

In the dynamic experiment, scheduling is carried out once in a fixed time step, that is, the time interval between the scheduling time tk-1 and the next scheduling time tk is a fixed time step. In this dynamic experiment, the time step is set to 60 s. The path planning algorithm uses a two-stage path planning algorithm.

The experiment compares three order scheduling algorithms, namely, the time-consuming greedy allocation strategy for the starting point of a new order, the order combining strategy considering the route of the delivery clerk without penalty, and the order combining strategy considering the route of the delivery clerk and taking the route length and overdue situation as punishment; The path planning algorithm uses a two-stage path planning algorithm. The experimental results are shown in Table 7. The performance of the first two scheduling strategies is poor. After taking into account the length of the route and the penalties for overdue situations, the average number of overdue orders is 15.8, the average overdue time is only 85.8 s, and the average service time is 1502.4 s. Each index is 82%, 98%, and 76% higher than that of the strategy without penalty, and the performance is greatly improved.

Table 2. Comparison of running results of various algorithm settings in dynamic experiment 2

Algorithm settings	Average overdue quantity (order)	Average overdue time (seconds)	Average service time (seconds)
Greedy allocation strategy for new order starting time	89.8(±3.1) 89.8 (± 3.1)	4569.0(±753.4) 4569.0 (± 753.4)	6336.2(±772.7) 6336.2 (± 772.7)
Non punitive order combining strategy considering the route of delivery staff	89.6(±3.2) 89.6 (± 3.2)	4467.6(±726.5) 4467.6 (± 726.5)	6233.0(±746.2) 6233.0 (± 746.2)
Ordering strategy with penalty and considering the route of delivery staff	15.8(±5.6) 15.8 (± 5.6)	85.8(±49.8) 85.8 (± 49.8)	1502.4(±100.4) 1502.4(± 100.4)

5.3 Influence of Super Parameters in Problem Scenarios

In the above static and dynamic problem solving, in order to analyze the performance of the scheduling algorithm and path planning algorithm, the super parameters of the problem are fixed. In this paper, factors such as the proportion of orders to the number of delivery staff will have an impact on the order delivery. This section analyzes the impact of the problem scenario super parameters on the algorithm performance.

A. *The proportion between the order and the number of delivery staff*

This section analyzes the impact of fixed order quantity on delivery efficiency with the increase of the number of delivery staff. The experiment in this section samples several days of simulated data from the original data of the real data of a business district, where the total number of orders per day is fixed, while the number of outsourcers changes. The experiment was divided into two groups. One group had 300 orders in total, simulating the situation of less distribution pressure in a day. The number of delivery staff increased from 5 to 65 in steps of 5; The total number of orders in the other group is 1000, and the number of delivery staff is increased from 5 to 115 in steps of 5.

The order scheduling in this section uses the order combining strategy that considers the route of the delivery clerk and takes the route length and overdue situation as punishment; The path planning algorithm is a two-stage path planning algorithm. The experimental results are shown in Fig. 6. The green, orange and red broken lines respectively represent the changes of the average overdue quantity, the average overdue time, and the average service time with the increase of the number of delivery staff. The solid line represents the total order quantity of 300, and the dotted line represents the total order quantity of 1000. According to the experimental results, when the total number of orders is 300 and the number of delivery staff reaches about 25, the average overdue time and average service time decrease slowly, and the average overdue number gradually decreases; When the total number of orders is 1000, and the number of delivery staff reaches about 45, the optimization of the average time related indicators is slow, and the average number of overdue still shows a significant downward trend. If the goal of the platform is only to minimize the time related indicators, then after the number of delivery clerks reaches the corresponding number in the above two cases, increasing the number of delivery clerks has no obvious effect on optimizing the time indicators, and there is no need to add more delivery clerks.

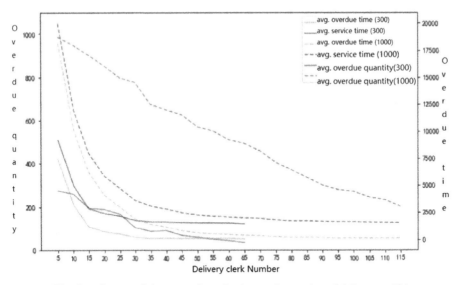

Fig. 6. Influence of the proportion of orders to the number of delivery staff 1

The above experiment can be applied under the influence of prevention and control measures such as closing the city during the epidemic period. In order to ensure the immediate distribution management of materials for the basic living needs of residents, the number of distribution personnel can be reasonably arranged according to the demand of each business district in the city, which can meet the living needs of residents to the greatest extent in a special period while reducing the flow of personnel, so as to reduce the possibility of epidemic transmission.

B. *Order reallocation*

Order reallocation is to reallocate the orders that will be overdue but have not yet started to be executed according to the current route planning. The specific method is to find out the orders that have not been executed by the pickup node in the path planning of each delivery clerk before the order allocation at the current scheduling time tk, and judge whether the action time of the delivery node is less than the expected delivery time (whether it will be overdue) And whether it is not less than the current dispatching time (whether it is executed after the current dispatching time). If it meets the requirements, it means that the order is about to be overdue but not executed, and can also be "saved". Then reset the order status, wait for redistribution, and update the route of the delivery clerk. Since this paper considers the delivery platform and the delivery staff as a whole, the discussion of order redistribution in this paper does not consider the rights and interests of the delivery staff.

This section explores the impact of order redistribution on the performance of the algorithm. Order scheduling is a combination strategy that considers the route of the delivery clerk and takes the route length and overdue situation as punishment; The path planning algorithm is a two-stage path planning algorithm. The experimental data is described as follows: the simulation data is extracted from the original data of a business district. The experimental data is divided into 8 groups, each group contains 50 days of data. The number of takeout workers and the number of orders in the same group are fixed for each day of data. Each group calculates the average result of 50 days. The data of the first group of 50 days includes 200 orders and 5 delivery agents, the second group includes 200 orders and 10 delivery agents, and so on. The eighth group includes 200 orders and 40 delivery agents. The proportion of orders to delivery agents gradually decreases, and the distribution pressure decreases.

The experimental results are shown in Fig. 7. Each point in the figure represents the percentage of the same data group when the order redistribution ratio is used but the improvement is not used. For example, the coordinate of the first point of the average overdue time is (40, 2.1%), which means that the average overdue time is 69.10 s after the order redistribution algorithm is improved on the group of 40 delivery staff, and the average overdue time is 67.67 s when the order redistribution algorithm is not used, then the percentage of improvement is (69.10–67.67)/67.67, about 2.1%.Fig. 7 Improvement effect of order reallocation 2.

The results show that with the decrease of the number of delivery staff, the distribution pressure gradually increases, and the improvement effect of order redistribution first increases and then decreases. When the number of delivery staff is large, the distribution pressure is small, and the algorithm performs well without improvement; When the number of outsourcers is small, the distribution pressure is high, and even if the order reallocation is used, the improvement effect is small. Therefore, if the distribution pressure is too large or too small, there is no need to waste computing resources for reallocation. When the proportion of orders to the number of delivery staff is between 8:1 and 20:1, the average overdue time and the number of overdue staff can be improved

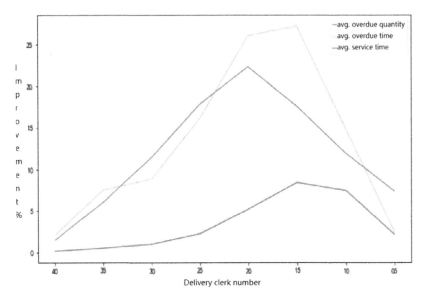

Fig. 7. Improvement effect of order reallocation 2

by more than 10%. The delivery platform can consider order redistribution to improve the overall delivery efficiency.

6   Conclusion and Future Work

Based on the existing literature, this paper analyzes the characteristics of the delivery order scheduling and distribution problem, and concludes the delivery order scheduling and distribution problem as a 1–1 pickup and delivery problem with soft time windows, vehicle capacity constraints, and dynamic order characteristics. Taking the average overdue time as the goal, this paper designs a combined order scheduling strategy that considers the delivery clerk's route, while taking the route length and overdue situation as punishment, And the two-stage path planning algorithm based on the nearest neighbor initial planning and tabu search. The research contents of this paper are summarized as follows.

In the context of the gradual development of the takeout industry and the fact that takeout has become the dining choice of most urban residents, the efficiency of delivery of takeout and user satisfaction have become hot issues of concern. By comparing the similarities and differences between takeout business and urban express and carpool business, this paper concludes that takeout distribution has the characteristics of small vehicle capacity, specific and decentralized delivery points for each order, high timeliness requirements, high and low daily flow peaks, etc.; At the same time, combined with previous research on VRP and PDP, the characteristics of delivery scheduling and distribution are analyzed, and it is classified as 1–1 pickup and delivery problem with soft time window, capacity limit and dynamic order characteristics.

In combination with the actual takeout business scenario, this paper describes the elements of the issue of order scheduling and distribution, and describes the task, that is, under the premise of meeting the maximum backorder quantity of the takeout staff, the order of taking and delivering meals and many other constraints, the new orders in each scheduling period are allocated and the specific path of the takeout staff is planned, with the goal of minimizing the average overdue time.

On the basis of existing research, corresponding order scheduling and path planning algorithms are proposed for this problem. In this paper, the order scheduling algorithm is proposed to consider the order consolidation strategy of the delivery clerk route. Specifically, when allocating orders, compare whether there are orders in the original route of each delivery clerk that are close enough to the starting point or the end point of the new order. If there are orders, it can achieve "incidental" delivery of new orders. In addition, it is necessary to punish the delivery clerk who has too long the original route and too long the overdue time of the original order. The two-stage path planning algorithm proposed in this paper is an improved two-stage algorithm based on the nearest neighbor initial planning and tabu search, The real-time exit mechanism of the tabu search algorithm is implemented based on the actual problem constraints in this paper to avoid invalid search without feasible neighborhood solutions.

The static problem experiment of this paper designed a scenario where one delivery clerk delivered five orders. The first come first service algorithm, the greedy initial planning of the nearest neighbor, and the path and route map improved by tabu search after the initial planning were visualized. The performance of the path planning algorithm proposed in this paper was verified by comparing the average overdue time. In the dynamic problem experiment, we designed a scenario of 20 delivery agents delivering 100 dynamic orders, compared the time consuming greed of the starting point of the new order with the order combining strategy considering the route of the delivery agents, and the results showed that each index of the strategy considering whether there are orders with the same starting point or the same destination in the original route of the delivery agents, and taking the route length and overdue situation as punishment was nearly twice as much as the former. This paper also discusses how many delivery staff can best meet the needs of users when the number of orders is fixed, which can provide a reference for the management of material delivery staff when taking epidemic prevention measures such as closing the city during the COVID-19 epidemic; In addition, this experiment gives the appropriate range for order rescheduling to improve delivery efficiency when the proportion of orders to the number of delivery staff is within.

This paper proposes corresponding order scheduling and path optimization algorithms for delivery scheduling and distribution problems with soft time windows, vehicle capacity constraints, dynamic orders, 1–1 pickup and delivery characteristics. The proposed algorithms have certain theoretical and practical value, but there are still many areas for optimization in the actual situation of this problem and the direction of algorithm improvement. The improvement directions that can be further studied in the future are as follows: (1) When planning the path, The distance between nodes is the spherical distance of the calculated longitude and latitude coordinate points. The actual road network in real life, as well as uncertain factors such as traffic status, bad weather, delayed meal delivery in restaurants, and delayed meal collection by customers are not taken into account, resulting in a gap between the order completion and the actual situation; (2) The order scheduling algorithm proposed in this paper is based on the business district as a unit. Each business district is independent of each other. The order distribution of the delivery staff is carried out in a fixed business district. In fact, there may be cross business district order distribution, and the joint scheduling of cross business districts can be considered subsequently; (3) In reality, the goal of the takeout business is not only to reduce the overdue situation, but also to consider the cost and income issues. In the future, we can consider the distribution cost, overtime compensation, the rights and interests of the takeout staff and other issues to model; (4) The path planning algorithm in this paper mainly uses greedy and meta heuristic algorithms, and does not consider the impact of future demand, traffic conditions, etc. on the solution at this stage. In the future, we can consider the use of reinforcement learning and neural network correlation algorithms to solve the order adjustment and distribution problems that are more in line with the real scene.

References

1. Na, D., Zhang, J.: Study on assign mode of o2o takeaway order delivery tasks. Shanghai Manag. Sci. **40**(01), 63–66 (2018)
2. Ma, S., Zheng, Y., Wolfson, O.: Real-time city-scale taxi ridesharing. IEEE Trans. Knowl. Data Eng. **27**(7), 1782–1795 (2014)
3. Asghari, M., Deng, D., Shahabi, C., et al.: Price-aware real-time ride-sharing at scale: an auction-based approach. In: Proceedings of the 24th ACM SIGSPATIAL International Conference on Advances in Geographic Information Systems, pp. 1–10 (2016)
4. Lin. Q., Deng, L., Sun, J., et al.: Optimal demand-aware ride-sharing routing. In: IEEE INFOCOM 2018-IEEE Conference on Computer Communications, pp. 2699–2707. IEEE (2018)
5. Jiang, W., Dominguez, C.R., Zhang, P., et al.: Large-scale nationwide ridesharing system: A case study of Chunyun. Int. J. Trans. Sci. Technol. **7**(1), 45–59 (2018)
6. Zheng Wang, Tingyu Li, Caifan Yue. Online scheduling for the instant delivery problem in a city based on multiple prediction scenarios. 系统工程理论与实践**38**(12), 3197–3211 (2018)
7. Chen, D., Chen, T., Jiang, S., Zhang, C., Wang, C., Lu, M.: Optimal model of chicken distribution vehicle scheduling based on order clustering. Smart Agricul **2**(4), 137–148 (2020)
8. Zhang, Y., Liu, Y., Li, G., et al.: Route prediction for instant delivery. Proceedings of the ACM Intera. Mobile, Wearable Ubiquitous Technol. **3**(3), 1–25 (2019)
9. Jahanshahi, H., Bozanta, A., Cevik, M., et al.: A deep reinforcement learning approach for the meal delivery problem. Knowl.-Based Syst. **243**, 108489 (2022)
10. Guo, B., Wang, S., Ding, Y., et al.: concurrent order dispatch for instant delivery with time-constrained actor-critic reinforcement learning. In: 2021 IEEE Real-Time Systems Symposium (RTSS), pp. 176–187. IEEE (2021)
11. Bozanta, A., Cevik, M., Kavaklioglu, C., et al.: Courier routing and assignment for food delivery service using reinforcement learning. Comput. Ind. Eng. **164**, 107871 (2022)
12. Zhou, M., Jin, J., Zhang, W., et al.: Multi-agent reinforcement learning for order-dispatching via order-vehicle distribution matching. In: Proceedings of the 28th ACM International Conference on Information and Knowledge Management, pp. 2645–2653 (2019)
13. Xu, Z., Li, Z., Guan, Q., et al.: Large-scale order dispatch in on-demand ride-hailing platforms: A learning and planning approach. In: Proceedings of the 24th ACM SIGKDD International Conference on Knowledge Discovery & Data Mining, pp. 905–913 (2018)

14. Dantzig, G.B., Ramser, J.H.: The truck dispatching problem. Manage. Sci. **6**(1), 80–91 (1959)
15. Battarra, M., Cordeau, J.F., Iori. M.: Chapter 6: pickup-and-delivery problems for goods transportation. In: Vehicle Routing: Problems, Methods, and Applications, Second Edition. Society for Industrial and Applied Mathematics, pp. 161–191 (2014)
16. Angelelli, E., Mansini, R.: The vehicle routing problem with time windows and simultaneous pick-up and delivery. In: Quantitative Approaches to Distribution Logistics and Supply Chain Management, pp. 249-267. Springer, Berlin (2002). https://doi.org/10.1007/978-3-642-56183-2_15
17. Qu, Y., Bard, J.F.: The heterogeneous pickup and delivery problem with configurable vehicle capacity. Trans. Res. Part C: Emerging Technol. **32**, 1–20 (2013)
18. Nagy, G., Salhi, S.: Heuristic algorithms for single and multiple depot vehicle routing problems with pickups and deliveries. Eur. J. Oper. Res. **162**(1), 126–141 (2005)
19. Wang, C., Qiu, Y.: Vehicle routing problem with stochastic demands and simultaneous delivery and pickup based on the cross-entropy method. In: Advances in Automation and Robotics, vol. 2, pp. 55-60. Springer, Berlin, Heidelberg (2011). https://doi.org/10.1007/978-3-642-25646-2_7
20. Pang, Y., Luo, H., Xing, L., Ren, T.: A survey of vehicle routing problems and solution methods. Control Theory Appl. **36**(10), 1573–1584 (2019)
21. Kalantari, B., Hill, A.V., Arora, S.R.: An algorithm for the traveling salesman problem with pickup and delivery customers. Eur. J. Oper. Res. **22**(3), 377–386 (1985)
22. Desrosiers, J., Dumas, Y., Soumis, F.: A dynamic programming solution of the large-scale single-vehicle dial-a-ride problem with time windows. Am. J. Math. Manag. Sci. **6**(3–4), 301–325 (1986)
23. Gajpal, Y., Abad, P.: Saving-based algorithms for vehicle routing problem with simultaneous pickup and delivery. J. Operat. Res. Soc. **61**(10), 1498–1509 (2010)
24. Jun, Y., Kim, B.I.: New best solutions to VRPSPD benchmark problems by a perturbation based algorithm. Expert Syst. Appl. **39**(5), 5641–5648 (2012)
25. Zheng, H., Wang, S., Cha, Y., et al.: A two-stage fast heuristic for food delivery route planning problem. In: Informs annual meeting, Seattle, Washington, USA (2019)
26. Chen, W., Dai, S.: A survey of algorithms for vehicle routing problems. J. Chuzhou Univ. **03**, 19–25 (2007)
27. Chen, J.F., Wu, T.H.: Vehicle routing problem with simultaneous deliveries and pickups. J. Operat. Res. Soc. **57**(5), 579–587 (2006)
28. Hosny, M.I., Mumford, C.L.: The single vehicle pickup and delivery problem with time windows: intelligent operators for heuristic and metaheuristic algorithms. J. Heuristics **16**(3), 417–439 (2010)
29. Tasan, A.S., Gen, M.: A genetic algorithm based approach to vehicle routing problem with simultaneous pick-up and deliveries. Comput. Ind. Eng. **62**(3), 755–761 (2012)
30. Demir, E., Bektaş, T., Laporte, G.: An adaptive large neighborhood search heuristic for the pollution-routing problem. Eur. J. Oper. Res. **223**(2), 346–359 (2012)
31. Su, X., Qin, H., Wang, K.: Tabu search algorithm for vehicle routing problems with time windows and multiple delivery personnel. J. Chongqing Normal Univ. (Nat. Sci.), **37**(01), 22–30 (2020)

Research on Gesture Recognition Based on Multialgorithm Fusion

Yuanyuan Zhu, Yanqing Wang$^{(\boxtimes)}$, Xiaofeng Gao, and Lihan Liu

Nanjing Xiaozhuang University, NanJing 211171, China
wyq0325@126.com

Abstract. Aiming at people with hearing and speaking obstacles, this paper proposes a multialgorithm fusion for gesture recognition. This paper aims to more clearly distinguish easily confused gestures in gesture recognition and improve gesture recognition accuracy by integrating lip-reading recognition. For gesture recognition, this paper first performs skin color processing and segmentation on the hand area of the collected video sequence and detects the hand feature points by calling the hand key point model. The extracted gesture features are trained and recognized by the support vector machine algorithm. For lip reading recognition, this paper first uses the AdaBoost algorithm to detect and track key points on the collected video sequence, locate the lips, extract the key points of the lips through a convolutional neural network, and input the extracted key point feature sequence into BiLSTM to extract semantic information. The fusion of gesture recognition and lip reading recognition algorithms using the YOLOV5 model can effectively improve the accuracy of gesture recognition. Through experimental verification, the recognition rate can be increased from 89.4% to 94.3%.

Keywords: Gesture recognition · Lip reading recognition · Mediapipe hand key point · HSV color space · Convolutional neural networks

1 Introduction

With the rapid development of artificial intelligence in the field of scientific research, sign language recognition has become an important component in the field of intelligent human-computer interaction. Sign language recognition technology has greatly facilitated the daily life of people with hearing and speech impairments and has received widespread attention from the academic community. According to the latest data from the WHO, the number of people with hearing impairment-related diseases worldwide has shown a significant upwards trend in recent years. Using gesture recognition and lip reading technology can effectively assist people with hearing impairment in solving communication problems. For people with hearing and speech impairments, obtaining information and communicating with the outside world through movement and lip reading is an effective means [1]. Therefore, lip-reading recognition technology can be used as an auxiliary means of gesture recognition, improving the accuracy of gesture recognition by visually supplementing the information of auditory channels.

Domestic and foreign research on sign language recognition technology and lip-reading recognition technology started early and has achieved a large number of scientific research results. Lee GC and others proposed a Kinect-based Taiwanese sign language recognition system that extracted three main features from gestures, used HMM to determine the direction of the hand movement trajectory and trained SVM to recognize hand shapes. The recognition rate of the system reached 85.14% [2]. Microsoft's Kinect, a somatosensory interaction device, has the advantages of flexible operation and simplicity. This new type of human-computer interaction device, Kinect, has quickly become a current research hotspot in the field of Chinese sign language recognition [3]. Li, TZ et al. proposed a static visual interactive gesture recognition method, which consists of real-time capture by a color camera and gesture extraction by FHOG features. The system achieved an average recognition rate of 95.31%, a rejection rate of 9.37% and an overall recognition efficiency of 90.63% [4]. Yang, XZ et al. proposed a recognition method that can be used for real-time dynamic gesture recognition and search for valid instantaneous gestures in dynamic gestures. The method outperforms the sliding time window algorithm in terms of recognition accuracy and efficiency by filtering invalid gesture data and building valid gesture features [5]. Muhamad Amirul Haq et al. used spatiotemporal convolution and SE-ResNet-18 to extract lip movement features. The back-end module used a custom BG-CL network. The model achieved an accuracy rate of 94.2% in the DMCLR data set [6]. Huijuan Wang et al. proposed a lip-reading method based on a 3D convolutional visual transformer (3DCvT), which combines visual transformers and 3D convolutions to extract spatiotemporal features of continuous images and extract local and global features from continuous images. The extracted features are then sent to a bidirectional gated recurrent unit (BiGRU) for sequence modelling, achieving state-of-the-art performance [7].

With the widespread application of computer technology and the gradual maturity of image processing and pattern recognition technology, gesture recognition and lip reading technology have also become a comprehensive product of multiple fields and disciplines. However, gesture recognition and lip reading technology are highly challenging due to many factors. For gesture recognition, whether the positioning of image sequences is accurate, whether image segmentation is accurate, and whether using only skin color detection algorithms can accurately distinguish nongesture regions are all issues that need to be addressed. To solve the problem of inaccurate gesture recognition algorithms mentioned above, this paper proposes a sign language recognition technology based on key point extraction and skin color detection. For gestures that are easily confused and indistinguishable, this paper incorporates lip-reading recognition technology into gesture recognition technology using lip reading recognition technology based on a convolutional neural network. Lip-reading recognition technology serves as a means of auxiliary gesture recognition technology to improve the progressiveness and accuracy of sign language recognition technology.

2 Hand Gesture Recognition Based on Key Point Extraction and Skin Color Detection

2.1 Process Framework for Gesture Recognition

This paper first performs skin color processing on the hand area of the collected data set and then combines the key point model of the hand to detect 21 feature points of the hand and uses the skin color segmentation gesture for image segmentation. Then, the morphological features of the hand region and the key point connection skeleton are extracted. Finally, the extracted gesture feature data set is trained and recognized by the support vector machine algorithm. The workflow diagram of gesture recognition is shown in Fig. 1.

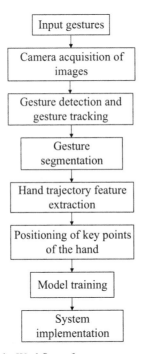

Fig. 1. Workflow of gesture recognition

2.2 Gesture Processing

To improve the recognition rate of hand gestures, image preprocessing is needed. Unpreprocessed images may be disturbed by various noises, such as complex backgrounds and uneven lighting. Therefore, preprocessing of gesture images is necessary. Image smoothing is performed on the input image in an effort to eliminate the interference of other factors, enhance the image information, and facilitate skin tone segmentation.

2.3 Segmentation Algorithm of Hand Region

Gesture changes are complex, but skin color does not change due to changes in hand position, so skin color can be used as a feature for gesture detection and gesture segmentation. Based on multiple experiments, it can be found that there are many interferences when studying gestures based on skin color, such as the degree of clutter in the background and changes in light, which have a significant impact on gesture detection and gesture segmentation. Therefore, we need to select an appropriate color space to convert the obtained image to a color space that can be less disturbed. This can minimize various interferences in the image recognition process and achieve better results in segmenting gestures.

To remove the interference of other factors, this paper has to split the target gesture from the image before continuing the operation. Splitting gestures are a crucial part of the implementation of the system, and their success directly affects the final recognition result. Segmenting gestures by HSV color space skin tones is one of the most common methods, and the process is shown in Fig. 2.

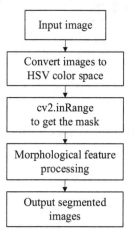

Fig. 2. Process of gesture segmentation

This paper uses HSV color space to extract gesture regions. The HSV color space threshold is mainly extracted by using the Trackbar adjustment threshold and cv2.inRange to generate a mask. Skin color detection solves the problem of how to identify pixels and regions with human skin color from images or videos and is a key technology in computer vision applications such as gesture recognition and human-computer interaction. A common skin color detection algorithm first converts an image into pixels located in a color space, then divides skin color and nonskin color spatial areas in that color space by setting a threshold, and finally identifies skin color pixels from the image by determining whether the pixel points are in the skin color area in the color space [8]. The conversion Fig. 3 is shown below.

The results of skin color detection are shown in Fig. 4, the results of HSV color space conversion are shown in Fig. 5, and the results of masking are shown in Fig. 6.

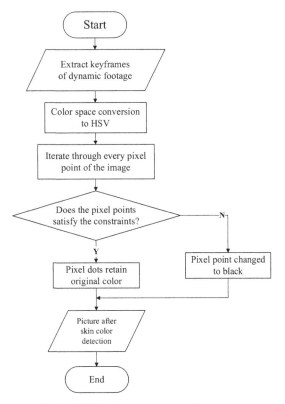

Fig. 3. HSV color space conversion process

Fig. 4. Results of the skin color test   **Fig. 5.** Results of HSV color space conversion

Fig. 6. Results of masking

2.4 Morphological Feature Processing

Even with high-performance skin color detection, the obtained gesture image still has defects and may show some black dots. To remove these traces, it is necessary to further optimize the gesture image through image morphology processing. After using morphological processing, the structure of the original image remains unchanged, and most of the defects and noise can be eliminated. After morphological processing, a clearer image can be obtained. The two main operations of morphological processing are expansion and erosion. Erosion and expansion are applied to highlight prominent areas. Figure 7 shows the results of morphological processing.

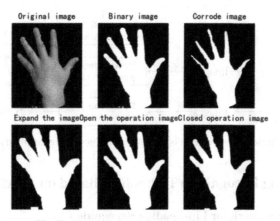

Fig. 7. Results of morphological processing

2.5 A Method of Hand Key Point Extraction Based on Mediapipes

Hand key point extraction is mainly based on Mediapipe, an open-source project of Google that can provide open-source, cross-platform common machine learning solutions. Mediapipe provides high-fidelity tracking of hands and fingers by inferring twenty-one 3D hand keys within a single frame based on machine learning. Through the positioning of 21 key points, the hand position can be located more accurately and quickly, and the hand area can be accurately divided. Figure 8 shows the 21 feature points marked on the key points of locating hands. Hand key points combined with skin color segmentation features have a resistance effect on the occlusion of gestures, and combined with gesture contours, they can accurately represent gestures. Hand gesture recognition based on MediaPipe includes a palm detection model, hand key part model and hand gesture recognizer [9]. The architecture of the Mediapipe gesture recognition system is shown in Fig. 9.

Fig. 8. 21 feature points for hand key point labelling

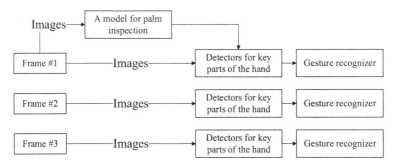

Fig. 9. Architecture diagram of the Mediapipe gesture recognition system

3 Lip-Reading Recognition Technology Based on AlexNet-BiLSTM

3.1 Process Framework for Lip-Reading Recognition

Lip-reading recognition mainly includes 4 steps: facial key point detection and tracking, lip region extraction, spatiotemporal feature extraction, and classification and decoding [10]. A schematic diagram of the lip-reading recognition steps is shown in Fig. 10.

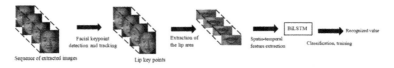

Fig. 10. Schematic diagram of the lip reading recognition steps

3.2 Detection and Tracking of Key Points of Faces

To achieve accurate positioning of facial feature points in sequence images for subsequent accurate extraction of lip area images, this paper uses the AdaBoost algorithm for face key point detection and tracking. The AdaBoost algorithm selects the optimal classifier through the feedback of weak classifiers during the training process and optimizes and combines the iteratively generated weak classifiers for face detection. The flowchart of the AdaBoost algorithm for face detection is shown in Fig. 11.

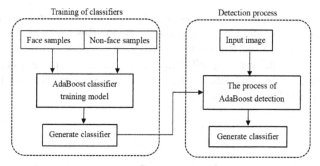

Fig. 11. Flowchart of the AdaBoost algorithm for face detection

3.3 Extraction of Lip Key Points

On the basis of face key point detection, lip region extraction is performed using the region where the key points of the lip contour are located. The main purpose of face key point detection is to provide geometric location information for key point extraction in the lip region. In this paper, an AlexNet convolutional neural network is used to extract the spatial feature extraction of the lip region picture and complete the localization detection of lip key points. The AlexNet network model for key point detection is shown in Fig. 12.

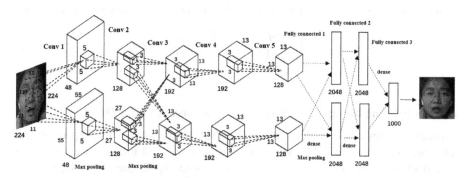

Fig. 12. AlexNet network model for key point detection

3.4 Spatiotemporal Feature Extraction

Spatiotemporal feature extraction is the core research of lip reading technology, through which precise key points can be extracted from the video data set for subsequent prediction of recognition values. With the rapid development of deep learning technology, the use of convolutional neural networks for lip reading feature extraction has gradually become mainstream. Convolutional neural networks are widely used in life and have made remarkable contributions in the fields of face recognition, sentiment analysis [11], and handwritten text data [12].

To fully utilize the dynamic information of the lip area and ensure effective learning of the interframe information in the lip-reading region. This paper adopts time feature extraction based on BiLSTM. BiLSTM has both forward and backwards information calculation directions, which can fully utilize lip-reading features, achieve effective learning of temporal features, and make the prediction results more accurate. The structure and principle of BiLSTM are shown in Fig. 13.

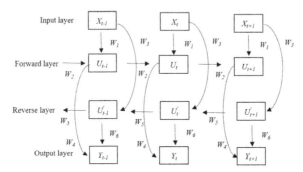

Fig. 13. Structure and principle of BiLSTM

The processing equations of the sequence are expressed as follows: U_t and U'_{t-1} are the forward and backwards values at moment t; X is the input variable in the input layer; f and g are the activation functions; and W_1-W_6 are the internetwork weight matrices.

$$\begin{cases} U_t = f(W_1 X_t + W_2 U_{t-1}) \\ U'_t = f(W_3 X_t + W_5 U'_{t+1}) \\ Y_t = g(W_4 U_t + W_6 U'_t) \end{cases} \tag{1}$$

3.5 Lip-Reading Recognition Process Based on AlexNet-BiLSTM

After detecting lip key points through AlexNet, the lip region images in the lip movement video are input into the BiLSTM network for analysis, and the model construction for extracting temporal features and recognition of the lip-reading video sequence is performed. Figure 14 shows the lip reading recognition frame diagram used in this paper that fuses AlexNet with BiLSTM.

The whole lipreading technology can be divided into the following four parts: the first part is the extraction and tracking of face key points. The AdaBoost algorithm is used to detect and track the key points of the face, which lays the foundation for further extraction of the lip region. The second part is lip key extraction. After the lip region image sequence is extracted, the AlexNet convolutional neural network is used to extract the spatial features of the lip region image, and the location detection of the lip key points is completed. In the third part, BiLSTM is used to extract the temporal features of the lip region. The lip region after key point extraction is transmitted to BiLSTM to extract temporal features. In the fourth part, the time series features learned by the BiLSTM network are trained, and the maximum probability value is selected as the prediction recognition result.

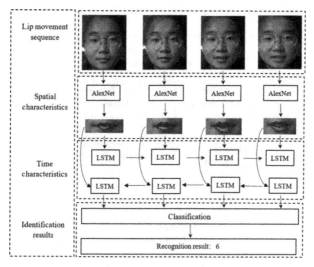

Fig. 14. Framework diagram of lip-reading recognition based on Alexnet-BiLSTM

4 Algorithm Fusion

In this paper, the YOLOV5 model is used to fuse gesture recognition and lip-reading recognition. The gesture recognition and lip-reading recognition models are loaded into the YOLOV5 model at the same time, and the input of the YOLOV5 model is input into the gesture recognition model and lip-reading recognition model to obtain the output of the two models. Then, the fusion classifier in YOLOV5 is used for processing. The final output result is the recognition result considering both gesture and lip-reading information. The fusion process is shown in Fig. 15.

5 Experimental Analysis

This paper uses a video data set of gestures and lip reading collected by the author in a normal environment. All of the data in this data set were collected under natural light indoors without any special lighting. The device that collected the images was a computer camera. There were a total of 15 types of sign language gesture video sequences (0–10 digital sign language gestures, "Family", "Hello", "Have a meal", and "Thank you"). The video sequence included gestures and lip reading. The data set consists of 450 segments, with 30 segments for each gesture. The obtained data set was expanded by 5 testers outputting gestures and corresponding lip reading, and each image segment was processed and saved. The training set and test set were distributed in a 2:1 ratio. The test set is 150 segments, and the training set is 300 segments. Figure 16 shows the collected video data set.

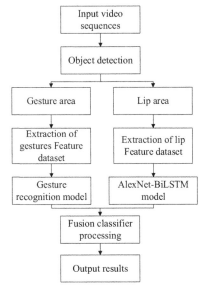

Fig. 15. Fusion process of gesture recognition and lip reading recognition

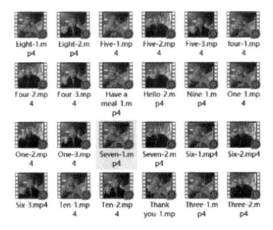

Fig. 16. Video data set

5.1 Experimental Results

In this paper, by fusing deep learning-based lip-reading recognition into sign language recognition, the fusion of the two techniques can resist the influence of confusing gestures and the surrounding environment on the recognition results. An illustration of the sign language is shown in Fig. 17. The final experimental results are shown in Figs. 18, 19, 20, 21, 22, 23, 24 and 25.

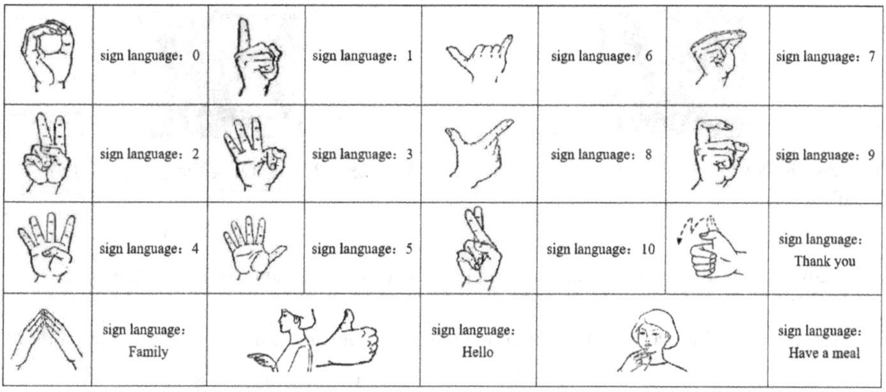

	sign language: 0		sign language: 1		sign language: 6		sign language: 7
	sign language: 2		sign language: 3		sign language: 8		sign language: 9
	sign language: 4		sign language: 5		sign language: 10		sign language: Thank you
	sign language: Family		sign language: Hello				sign language: Have a meal

Fig. 17. Illustration of sign language

Fig. 18. Recognizing the number "0"

Fig. 19. Recognizing the number "7"

Fig. 20. Recognizing number "9"

Fig. 21. Recognizing the number "10"

Fig. 22. Recognizing the gesture "Thank you"

Fig. 23. Recognizing the gesture "Family"

5.2　Analysis of Experimental Results

In this paper, the accuracy of gesture recognition is improved by fusing deep learning-based lip-reading recognition to sign language recognition through interception of image sequences for 0–10 numbers, "Family", "Hello", "Have a meal", and "Thank you" sign language sign recognition. Table 1 and Table 2 are the accuracy comparison results before and after the fusion lip reading recognition technology.

Fig. 24. Recognizing the gesture "Have a meal"

Fig. 25. Recognizing the gesture "Hello"

Table 1. Gesture recognition based on key point extraction and skin color detection

Test Set (piece)	Correct identification of the number	Gesture (kind)	Recognition rate(%)
1000	894	15	89.4

Table 2. Sign language recognition based on the fusion of AlexNet-BiLSTM lip reading recognition and gesture recognition (the method proposed in this paper)

Test Set (piece)	Correct identification of the number	Gesture (kind)	Recognition rate(%)
1000	943	15	94.3

By comparing the recognition results of 15 gestures before and after fusion, it can be seen that the accuracy of sign language recognition based on AlexNet-BiLSTM lip-reading recognition fusion is higher than that based on key point extraction and skin color detection. In particular, the words "Thank you" and "Hello", the sign language "2" and the sign language "10" are easily confused in the process of sign recognition. Figure 26 shows the extraction sequence of the lip region during the "Thank you" gesture recognition process. Figure 27 shows the extraction sequence of lip language regions during the "Hello" gesture recognition process. Figure 28 shows the extraction sequence of lip language regions during the "2" gesture recognition process. Figure 29 shows the extraction sequence of lip language regions during the "10" gesture recognition process. The fusion of lip reading recognition can improve the recognition rate of similar gestures.

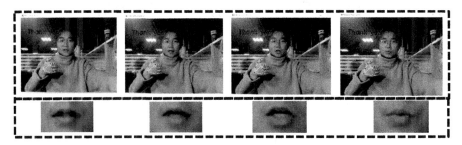

Fig. 26. "Thank you" gesture recognition process lip region extraction sequence

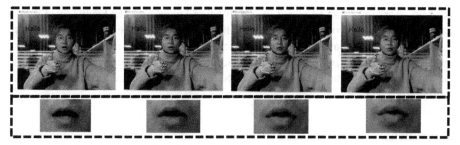

Fig. 27. "Hello" gesture recognition process lip region extraction sequence

Fig. 28. "2" gesture recognition process lip region extraction sequence

Fig. 29. "10" gesture recognition process lip region extraction sequence

6 Conclusion and Outlook

This paper combines sign language recognition based on key point extraction and skin color detection and lip reading recognition based on AlexNet-BiLSTM, processes the collected data set, combines the detection of hand key points and skin color segmentation recognition, and integrates lip reading recognition into gesture recognition. Finally, the recognition accuracy is enhanced by training and recognition of the model. Experiments have shown that the method can solve the problem of low recognition rate and low accuracy of confusing gestures and can help people in need to a certain extent. In addition to relying on hand movements, hand detail features and nonmanual features such as lip shape and eyes need to be finely modelled [13]. In subsequent research, recognition of facial emotions, etc., can be added to identify sign language more accurately. To promote sign language recognition technology, it is necessary to cope with the recognition of complex environments such as light, changes in perspective, and hand obstruction. Expanding the data set, enhancing the number of gesture recognitions while improving recognition accuracy, and expanding the scope of application are the directions of future efforts.

References

1. Li, G., Wang, M., Lin, L.: Chinese visual speech database for the disabled. Chin. J. Biomed. Eng. **112**(03): 355–360+388 (2007)
2. Lee, G.C., Yeh, F., Hsiao, Y.: Kinect-based Taiwanese sign-language recognition system. Multimedia Tools Appli. **75**(1), 261–279 (2016)
3. Qin, M.: A review of sign language recognition research. Software Guide **20**(02), 250–252 (2021)
4. Li, T.Z., Qin, Q., Chen, Z.Y.: Interaction method based on visual gesture recognition. Second Target Recogn. Artifi. Intell. Summit Forum, 11427 (2020)
5. Yang, X.Z., et al.: An approach to dynamic gesture recognition based on instantaneous posture. In: IEEE 7th International Conference on Virtual Reality (ICVR 2021), pp. 90–95 (2021)
6. Mendes Junior, J.J.A., et al.: Feature selection and dimensionality reduction: an extensive comparison in hand gesture classification by sEMG in eight channels armband approach. Biomed. Signal Proc. Control 59, 101920 (2020)
7. Wang, H., Pu, G., Chen, T.: A lip reading method based on 3d convolutional vision transformer. IEEE Access **10**, 77205–77212 (2022)
8. Wang, S., Li, Y.: Research on skin color detection algorithm based on different color spaces. Inform. Recording Mater. **23**(03), 157–159 (2022)
9. Wang, R.: Gesture recognition based on MediaPipe for excavator teleoperation control. Civil Construction Eng. Inform. Technol. **14**(04), 9–169 (2022)
10. Chen, X.: Research progress and prospect of lip reading. Acta Autom. Sinica **46**(11), 2275–2301 (2020)
11. Huang, M.H., et al.: Lexicon-based sentiment convolutional neural networks for online review analysis. IEEE Trans. Affect. Comput. **13**(3), 1337–1348 (2022)
12. Espana-Boquera, S., et al.: Improving offline handwritten text recognition with hybrid HMM/ANN models. IEEE Trans. Pattern Anal. Mach. Intell. **33**(4), 767–779 (2011)
13. Tao, T., Liu, T.: A review of sign language recognition technology based on sign language expression content and expression features. J. Elect. Inform., 1–19 (2022)

Design of Fitness Movement Detection and Counting System Based on MediaPipe

Yinan Chen[✉] and Xia Liu

Sanya Aviation and Tourism College, Sanya 572000, Hainan, China
305780161@qq.com

Abstract. With the rapid advancements in artificial intelligence and computer vision technology, the field of visual-based human pose detection has emerged as a highly sought-after research area in recent years. The identification of human poses has practical applications in diverse domains, ranging from motion-sensing games for human-computer interaction to activity prediction and medical rehabilitation. The present study is focused on the utilization of human pose detection for fitness movement counting. The ultimate aim of the system design is to accurately detect the skeletal key points of each body part in the image and subsequently connect them to form a human pose skeleton, which serves as a vital representation of the characteristics of human motion, particularly in the context of video data, where multiple human poses can be linked to form a certain movement trajectory. By judging the trajectory and angle changes, the system can determine whether people's fitness movements are correct and help them improve their fitness effectiveness. Hence, an increasing number of researchers are investing time and effort in this field. One common approach for human pose detection is OpenPose, but this model has a large and complex structure and low detection accuracy. Therefore, this fitness movement detection and counting system uses a lightweight MediaPipe model and improves it to enhance the algorithm's accuracy and recognition speed. The specific work in this paper includes three main points: (1) a suitable network structure to detect human skeletal points; (2) the appropriate skeletal structure for fitness movements through experiments to obtain accurate results; and (3) a Qt interface for human-computer interaction.

Keywords: Human Pose Detection · Fitness Movement Counting · MediaPipe · Skeletal Nodes

1 Introduction

1.1 Research Background

In recent times, the maturation of artificial intelligence technology has witnessed the proliferation of various AI application technologies that have been widely deployed across diverse domains of human existence [1]. Of particular significance in the realm of AI technology is human body posture recognition technology, which holds unparalleled importance for the future growth and development of intelligent technology in our country. Since the 1970s, China has been at the forefront of research on human body posture recognition, which has significantly catalyzed the advancement of AI and laid a robust foundation for the modernization of AI technology. At present, human body posture recognition can be applied in many fixed or standard scenes, but human body posture recognition in more complex environments is still in the theoretical stage and has not been used in designed testing environments. With the high-speed development of science and technology in China in recent years, people have higher demands for technology in their daily lives. Human body posture estimation has gradually penetrated all aspects of life and has very broad development prospects. Nonetheless, the detection of human key points is susceptible to the influence of diverse factors, including but not limited to clothing, lighting, and object occlusions, thereby impeding the attainment of a real-time, efficient, and accurate human body posture estimation algorithm. As such, there exists a considerable gap that needs to be bridged to achieve a robust and reliable algorithm that can surmount the challenges posed by these factors and deliver optimal results [2].

With the changing consumer attitudes of increasingly more Chinese people, especially the improvement in consumption levels, people have become more concerned about health, especially the physical fitness of young people. In addition, the participation rate in sports has increased. According to official data, between 2022 and 2023, there were 400 million people in China who frequently participated in sports, with a sports industry scale of approximately 1.5 trillion yuan. It is expected that these two indicators will likely reach 435 million and 3 trillion yuan by 2023. Based on this, a large number of sports and fitness applications have emerged both domestically and internationally, with rapid development momentum and a surge in the number of users [4]. An increasing number of people have gained a deeper understanding of sports and fitness through the widespread use of mobile fitness apps and have further insights into them. Now, without relying on fitness centers and coaches and without having to set aside special time to go to specific locations for scientific fitness, fitness has diversified into various forms that can be done at any time and anywhere and can effectively record the results and achievements of fitness. However, there are still many problems with current fitness-assistant systems, such as inaccurate counting and limited functionality [5].

Based on this, this article will focus on achieving accurate counting of fitness movements and enriching the number of recognized fitness movements to better help people perform standardized fitness movements.

1.2 The Main Work and Innovative Points

This article designs and implements a fitness detection and counting system based on the MediaPipe algorithm, which detects and counts fitness actions and evaluates their accuracy [3]. When the action is not correct, counting is not performed. The main work includes the following:

(1) After conducting an extensive review of a vast corpus of literature, the hardware and software design of the machine vision-based fitness detection and counting system was duly finalized. The culmination of this comprehensive review has facilitated the

meticulous and systematic construction of an innovative and advanced system that harnesses the power of machine vision to enable efficient and accurate detection and counting of fitness movements. According to the system requirements, the algorithm architecture is designed, the hardware platform of the machine vision detection system is built, the algorithm is written in Python, and the manipulation interface is written in Qt to improve the man-machine interface.

(2) Suitable action detection is designed for the acquired body skeleton points, and appropriate action skeleton angles are determined through experiments. The appropriate angles ensure the accuracy of the system's detection.

(3) In the primary MediaPipe algorithm, a deep separable convolution approach has been adopted in lieu of the conventional convolution technique. This innovative convolution strategy enables the efficient decomposition of the spatial and channel dimensions of the input data, thereby achieving significant improvements in both computational efficiency and model accuracy. Compared with traditional convolution, deep separable convolution has less parameter quantity and less computation workload. The accuracy of deep separable convolution is slightly lower than that of traditional convolution, so dilated convolution is introduced to expand the receptive field of each layer feature map to compensate for the deficiency. Finally, a smaller model and faster speed are achieved.

(4) Finally, research on the Qt interface is conducted. The corresponding interaction interface is designed to ensure the usability of the system.

2 Theoretical Foundation

2.1 Classification Algorithms of Traditional Machine Learning

2.1.1 Support Vector Machine (SVM)

The support vector machine (SVM) method is one of the traditional classification methods in machine learning [6]. To explain it vividly, it maps the collected data into a two-dimensional storage space and uses a two-dimensional line or plane to divide the data, dividing the mapped data into two categories. Before mapping the data, SVM needs to label the data. The classification method of SVM is generally used in binary-classification problems, but some scholars use the multiple vector machine approach to achieve multiclassification functions. The SVM method also provides valuable reference opinions for future classification methods and lays a certain foundation.

2.1.2 K-Nearest Neighbor Algorithm

The K-nearest neighbor algorithm is generally abbreviated as the KNN algorithm. The essence of this algorithm is similar to the traditional Chinese culture of "close to red and black." Simply put, the KNN algorithm is also used for classification of data points in a two-dimensional plane. It determines the number of neighboring points by selecting the K value and then uses the voting principle to determine the category of the selected point [7]. The KNN method is also applied to the classification of objects, but this algorithm requires strict preprocessing of the data; otherwise, the classification effect is not particularly good. In addition, the KNN algorithm does not perform well on global data compared to local data and is more suitable for combination with other global algorithms.

2.1.3 The Random Forest Algorithm

The fundamental principle underlying the random forest algorithm is that of voting, wherein a decision is made based on the collective decision of multiple constituent elements. In this instance, the voting principle is predicated on the decision tree of the algorithm, whereby the system generates individual decision classifications based on each decision tree and subsequently selects the decision tree with the most votes as the final output of the system [8]. This intricate implementation process entails the input of data into the system, followed by the generation of multiple decision trees that are subsequently harnessed to produce a collective decision that is reflective of the best outcome. It is obvious that the random forest method is designed to be very simple and does not require much training time. However, the random forest method is prone to overfitting in many complex problems, which is one of the disadvantages of the random forest method.

2.2 Mainstream Neural Network Methods

2.2.1 AlexNet Network Architecture

The AlexNet network is a highly accurate network architecture that was first proposed in the 2012 ImageNet competition. AlexNet achieved higher accuracy and won first place in the competition. It can be said that the presentation of AlexNet was a groundbreaking work that marked the true development of convolutional neural networks both domestic and international. The fundamental network structure of AlexNet is characterized by a multifaceted composition of convolutional layers, pooling layers, and fully connected layers, all of which serve to bolster the overall functionality and efficacy of the network. This intricate network architecture is comprehensively illustrated in Fig. 1, which

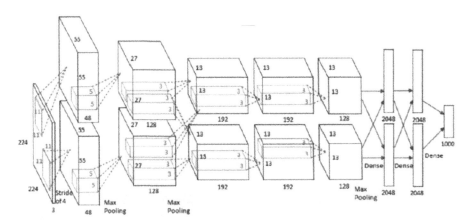

Fig. 1. The network architecture diagram of AlexNet.

provides a detailed graphical depiction of the various constituent elements and their respective roles within the network.

In the year following AlexNet's championship victory in the competition, the ZFNet network model was proposed, which improved the recognition success rate by more than 6% compared to the AlexNet network. However, this method is actually an improvement on the network architecture of AlexNet, and the main network has not been significantly modified. The original convolution kernel size of the convolutional layer was changed to 7*7, and the stride of the last two convolutional layers was changed to 2. In addition, the number of convolutional kernels of the last two layers was increased to better preserve the feature information of the input images.

The VGG network architecture represents another notable structural paradigm that has been derived from the foundational blueprint of the AlexNet architecture. In broad terms, the VGG network is characterized by four distinct models, denoted as VGG11, VGG13, VGG16, and VGG19, each of which possesses a unique network structure and configuration. The comprehensive network architecture of the VGG model is depicted in Fig. 2, which provides an intricate graphical representation of the various constituent layers and their respective roles within the network. Specifically, the VGG16 model, which is widely hailed as the quintessential embodiment of the VGG network, comprises a total of 13 convolutional layers, four pooling layers, and three fully connected layers, which collectively serve to optimize the efficiency and accuracy of the network.

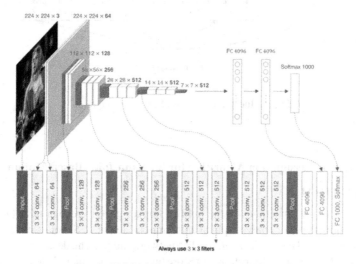

Fig. 2. VGG16 Network Structure Diagram.

2.2.2 GoogLeNet Network Architecture

The GoogLeNet network model represents a path-breaking paradigm shift in the realm of network architecture, having emerged as the undisputed winner of the 2015 competition. Propounded by Google, this novel network model introduced a radical Inception structure that revolutionized the preservation of feature information within input data. This pioneering network architecture, as illustrated in Fig. 3, has effectively surmounted the vexing convolutional structure predicament that plagued the VGG network model, wherein the output of the previous layer exerted an undue influence on the subsequent convolutional layers, leading to an irreversibly negative impact on the final results. Comprising three distinct convolutional layers, the GoogLeNet architecture has deliberately dispensed with the first two fully connected layers of the AlexNet network and leverages average pooling to reduce the parameters required for data processing. This innovative network model also incorporates a classifier that effectively resolves the issue of gradient disappearance that was endemic to the original model.

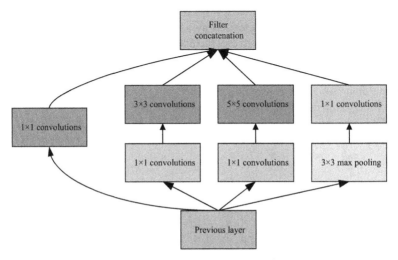

Fig. 3. Inception Network Structure Diagram.

2.3 A Behavior Recognition Algorithm for Skeleton Joint-Based Motion

2.3.1 AGC-LSTM

There are still many challenges in the field of human posture recognition. One major challenge is how to better analyse spatial and temporal features. Scholars have proposed a new network model called the AGC-LSTM network model. The design of the network model generally consists of four steps: (1) Store the collected human skeletal information in a spatial matrix, with each joint information placed in a row-column feature matrix; (2) Use LSTM technology to compare the differences between two frames of data in the collected data stream; (3) Use the AGC-LSTM's three concatenated layers to simulate and process the spatial and temporal feature points of the collected data, and then use global feature information to determine human actions; and (4) To better improve the network model's ability to receive advanced spatial and temporal semantic features, time features are added on the original basis, and a new time average pooling method is used. The overall network model architecture is shown in Fig. 4.

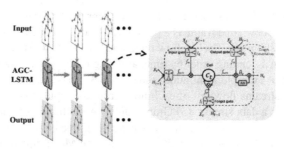

Fig. 4. Schematic Diagram of the AGC-LSTM Network.

2.3.2 DPRL

DPRL is a network model framework designed for human skeleton joint recognition. It performs well in recognizing dynamic human postures from video streams by filtering out useless feature information and retaining useful information frames. For instance, for a short 10-s video clip, measured by the nearly universal standard of 24 frames per second, the database gradually forms 240 frames of data. The specific process for both theory and practice takes into account another influencing factor, one of which is the ability of the currently selected framework to discriminate [9]. One is to choose the relationship with the entire posture sequence. The final selected frames are regarded as representative of the information related to the entire online video and are used for the final visual recognition of human posture. In addition [20], to consider the internal topological structure of the human posture east–west network, a basic structural model of the graph is designed for joint-related information and the connections between joints. The equilateral triangles of the graph represent the skin joints of the human body, the coordinate position information of each joint is stored in a file, and the connection between joints is represented by the adjacency matrix. Then, the neural network model of the graph is selected to learn the relationships between joints. Finally, a deep learning model that can display gestures but cannot recognize gestures was used to test the specific method, which proved its regularity.

2.3.3 MediaPipe

MediaPipe is an advanced multimedia machine learning model application framework that has been meticulously developed and generously open-sourced by the distinguished research team at Google. This pioneering framework has already been seamlessly integrated across a wide spectrum of key products and services within the Google ecosystem, including but not limited to YouTube, Google Lens, ARCore, Google Home, and Nest, attesting to the broad-based utility and versatility of MediaPipe. MediaPipe has a wide range of applications, including object detection, portrait segmentation, hair segmentation, face detection, hand detection, motion tracking, and more.

Boasting a highly sophisticated core architecture that has been meticulously implemented in C++, the MediaPipe framework offers seamless support for a diverse range of programming languages, including but not limited to Java and Objective-C. The main concepts of MediaPipe include the following: (1) Graph: A directed graph, output by display. (2) Packet: The most basic unit of data, representing data at a specific time node. In the figure above, one frame is a packet. (3) Node: A node in the graph can be a computing unit or subgraph. Each yellow rectangle is a node. (4) Stream: Multiple data packets arranged in ascending order by time, allowing for only one data packet at a specific timestamp. For example, continuous video frames can be seen as a continuous data stream. The data streams flow out of each node from the input stream, and the yellow line in the figure is a data stream. (5) GraphConfig: A configuration that describes the topology and functionality of the graph, corresponding to a configuration in the figure above. (6) Calculator: A C++ class that implements the MediaPipe protocol to handle data packets. It serves as a node, and the yellow box in the figure represents a calculator. (7) Subgraph: Additionally, a node, a subgraph contains a complete graph. The yellow box in the figure can be a subgraph.

In human detection, the backbone network used by MediaPipe is BlazePose, a detection network proposed by Google. It uses a face detector instead of a body detector to detect humans. Research has found that using a body detector can be easily affected by crowded scenes where people are heavily occluded. In cases with severe occlusion, the trust value of the body detector may not be very high. However, the face of a person is less likely to be occluded than the whole body, and the response values of this part are often the highest in neural networks. The specific detection points are shown in Fig. 5.

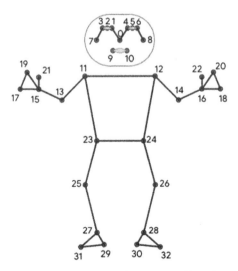

Fig. 5. The 33 Human Key Points Detected by BlazePose.

The BlazePose network consists of two parts, namely, the keypoint detection part and the keypoint regression part. What sets this network apart is that during the training phase, keypoint detection and regression are trained together. During testing, the detection part is removed, and only the regression part is retained to speed up the process. The specific architecture of this network is shown in Fig. 6.

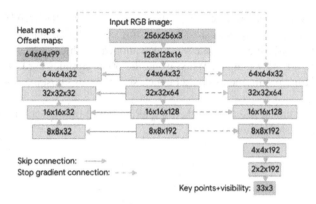

Fig. 6. BlazePose Network Architecture Diagram.

The comprehensive network architecture adopted in this study is characterized by a fully convolutional network that effectively detects heatmaps, offsets, and regressions separately. During the training stage, the algorithm effectively utilizes heatmap and offset losses while strategically deleting the corresponding output layers from the model prior to running inference. This approach enables the optimal utilization of heatmaps for supervising lightweight embedding, which is subsequently leveraged for regression encoder networks [10, 17]. Notably, this approach has been partially inspired by the groundbreaking stacked hourglass method proposed by Newell et al. Nonetheless, in our case, we have ingeniously stacked a miniature encoder-decoder heatmap-based network [16], along with a subsequent regression encoder network. Within the network architecture, images are processed with 3×3 convolutions. Owing to the potential for a considerable slowdown in program execution that may arise due to the large number of convolutions [19], the system design has been carefully optimized to incorporate dilated convolutions, which significantly improve the operational speed of the network.

3   Technical Scheme

The system consists of two parts: acquisition and processing, where the acquisition part can be divided into two types, video streams and cameras. Considering the system deployment requirements for performance and the difficulty of meeting the performance requirements on mobile devices, a video solution is used, enabling users to capture images and then replay and detect them at their convenience [11].

This solution consists of both hardware and software components [18]. The software solution includes system selection and deployment, while the hardware solution includes the impact of different PC processing powers on the system.

On the software side, this design considers two common deep learning frameworks, TensorFlow and PyTorch. Because TensorFlow has issues with version incompatibility and documentation confusion, Pytorch was chosen as the underlying framework for the system. The corresponding buttons and interface were developed using Qt [12].

On the hardware side, we tested the system using no GPU, NVIDIA1650 graphics cards, and NVIDIA3060 graphics cards. After testing, it was found that the graphics card had a significant impact: the frame rate was low without a graphics card, and action could not be judged in a timely manner. When using the 3060 graphics card, the effect was better, and real-time requirements could be met, so 3060 was chosen as the computing unit.

The overall technical solution for the MediaPipe-based fitness action detection and counting system is shown in Fig. 7.

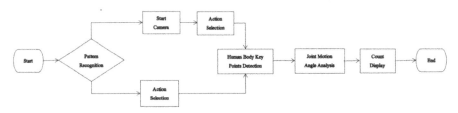

Fig. 7. Overall Design Flowchart of The Solution.

4   System Implementation

4.1   Introduction to Experimental Equipment

The experimental equipment used in this article is a Lenovo Y9000P laptop with an NVIDIA GEFORCE RTX3060 graphics card. The experiments were conducted using Python language development based on the PyTorch framework. The specific experimental environment is shown in Table 1.

Table 1. Experimental Environment Configuration.

Operating System	Windows11
System Type	64-bit
GPU	NVIDIA GEFORCE RTX3060
RAM	16G
VRAM	6G
Programming Language	Python3.7
Compiler	Pycharm
Deep Learning Framework	Pytorch
GPU Acceleration Tool	CUDA11.6 + CUDNN8.3

4.2   Introduction to the Dataset

The training uses the Leeds Sports Pose (LSP) dataset, which is a sports pose dataset categorized into various sports, such as athletics, badminton, gymnastics, baseball, parkour, volleyball, soccer, and tennis, and contains approximately 2000 annotated poses. The images are sourced from sports personnel in Flickr [14].

Each image is a 3-channel color image with pixel rows ranging from 64 to 202 and columns ranging from 57 to 202. Each image is annotated with 14 joint positions, with left and right joints always marked relative to the person's center. See Fig. 8 for specific images.

Fig. 8. BlazePose Network Architecture Diagram.

4.3   Training Process

The hyperparameter settings for the training process in this article are shown in Table 2. Due to computer power limitations, the model training was conducted on a rented server platform.

The final training results of the server are shown in Fig. 9 and Fig. 10.

The results from the model running on the server show that the average accuracy is above 90%, and the recognition of human posture is very good on the server, accurately identifying the human posture in the image.

4.4   Introduction to Evaluation Metrics

To better evaluate the performance of human pose recognition algorithms, four metrics have been selected as evaluation indicators for algorithm performance, namely, parameter size, model memory size, recognition accuracy, and inference latency. These metrics can effectively reflect the performance of an algorithm [13]. Parameter size represents the number of parameters in a human pose recognition algorithm, while model size represents the amount of computer storage required for a trained human pose model. Recognition accuracy reflects the intrinsic quality of a recognition algorithm, and inference latency represents the running time of a model processing an image. Below is a detailed explanation of these evaluation metrics [15].

Table 2. Table of Training Parameter Settings.

Training Parameters	Parameter value
Initial learning rate	0.01
Cyclic learning rate	0.2
Momentum	0.937
Weight decay	0.0005
Warm-up learning	3
Warm-up initial learning rate	0.1
Warm-up learning momentum	0.8
Training batch	16
Number of threads	4
Image size	1000×350–1024×768

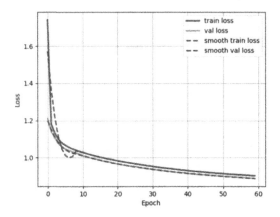

Fig. 9. Loss Curve Graph

(1) Precision

Precision, a crucial metric of model performance, is precisely defined as the ratio of true positives to the total number of samples that have been predicted as positive. By quantifying the degree of false detections, precision serves as a powerful indicator of model accuracy and efficacy in the context of predictive modelling. The formula for calculating precision is as follows:

$$Precision = \frac{TP}{TP + FP} \tag{1}$$

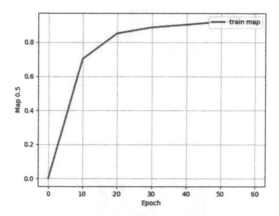

Fig. 10. MaP Variation Chart

(2) Recall Rate

The recall rate, a pivotal metric of model performance, is precisely defined as the ratio of samples with positive predicted values to the total number of samples that have true values equal to true. By providing a quantitative measure of missed detections, the recall rate serves as a powerful indicator of model sensitivity and performance in the context of predictive modelling, and its calculation formula is as follows:

$$Recall = \frac{TP}{TP + FN} \tag{2}$$

(3) Average Precision

Average precision (AP) is a measure that combines precision and recall. The calculation formula is as follows:

$$AP = \int_0^1 Rd(P) \tag{3}$$

(4) MAP

In recent years, the field of object detection has mostly moved beyond simple binary classification tasks, and the accuracy of complex multiclass tasks is represented using the original AP value averaged, known as mAP. The calculation formula for mAP is shown in Eq. 4 below:

$$mAP = \frac{\sum_{j=1}^{C} AP_j}{C} \tag{4}$$

5 Experimental Results

In the experiment, three movements were selected for testing and counting. The experimental setup is shown in Fig. 11, Fig. 12 and Fig. 13.

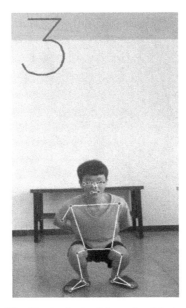

Fig. 11. Analysis of Squatting Motion

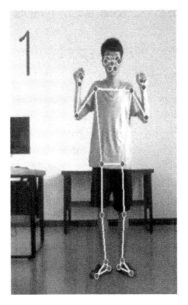

Fig. 12. Analysis of Pull-up Motion

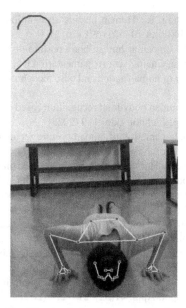

Fig. 13. Analysis of Press-up Motion

It can be seen that real-time monitoring by the camera has a good detection effect on movements, and the system does not provide feedback when the movement is not performed correctly.

References

1. Liu, Y., Xiong, X., Lin, P.: Warning of dangerous operations of port container crane operators based on active human posture recognition. Cranes Transport Mach. **24**, 15–20 (2022)
2. Sheng, Y., Wang, J.: Research on human body posture recognition based on computer vision. Mod. Inform. Technol. **6**(16), 87–91+95 (2022)
3. Jin, W., Meng, J., Huang, Y.: Medical human body posture recognition method based on CNN and high-speed communication technology. J. Microcomput. Appl. **38**(07), 20–22+26 (2022)
4. Han, K., Huang, Z.: Falling behavior recognition method based on dynamic characteristics of human body posture. J. Hunan Univ. (Natl. Sci.) **47**(12), 69–76 (2020)
5. Qian, Y., Shen, Y.: Hybrid of pose feature and depth feature for action recognition in static image. Acta Automatica Sinica **45**(03), 626–636 (2019)
6. Zheng, X., Peng, X., Wang, J.: Human action recognition based on pose spatio-temporal features. J. Comput. Aided Design Comput. Graph. **30**(09), 1615–1624 (2018)
7. Ge, D.: Application of human body posture recognition technology in intelligent control system of smart substation. Electric Eng. **13**, 132–134 (2022)
8. Chai, D., Xu, C., He, J.: Inception neural network for human activity recognition using wearable sensor. J. Commun. **38**(Z2), 122–128 (2017)
9. Wei, X.: Human body posture recognition algorithm based on inertial sensors. Intell. Comput. Appl. **12**(06), 97–101+105 (2022)
10. Liu, J., Liu, Y., Jia, X.: Research on human pose visual recognition algorithm based on model constraints **41**(04), 208–217 (2020)

11. Zheng, L., Huang, X., Liang, R.: Human posture recognition method based on SVM. J. Zhejiang Univ. Technol. **40**(06), 670–675+691 (2012)
12. Sun, J., Han, S., Shen, Z.: Binocular human body posture distance positioning recognition based on double convolution Chain. Acta Armamentarii **43**(11), 2846–2854 (2022)
13. Huang, G., Li, Y.: A survey of human action and pose recognition. Comput. Knowl. Technol. **9**(01), 133–135 (2013)
14. Sun, Z., Li, H., Ye, J.: 3D human body joint recognition based on weakly supervised transfer network. J. Jilin Univ. (Eng. Technol. Edn) 1–9 (2023)
15. Yang, G.: Design and implementation of human posture recognition system based on deep learning. China's New Technol. New Prod. **07**, 22–24 (2022)
16. Duan, J., Liang, M., Wang, R.: Human pose recognition based on human bone point detection and multi-layer perceptron. Electron. Measure. Technol. **43**(12), 168–172 (2020)
17. Guo, X., Zhang, L.: Research and implementation of human body posture feature selection method. Comput. Eng. **37**(04), 184–186 (2011)
18. Li, Y.: Research on human posture recognition based on acceleration sensor. Electron. Compon. Inform. Technol. **6**(03), 1–3+6 (2022)
19. Zhou, K.: Fitness motion recognition system based on deep learning. Indust. Control Comput. **34**(06), 37–39 (2021)
20. Ma, Z., Lin, Y., Wang, Z.: Human posture recognition and fighting behavior monitoring in closed environments. Comput. Appl. **41**(S2), 214–220 (2021)

Data-Driven Smart City/Planet

.

MetaCity: An Edge Emulator with the Feature of Realistic Geospatial Support for Urban Computing

Lin Wu, Guogui Yang, Ying Qin, Baokang Zhao, Xue Ouyang[✉],
and Huan Zhou[✉]

National University of Defense Technology, Changsha 410073, China
{wulin20,ggyang,yingqin,bkzhao,ouyangxue08,huanzhou}@nudt.edu.cn

Abstract. The edge computing paradigm is an important supplement to the traditional cloud computing paradigm in current IoT application scenarios. However, edge computing is highly related to a specific application scenario, in which the mobility of edge devices and the geographical distribution of edge infrastructure are strongly correlated. However, it is expensive to deploy the solution in the real world and most current edge computing emulators lack realistic scenario support and mobility support. Therefore, it is challenging to evaluate whether an edge infrastructure deployment solution can satisfy the QoS (Quality-of-Service) requirement of an edge application in a cost-effective manner. In this paper, we propose and implement an edge emulator, MetaCity, which is able to effectively enforce edge computing policies and construct realistic application scenarios. MetaCity can leverage geographical data to establish an emulation environment according to the realistic infrastructure deployment strategy and emulate the mobility process of edge devices based on the actual urban road network. MetaCity can also provide an extensible network QoS monitoring module that supports the concurrent execution of various QoS monitoring in an emulated environment. In addition, MetaCity provides a user-friendly web-based graphical user interface instead of text-based configuration files. For evaluation, three smart transportation-based experiments are conducted to validate the functionality, scalability, and emulation accuracy of MetaCity.

Keywords: edge computing emulator · mobility emulation · realistic geospatial support · urban computing

1 Introduction

With the rapid growth of IoT (Internet of Things) technology, an increasing number of smart devices, such as smart cameras, smart watches, smart home gadgets, and smart cars are interconnected with the network, realizing the connectivity of all things. Meanwhile, these devices generate considerable data and requests. Traditional Cloud computing techniques are not sufficient to process

© The Author(s), under exclusive license to Springer Nature Singapore Pte Ltd. 2023
Z. Yu et al. (Eds.): ICPCSEE 2023, CCIS 1880, pp. 95–115, 2023.
https://doi.org/10.1007/978-981-99-5971-6_7

these data because of its centralized manner, which causes high transmission latency, high bandwidth consumption and energy consumption in data centers. Edge computing is therefore proposed to enable the network edge with computing capabilities, usually in the form of base stations (or named as edge clouds). By properly arranging the position of edge clouds, most of the data and requests can be processed closely without long-distance network transmission, and are therefore able to satisfy the application QoS, which is necessary for edge application developers and operators. In other words, to assure the application QoS in a specific scenario, for example, to operate the video streaming application for mobile devices in vehicles, the locations of edge clouds along roads are important. However, in most cases, it is not feasible to directly conduct the placement evaluation in the real world, because of the huge overhead, both time-wise and resource-wise.

Instead of the actual edge cloud placement evaluation, another important method is through the use of simulators (or emulators) to evaluate the quality of the location selection. However, existing simulators are faced with the following challenges.

- **Inaccurate simulations of real-world scenarios.** Most existing edge simulators provide only abstract network topology construction. Users need to build edge computing network topology using elements such as cloud center nodes, edge cloud nodes, and edge device nodes and customize the network topology's latency and bandwidth. Since the user-defined network topology may not be closely related to the realistic scenario, it is unreliable for the user to estimate the network QoS metrics of the wireless network environment. In this case, the simulated environment constructed by the edge computing simulator can easily be out of realistic scenarios.
- **Lack of complete support for mobility.** Edge devices in edge computing scenarios typically have high mobility and different movement ways, such as flying, walking and driving, etc. However, most existing edge simulators only support fixed network topology but do not support simulated edge device mobility to dynamically change the network topology in real-time. Although some edge simulators can support the mobility of edge devices, the movement paths of edge devices are either random or imaginative and unrealistic. The route and speed of edge device movement cannot usually be customized for these edge simulators that support mobility simulation.
- **Lack of real application support.** Most existing edge simulators are only based on various network models, where edge devices, edge clouds, links, and even network protocol stacks in an edge computing network environment are modelled and simulated through various models. In other words, all behaviors of devices and networks are based on models for network simulation, so the simulated environment constructed by the edge simulator cannot run actual network protocols. Since the simulation environment does not support real network protocols or provide a base runtime environment library, the edge simulator cannot run real applications. Another type of simulation tool emulates the network link but without operating system complete support. The

performance of real applications is therefore hardly being estimated through current simulators, even emulators. To clarify the concept for this paper, we name the former as the "simulator", which is all based on models. We name the latter as the "emulator", some parts of which are real components.

To address the above problems, this paper introduces real urban geospatial datasets provided by the OpenStreetMap service to build a realistic scenario-based edge computing deployment environment. Second, real urban road network data are utilized to plan and emulate close-to-realistic movement paths of edge devices. Then, the host nodes are emulated based on the OS virtualization technique to realize the requirement of running local applications. Finally, we propose and implement a hybrid network edge computing emulator, MetaCity[1], which supports both wireless and wired network environments, and provides features including the simulation of geographically distributed infrastructure, the mobility of devices, and the QoS monitoring of native applications in edge computing environments. Compared to other edge computing emulators that are difficult to use, MetaCity enables developers and operators to evaluate their products and strategies in a cost-effective, flexible, and highly realistic manner. In addition, MetaCity provides a web application with a user-friendly interface that is simple for developers and regular users, and contains a front-end framework that enables developers to customize the web application for edge computing emulators to their own requirements.

The rest of the paper is organized as follows. Section 2 presents related work. Section 3 denotes the architecture and design of MetaCity. The experimental studies are presented in Sect. 4. Finally, Sect. 5 concludes the paper with future work.

2   Related Work

Emulator and simulator technologies have been developed for some time in the field of IoT. In 2019, the Cloud Computing and Distributed Systems (CLOUDS) Lab at the University of Melbourne in Australia conceived and developed CloudSim [1], the most well-known cloud computing simulator. From 2015 to 2017, a three-year period of significant development occurred in edge computing. Emulator-related technologies for edge computing have emerged only since 2017. To date, emulator and simulator technologies for edge computing are still in the stage of development: neither has a complete system and ecosystem been established nor does a product have a position of absolute dominance.

In industry and academia, there are various network emulators and simulators. Mininet [2] is a stable, lightweight process virtualization network emulator that operates in a Linux environment. It enables users to design customized network environments and execute actual applications and protocols by rapidly building hosts, switches, and links. Mininet enables the building of customized software defined networking (SDN) [3] because it supports OpenFlow [4] and

---

[1] https://github.com/wulin-nudt/MetaCity.

OpenvSwitch [5], as well as provides an interface to SDN controllers. MaxiNet [6] is a distributed extension of Mininet that enables large SDN emulation on multiple physical machines. Containernet [7] is an extension of Mininet that enables the emulation of host nodes as Docker [8] containers. Mininet-WiFI [9] is a wireless network extension of Mininet that supports the emulation of wireless SDN by adding virtual wireless workstations and wireless access points. CORE (Common Open Research Emulator) [10] is a real-time network emulator that rapidly enables users to construct custom common virtual network topologies, supports Docker, and provides a user-friendly desktop interface. OMNeT++/INET [11] is a modular component-based discrete event network simulator, and the INET framework is an open-source communication network simulation library containing various protocol models, including TCP protocol models, wireless and wired protocol models, etc. It is primarily used for simulating communication networks. OPNET [12] is a commercial network simulator that holds a high level of integrity and reliability, as well as more protocol model packages. SCORE [13] is used to simulate and evaluate the performance of cluster scheduling policies in a cloud computing network environment.

The edge computing simulator abstractly models the infrastructures of the edge computing architecture by extracting characteristics such as processing performance and storage capacity from the infrastructure and then modelling the behavior of these infrastructures using mathematical models or predefined rules. The edge computing simulator does not execute actual applications and policies; rather, it uses built-in models to analyse the features of applications and policies. Harshit Gupta proposed iFogSim [14], a famous edge computing simulator based on the cloud simulator CloudSim that adheres to the Sense-Process-Actuate model. iFogSim can evaluate scheduling policies based on various quality-of-service metrics, but it does not support device mobility, energy usage, or the heterogeneity of IoT devices. iFogSim2 [15] extends iFogSim to support device mobility, cluster management, and microservice deployment policies. IoTSim-Edge [16] is an extension of CloudSim that includes an edge controller module mainly consisting of EdgeDataCenter, EdgeBroker, and EdgeDevice. IoTSim-Edge supports a variety of features, such as network communication, device heterogeneity, edge device communication protocols, battery usage, and edge device mobility. In addition, there are various other edge computing simulators, some of which extend to the cloud computing simulator CloudSim, such as EdgeCloudSim [17], CloudSimSDN [18], PureEdgeSim [19], and MobFogSim [20], and some of which extend to the OMANet++/INET network simulator, such as ECSim++ [21], FogNetSim++ [22], and xFogSim [23]. Since the edge computing simulator is dependent on various edge computing models, the completeness and precision of the models will significantly impact the final simulation, and it does not actually execute the evaluated applications and rules; its results suffer from inaccuracies.

The edge computing emulator allocates (virtual or actual) hardware resources to the emulated edge computing infrastructure to establish an edge computing emulation environment in which applications or policies can run, and then runs the evaluated application and policy in this edge computing emulation environment and evaluates the performance of these policies or applications in real-time. Ruben Mayer proposed EmuFog [24], an edge computing emulator based on the MaxiNet network emulator. EmuFog focuses on the generation of edge computing network topology rather than on-device heterogeneity, device mobility, service migration, and energy consumption. FogBed [25] extends the MaxiNet network emulator to support distributed edge computing environment emulation. Emuedge [26] is the first edge computing emulator that utilizes both OS-level and full-system virtualization, providing an interface that supports integration with edge computing simulators and testbeds. MockFog [27] can inject faults at runtime to evaluate applications under various deployment and failure situations. MockFog 2.0 [28] offers automated experiment orchestration. openLEON [29] is a multi-access edge computing end-to-end emulator based on MiniNet and srsLTE [30], which can support the emulation of mobile network and data center. Fogify [31] is an emulation framework that models complex edge computing topologies comprised of heterogeneous resources, network capabilities, and QoS criteria, and supports fault injection and configuration modification at runtime. MACE [32] is based on CORE and is capable of supporting node mobility control. It allows users to plan deployment topology and observe and interact with nodes in real-time. On the other hand, CloudsStorm [33] is a network testbed providing real network links among datacenters based on cloud computing, which can be leveraged to evaluate cloud performance [34].

Table 1 shows a comparison of existing edge computing emulators with MetaCity in terms of features. Existing edge computing emulators do not provide comprehensive support for the various features of edge computing, the majority of edge computing emulators lack suitable user interfaces, and some edge computing emulators only target and solve problems for specific application scenarios.

Table 1. The Comparison of Edge Computing Emulators

Feature / Emulator	mobility	geographical	visualization	QoS monitoring
EmuFog [24]	✗	✗	✗	✗
FogBed [25]	✗	✗	✗	✓
Emuedge [26]	✓	✗	✗	✗
MockFog 2.0 [28]	✗	✗	✗	✗
openLEON [29]	✓	✗	✗	✗
Fogify [31]	✓	✗	✗	✓
MACE [32]	✓	✗	✗	✗
MetaCity(this work)	✓	✓	✓	✓

3   MetaCity Architecture and Design

This section demonstrates the architecture and supported features of MetaCity. MetaCity is a convenient, low-cost, and efficient edge computing emulator that is based on Mininet-WIFI and ContainerNet, two well-known network emulators, enabling the construction of wireless network environment emulation and Docker container-based edge computing infrastructure emulation. In recent years, with the rise of smart transportation and smart city, to enable MetaCity to emulate a more realistic edge computing environment that is suitable for these application scenarios, MetaCity uses OpenStreetMap [35], a third-party open-source, free, editable mapping service built using the collective power of the public. OpenStreetMap not only has comprehensive and accurate geospatial data, but the geospatial data are owned by anyone and can be used for any purpose. MetaCity employs the OSMnx [36] python package, a third-party python tool that enables users to get geospatial data from OpenStreetMap and to quickly process and visualize these geospatial data. Since the OSMnx python package is constructed according to a process-oriented paradigm, its extensibility is extremely poor; therefore, we built on top of it a more extensible geospatial data processing Python library that serves MetaCity.

To emulate a more realistic edge computing environment and to support use cases such as urban planning and the metaverse [37], the MetaCity design objectives can satisfy and support the following characteristic:

Realistic Geographic Distribution: Edge computing is a scenario that covers a broad range of people and radiation, typically a city's area. When deploying and installing edge computing infrastructure, geographic dimension elements such as distance and location between facilities exist, and MetaCity can completely support these realistic geographic distribution characteristics.

Complete Process Emulation of Edge Device Mobility: Edge device mobility is a continuous, non-static, and non-instantaneous change process; hence, a highly believable edge device mobility emulation is to conduct the complete process of mobility tracking. Moreover, edge device mobility paths should be actual rather than imaginary. MetaCity employs geospatial data to plan and predict device mobility paths and comprehensively analyse edge devices' quality of service changes during the mobility process.

Hybrid Network Emulation: The network environment of edge computing is typically not a simple network but rather a hybrid network architecture, such as 5G, WIFI, and Ethernet. MetaCity can emulate both wireless and wired networks to support as much diversity as possible in edge computing network environments.

User-Friendly: To reduce the user's threshold and give rapid visual feedback on emulation outcomes, MetaCity follows the trend of Internet development and develops the emulator as a web application to assist users in designing and building the desired edge computing emulation environment.

Customized Emulation Experiment Applications: MetaCity provides a front-end framework based on the back-end that developers can use to customize emulator web applications without familiarity with the front end.

Furthermore, MetaCity supports features such as infrastructure resource heterogeneity, communication link heterogeneity, extensibility, scalability, QoS monitoring, lightweight deployment, low cost, replication, and convenient repeatable experiments.

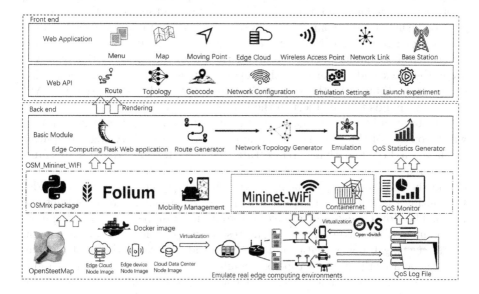

Fig. 1. The architecture overview of MetaCity

3.1   Overview of MetaCity

MetaCity employs mininet-wifi and containerernet as fundamental components to ensure the rapid prototype implementation of an edge computing emulator, and gradually expands the functionality based on these network emulators. MetaCity's underlying architecture assures that it can support hybrid network topology emulation and OS virtualization (Docker containers), allowing it to directly deploy and execute real applications and policies to achieve outcomes that are close to reality. MetaCity is a web application that enables users to construct the required network topology and rapidly test and evaluate their apps and policies in this emulated edge computing environment.

The MetaCity architecture is composed of five primary layers. Figure 1 demonstrates a high level overview of the MetaCity architecture's components and their relationships. Although MetaCity as a web application is primarily divided into front-end and back-end parts, the front-end content is entirely rendered by the back-end; hence, the front-end and back-end are not separated. As the interface to interact with users, the front end of MetaCity consists of two layers: a web application and web API. The web application layer provides several basic apps, such as map applications and edge computing infrastructure applications, which users can use to construct the required edge computing network topology and environment atop city spatial map application. Each application in the web application layer corresponds to one object in the back end, and developers can consult the related object's application programming interface (API) about development. The Web API layer provides the API interface to interact with the back end, through which the front end sends JSON data to the back end, such as edge computing network topology, infrastructure geolocation, moving point routing location parameters, network configuration, and emulation experiment configuration. After receiving the JSON data sent by the Web API layer, the back end can execute the corresponding edge computing emulation experiments. Similar to the web application layer, each API in the web API layer corresponds to one object in the back end.

The back end handles requests for emulation experiments from the front end, which consists of three layers: the basic module, the MetaCity edge computing emulator, and the underlying architecture. The basic module layer contains functional modules for external usage and interaction with the front end; users can obtain the corresponding MetaCity services by calling these modules. The Edge Computing Flask [38] Web application module generates and renders front-end pages. The Route Generator receives the route location parameters of the moving point and then constructs the shortest move path for that moving point. After receiving topology data from the front end, the Network Topology Generator generates the network topology emulation experiment Python executable file. The Emulation module is accountable for invoking the underlying edge computing emulator to execute the emulation experiment executable file, perform some preprocessing, and provide feedback. QoS Statistics Generator analyses the underlying QoS log files and presents the results as a visual Vega [39] JSON file. The MetaCity edge computing emulator uses mininet-wifi and containernet as its fundamental components. It also adopts and extends OSMnx and Folium to process geographical data and render leaf map pages. The mobility management module in MetaCity is primarily responsible for emulating the moving process of moving points. In addition, the QoS monitor is responsible for executing applications or commands in the emulated nodes to collect the required metric data.

In the underlying architecture, the OpenStreetMap service provides city geospatial data for map page rendering and route planning. Users use predefined Docker images to instantiate various infrastructures in the edge computing architecture, and these Docker images commonly contain edge cloud, edge device, and cloud data center images. OpenvSwitch [40] is a network virtualization technique that visualizes a router or switch within an edge computing architecture. In an emulated network environment, Linux Traffic Control [41] and Linux netem [42] control the QoS performance of network links. While monitoring QoS in the emulated environment, some QoS log files are generated. Eventually, a basic network emulation environment for edge computing was constructed.

3.2 Edge Device Mobility Emulation

The edge device layer usually contains high-mobility devices or facilities, such as smartphones, laptops, UAVs, and IoT vehicles, demanding a more realistic emulation of edge device mobility. MetaCity emulates edge device mobility through four steps: specifying edge device (moving point) routing location, planning edge device move paths, configuring edge device mobility, and emulating edge device mobility. The four steps are described as follows:

- **specifying routing location:** Users can specify several moving locations for edge devices on the city map, such as three moving point locations XYZ, which are depicted as edge devices moving from X to Y and then Y to Z. Due to the possibility that the user-specified moving location does not exist in the geospatial data, the back end uses scikit-learn [43] to construct a ball tree for haversine [44] nearest neighbor search, then searches for the road node closest to the moving location in the OpenStreetMap geospatial data and uses this node as the edge device moving location for the final emulation.
- **planning move paths:** Once the edge device routing location has been determined, it is required to plan each segment of a path for edge device moves, such as three moving point locations XYZ, and two shortest routes must be planned from X to Y and from Y to Z. There are two modes of path planning: drive and walk. The former considers one-way directionality, while the latter ignores it, resulting in significant disparities in path planning on the road network for the two modes.
- **configuring device mobility:** At the finish of path planning, the user should also configure the fundamental parameters of edge device movement, such as speed, wireless access point handover association mechanism, and propagation model. During emulation, the speed parameter impacts the total move time of the edge device and the moving route points' location. The primary wireless access point handover association mechanisms are least loaded first (llf) and strongest signal first(ssf). ssf focuses on the received signal strength indicator (rssi) between the edge devices and the wireless access point, whereas llf focuses on the workload of the wireless access point. The propagation model is the basis for calculating rssi values, and the default propagation model is the log-distance propagation loss model (logDistance) [45].
- **emulating mobility:** MetaCity creates two threads to emulate the movement process of edge devices. One thread simultaneously emulates the movement process of all edge devices, and the other thread sets the network QoS of all edge devices throughout the movement process. In the emulation of the edge device movement process, the location of the edge device movement per millisecond throughout the full path is calculated in advance. Then the location to be emulated is selected from the calculated location in units every 0.1 s. To optimize the calculation processes of emulation, the emulation's total time and move location differ from the actual movement of the edge device. When emulating network QoS changes during edge device movement, the network QoS performance of the edge device is adjusted in real-time according to

factors such as the edge device's distance from its associated wireless access point, the handover association mechanism, and the IEEE 802.11 standard.

3.3   QoS Monitor

The QoS monitoring framework consists of two components: a monitor specifying what QoS analysis tools to use and how to capture QoS, and a parser defining how to parse the collected QoS results. MetaCity has provided a module for monitoring basic network QoS that is highly extendable. Based on this network monitoring module, the network delay monitor, network bandwidth monitor, and their corresponding parser are implemented, the former for monitoring metrics such as network delay and packet loss rate, and the latter for monitoring metrics such as bandwidth. Before running the monitor, the corresponding Docker containers of the edge computing infrastructure should be pre-installed with relevant third-party network performance evaluation tools or custom tools (e.g., ping and iperf3), and the network monitor automatically runs these network performance evaluation tools to capture network QoS metrics in real-time.

A QoS monitor can simultaneously instantiate multiple instance objects, each of which is a separate thread, and these instance objects can be placed on edge computing infrastructures for QoS monitoring. We use pseudo-terminal (virtual terminal pseudo-tty) technology and IO multiplexing to extend the original edge computing nodes, thereby achieving the demand for an edge computing infrastructure to run many monitors simultaneously. Therefore, each monitor holds a private shell terminal to run the QoS evaluation tools in the computing infrastructure. Many monitor instances of the same type or different types can run concurrently on an edge computing infrastructure. These monitor instances run processes without interfering with each other, but together affect the consumption of edge computing infrastructure resources (e.g., CPU, network bandwidth resources for Docker containers).

To optimize the QoS monitoring process in the emulated edge computing environment, all QoS monitoring tasks are aggregated and classified, and those that can be executed together are merged. For example, the delay and packet loss monitoring tasks from node A to node B can be merged into one task (e.g., QoS monitor using Ping for network performance evaluation) rather than two tasks (threads) to execute this task independently. The control groups (Cgroups) mechanism provided by the Linux kernel is used to manage the allocation of resources to Docker containers and maintain the current operational status of the containers in real-time. Hence, the Cgroups mechanism can be used to achieve QoS monitoring for computations.

3.4   The Framework of Front End

The front-end framework is built on top of Flask, Folium's branca library, leaf.js [46], and Layui.js, with front-end code entirely generated by back-end rendering. Each JavaScript object, HTML element, and event-handling function in the front-end page corresponds to a python object in the back end, and the

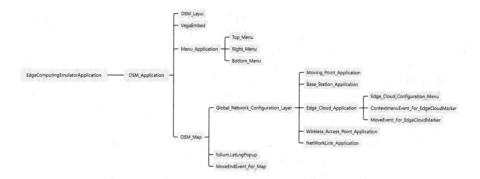

Fig. 2. An Overview of The Front End Framework

back-end objects define and establish the relationships between the front-end elements. Figure 2 gives an overview of the front-end framework's components and relationships. It is evident from the picture that the front-end framework is represented as a tree and that it is organized similarly to the HTML DOM (Document Object Model).

The back-end python object establishes their parent-child relationship through the add_child or add_to method, establishing a front-end framework tree; the object determines the rendering order of the parent and child nodes through the render method, hence determining the order in which the framework nodes are rendered when the front end framework tree is rendered into a front end HTML+JS page; the object defines the content of the Jinja2 template through the class variable _template, which decides the content and code that will eventually be presented on the front end page. Knowing that each node can define its parent-child rendering order, the actual rendering order of each node in the front-end framework tree is a mixture of preorder and postorder traversal.

The code for the front-end pages is embedded in the back-end objects to establish a front-end framework. Although this approach does not divide the front end and back end and makes maintenance more challenging, it can increase the code reuse rate. Back-end objects modify the generated front-end pages in a configurable manner, and the object relationships can be altered or created as needed. Therefore, users can use the front-end framework objects to build their own front-end page content and style styles without being too familiar with front-end development, and eventually realizing the customization required for emulator web applications or edge computing web applications.

3.5   Realistic Edge Computing Network Topology Construction

Utilizing realistic city geospatial maps, creating a more realistic edge computing network topology is achievable. First, specify the city where the edge computing infrastructure will be deployed or have the system automatically locate your location. On the city map, it is available to choose the building or geographic location where edge computing infrastructure, such as base stations, edge devices, and edge clouds, will be deployed to match the deployment in the real world. After determining the locations of the edge computing infrastructures and the links between them, the front end will record their geographical factors, including latitude, longitude, relative location, and distance. Although OpenStreetMap provides only 2-dimensional geospatial data, the deployment of infrastructure can support both 2-dimensional and 3-dimensional geospatial data. Hence, such a network topology of edge computing is realistic with fully incorporated geospatial elements.

In network topology construction, the front end page provides the fundamental elements of moving points (edge devices), WIFI points, edge clouds, base stations, and network links, allowing for the customization of simple to complicated edge computing network topologies. In the edge computing network topology, WIFI points and base stations serve as intermediary devices for all elements to connect and communicate, i.e., direct links cannot be constructed between other elements (e.g., edge cloud and edge device), which must instead use WIFI points or base stations to communicate. Network links establish connectivity between nodes and configure basic network QoS requirements such as bandwidth, latency, jitter, and packet loss rates for each node.

In implementing the underlying network topology for edge computing, the WIFI access point and base station are represented as an OpenvSwitch virtual switch, and the network link is represented as a virtual network card interface. Linux traffic control and Linux netem configure and set the network QoS of the link. The edge cloud and edge devices are represented as Docker containers, resulting in a hybrid network topology consisting of Docker containers with virtual switches and mac80211_hwsim [47] wireless network simulation. Nodes in the network topology rely on OpenvSwitch's flow tables or routing rules for indirect communication. In contrast, direct communication between nodes requires virtual ethernet device (veth) pairs to build point-to-point links and the configuration of the correct routing tables to communicate. Regarding computational realism, the computational resources of the edge infrastructure can be configured and limited utilizing container and Cgroups mechanisms. The computational performance of the edge infrastructure containers can be monitored within the relevant Cgroups directories.

4   Evaluation

This section conducts a study to evaluate the functionality of the MetaCity. First, a typical edge computing use case scenario is described to be emulated by MetaCity. Next, the network topology and experimental configuration of the emulated edge computing use case scenario are presented, together with the hardware configuration of the MetaCity-running virtual machine. Finally, we present the results of the execution of edge computing emulation experiments.

4.1   A Demo Use Case of Edges for Urban Computing

Smart transportation is one of the most typical application scenarios for edge computing. In smart transportation, the most common services include real-time traffic road condition push, road traffic data collection, vehicle location data collection, edge-side data processing and rapid feedback, and many other services with stringent real-time requirements. With the advent of autonomous driving technologies over the past several years, real-time system response requirements have increased dramatically. For instance, when a driverless car encounters an unexpected situation, it must immediately perform calculations at the edge and receive feedback before taking emergency measures and applying the brakes. If the driverless car sends the data to the cloud for processing, the driverless car may be involved in an accident before waiting for the response results.

With the rise of cities and the vast increase in vehicles on the road, a city creates enormous amounts of traffic data daily. Uploading this enormous volume of traffic data directly to the cloud for processing would consume valuable bandwidth and storage resources and produce high latency feedback, which is unsustainable. In edge computing application scenarios, data should be collaboratively processed by the cloud-edge to achieve real-time feedback and highly accurate results.

In this experiment, a simple edge computing network topology is built based on a smart transportation scenario. The QoS variations between a moving vehicle and the edge infrastructure will be emulated.

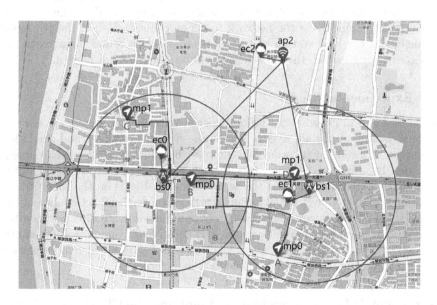

Fig. 3. A Simple Edge Computing Network Topology of Smart Transportation

4.2   Experimental Configuration

The MetaCity emulator runs on a WMware virtual machine running the ubuntu 20.04.4 operating system, with 4 cores at 2.8GHz, 4G of RAM, and 50GB disk. Since the emulation experiments do not currently consider computational QoS, the use of computational resources of the emulated edge computing infrastructure (docker containers) is unrestricted and uncontrolled.

Based on the previously described smart transportation scenario, a simple network topology for edge computing is constructed. As shown in Fig. 3, this edge computing network topology consists of two moving points (mp0 and mp1), two edge clouds (ec0 and ec1), one cloud data center (ec2), two base stations (bs0 and bs1), one wireless access point router (ap2), and five cable links. The moving points (edge devices or smart vehicles) are connected to the base station through a wireless network using wireless channel one and the IEEE 802.11g standard. Edge clouds, cloud data centers, and moving points will be virtualized as docker containers, allowing the execution of real applications and policies in the above network topology. Hence, Fig. 3 shows a basic hybrid network topology for edge computing.

Table 2 illustrates the emulation configuration parameters for the link, base station, and moving point in the above edge computing hybrid network topology. The base station and wireless access point router are essentially OpenvSwitch. When the fail mode configuration parameter is standalone, the OpenvSwitch turns into a standard MAC learning layer 2 switch when it cannot connect to the SDN controller. SDN does not cope well with rapid network topology changes caused by mobility, i.e., the flow table entries issued by the SDN controller are not updated as quickly as the network topology changes. Hence, the fail mode parameters of the base station and wireless access point router, along with the SDN controller to which they are linked, will be configured as needed.

4.3   Experimental Study

The back end will generate a Python experiment script with geographical elements according to the above edge computing network topology and emulation parameters. The MetaCity emulator enforces the edge computing environment emulation by directly executing this Python experiment script and recording the desired emulation results as the script executes. Based on the smart transportation scenario above, the experiment will evaluate the functionality of the MetaCity emulator in three aspects: (I) communication emulation between the moving point and the cloud data center; (II) end-to-end communication emulation between the moving point and edge clouds; and (III) concurrent emulation of multiple moving points and multiple QoS monitoring.

Experiment 1: Between the Moving Point and the Cloud Data Center
moving point mp0 from point A to point B takes approximately 976 m. The move speed is 10 m/s, and the total time should be 97.6 s. Due to MetaCity's optimization of the computation process of the emulated position points, the

Table 2. Emulation Parameters

element	parameter	value	element	parameter	value
Link(bs0,ap2)	delay	80 ms	MovingPoint(mp0)	speed	10 m/s
	band width	10 M		starting-point	A
	jitter	10 ms		end-point	B
	loss	5 %	BaseStation(bs0)	signal range	500 m
Link(bs1,ap2)	delay	40 ms		fail mode	standalone
	band width	20 M	BaseStation(bs1)	signal range	500 m
	jitter	5 ms		fail mode	standalone
	loss	10 %	AccessPoint(ap2)	signal range	10 m
Link(bs1,ec1)	delay	10 ms		fail mode	standalone
	band width	10 M	MovingPoint(mp1)	speed	12 m/s
	jitter	1 ms		starting-point	C
	loss	0 %		end-point	D
Link(bs0,ec0)	delay	20 ms	Link(ap2,ec2)	delay	5 ms
	band width	10 M		band width	10 M
	jitter	5 ms		jitter	1 ms
	loss	0 %		loss	0 %

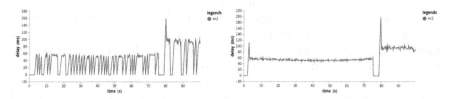

(a) Delay analysis of mp0 (non-zero packet (b) Delay analysis of mp0 (zero packet loss) loss)

Fig. 4. The Emulation Results between the Moving Point and the Cloud Data Center

actual total emulation time is slightly faster than the actual move time. Figure 4 shows the outcomes of the emulation for 100 s. Figure 4a or 4b depicts that the routing handover at moving point mp0 occurs at approximately 75 s. The delay of moving point mp0 varies around 53ms before the routing handover and around 95ms after the handover. As the wireless communication between moving point mp0 and the base station likewise has a delay variation of approximately ten milliseconds, the outcomes of the delay emulation are accurate. As depicted in Fig. 4a, the packet loss rate difference before and after the routing handover is calculated to be approximately 4.4%, and the emulation results for the packet loss rate are likewise primarily compatible with the parameter configuration.

After adjusting the packet loss rate of Link(bs0,ap2) and Link(bs1,ap2) to 0%, the emulation result is depicted in Fig. 4b.

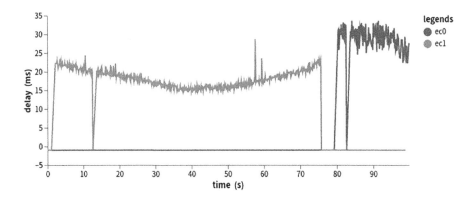

Fig. 5. The Emulation Results between the Moving Point and Edge Clouds

Experiment 2: Between the Moving Point and Edge Clouds. In the smart transportation scenario, smart cars send data through nearby base stations to the edge cloud for preprocessing instead of sending data directly to the cloud for processing. As described previously, this experiment emulates the communication between moving points and edge clouds. The hybrid network topology of edge computing described above supports indirect communication from the moving point to the edge cloud via the wireless access point router; therefore, Link(bs0,ap2) and Link(bs1,ap2) must be removed from this experiment, and only allow direct communication between the moving point and the edge cloud via the base station. Figure 5 depicts the results of the end-to-end communication emulation. The moving point mp0 is initially out of range of the signal from the base station bs0 and is, therefore, unable to establish communication with the edge cloud ec0. As mp0 leaves the signal range of bs1 and enters the signal range of bs0, it can naturally communicate with ec0 but not ec1. When mp0 is within the signal crossing range of bs1 and bs0, the wireless access point handover association mechanism selects the associated base station. The delay between mp0 and each edge cloud depicted in Figure 5 is mainly consistent with the configuration of the emulation parameters.

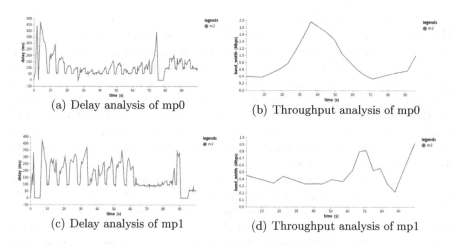

(a) Delay analysis of mp0

(b) Throughput analysis of mp0

(c) Delay analysis of mp1

(d) Throughput analysis of mp1

Fig. 6. The Result of Concurrent Emulation

Experiment 3: Concurrent Emulation of Multiple Moving Points and Multiple QoS Monitoring. In edge computing usage scenarios, the number of edge devices is usually enormous, and each edge device runs various services concurrently. This experiment investigates the scenario of concurrently emulating the mobility of several moving points (edge devices), running multiple QoS monitors per moving point, and setting the packet loss rate to 0 for all links in the configuration described above for better emulation results. Figure 6 shows the latency and bandwidth QoS monitoring results for each moving point mp0 and mp1 during concurrent moves. By comparing the results of Fig. 6a with those of Fig. 4b, it is evident that latency QoS monitoring and bandwidth QoS monitoring that runs concurrently will affect each other and compete for network resources. Emulating the concurrent running of numerous moving points and multiple QoS monitoring demands an adequate host hardware resource supply. When the physical machine running the MetaCity emulator has adequate hardware resources, it can emulate large-scale edge computing environments at a lower cost.

5 Conclusion

In this paper, we design and implement MetaCity, an edge computing emulator based on Mininet-WiFI and Containernet. It enables users to quickly build a hybrid network topology for edge computing that takes real geographic information into consideration, and is capable of executing actual edge computing applications and policies in a constructed emulation environment. Utilizing MetaCity to thoroughly test and evaluate these applications and policies can provide useful insights before deploying them to production environments.

MetaCity uses the city road network information provided by Open-StreetMap to build a realistic emulation environment where edge computing infrastructure is deployed and placed. To emulate the mobility characteristics, MetaCity uses real road network data to plan suitable and accurate paths for edge devices during mobility emulation. To track network QoS metrics during the emulation, MetaCity provides a scalable network QoS monitoring module that can perform various QoS monitoring concurrently. Four experiments are designed based on the smart transportation edge computing scenario to evaluate the functionality of the MetaCity, and the impact of the SDN controller's policy of issuing flow tables on edge computing mobility emulation is discussed. Meanwhile, MetaCity is functional and easy to use, and it provides a user-friendly interface as a web application.

In conclusion, MetaCity enables users to build edge computing emulation environments and evaluate applications and policies cost-effectively, scalably, efficiently, and realistically. In future work, two improvements are considered. First, we want to implement a computing QoS monitoring module utilizing Docker and control groups mechanisms, as MetaCity is based on Docker containers to emulate edge computing infrastructure and provides control over computing resources. Second, to improve the scalability of MetaCity to enable the construction of large-scale edge computing environments with distributed multi-physical machine emulation.

Acknowledgment. The work is supported by the National Natural Science Foundation of China under grant No. 62102434, No. 62002364 and No. U22B2005, and is partially supported by the Natural Science Foundation of Hunan Province under grant No. 2022JJ30667.

References

1. Calheiros, R.N., Ranjan, R., Beloglazov, A., De Rose, C.A., Buyya, R.: Cloudsim: a toolkit for modeling and simulation of cloud computing environments and evaluation of resource provisioning algorithms. Software: Pract. Exper. **41**(1), 23–50 (2011)
2. Lantz, B., Heller, B., McKeown, N.: A network in a laptop: rapid prototyping for software-defined networks. In: Proceedings of the 9th ACM SIGCOMM Workshop on Hot Topics in Networks, pp. 1–6 (2010)
3. De Oliveira, R.L.S., Schweitzer, C.M., Shinoda, A.A., Prete, L.R.: Using mininet for emulation and prototyping software-defined networks. In: IEEE Colombian Conference on Communications and Computing (COLCOM), pp. 1–6. IEEE (2014)
4. McKeown, N., et al.: Openflow: enabling innovation in campus networks. ACM SIGCOMM Comput. Commun. Rev. **38**(2), 69–74 (2008)
5. Pfaff, B., Pettit, J., Amidon, K., Casado, M., Koponen, T., Shenker, S.: Extending networking into the virtualization layer. In: Hotnets (2009)
6. Wette, P., Dräxler, M., Schwabe, A., Wallaschek, F., Zahraee, M.H., Karl, H., Maxinet: Distributed emulation of software-defined networks. In: IFIP Networking Conference, pp. 1–9. IEEE (2014)
7. Peuster, M., Karl, H., van Rossem, S.: Medicine: Rapid prototyping of production-ready network services in multi-pop environments. In: 2016 IEEE Conference on Network Function Virtualization and Software Defined Networks (NFV-SDN), pp. 148–153 (2016)

8. Merkel, D., et al.: Docker: lightweight linux containers for consistent development and deployment. Linux j **239**(2), 2 (2014)

9. Fontes, R.R., Afzal, S., Brito, S.H., Santos, M.A., Rothenberg, C.E.: Mininet-wifi: emulating software-defined wireless networks. In: 2015 11th International Conference on Network and Service Management (CNSM), pp. 384–389. IEEE (2015)

10. Ahrenholz, J., Danilov, C., Henderson, T.R., Kim, J.H.: Core: a real-time network emulator. In: MILCOM 2008–2008 IEEE Military Communications Conference, pp. 1–7. IEEE, 2008

11. Varga, A., Hornig, R.: An overview of the omnet++ simulation environment. In: 1st International ICST Conference on Simulation Tools and Techniques for Communications, Networks and Systems (2010)

12. Chang, X.: Network simulations with opnet. In: Proceedings of the 31st Conference on Winter simulation: Simulation–a Bridge to the Future, vol. 1, pp. 307–314 (1999)

13. Fernández-Cerero, D., Fernández-Montes, A., Jakóbik, A., Kołodziej, J., Toro, M.: Score: simulator for cloud optimization of resources and energy consumption. Simul. Model. Pract. Theory **82**, 160–173 (2018)

14. Gupta, H., Vahid Dastjerdi, A., Ghosh, S.K., Buyya, R.: IFOGSIM: a toolkit for modeling and simulation of resource management techniques in the internet of things, edge and fog computing environments, Software: Pract. Exper. **47**(9), 1275–1296 (2017)

15. Mahmud, R., Pallewatta, S., Goudarzi, M., Buyya, R.: ifogsim2: an extended ifogsim simulator for mobility, clustering, and microservice management in edge and fog computing environments. J. Syst. Softw. **190**, 111351 (2022)

16. Jha, D.N., et al., IOTSIM-edge: a simulation framework for modeling the behavior of internet of things and edge computing environments. Software: Pract. Exper. **50**(6), 844–867 (2020)

17. Sonmez, C., Ozgovde, A., Ersoy, C.: Edgecloudsim: an environment for performance evaluation of edge computing systems. Trans. Emerg. Telecommun. Technol. **29**(11), e3493 (2018)

18. Son, J., Dastjerdi, A.V., Calheiros, R.N., Ji, X., Yoon, Y., Buyya, R.: Cloudsimsdn: modeling and simulation of software-defined cloud data centers. In: 15th IEEE/ACM International Symposium on Cluster, Cloud and Grid Computing, pp. 475–484. IEEE (2015)

19. Mechalikh, C., Taktak, H., Moussa, F.: Pureedgesim: a simulation toolkit for performance evaluation of cloud, fog, and pure edge computing environments In: International Conference on High Performance Computing and Simulation (HPCS), pp. 700–707. IEEE (2019)

20. Puliafito, C., et al.: Mobfogsim: simulation of mobility and migration for fog computing. Simul. Model. Pract. Theory **101**, 102062 (2020)

21. Nguyen, T.-D., Huh, E.-N.: Ecsim++: an INET-based simulation tool for modeling and control in edge cloud computing. In: 2018 IEEE International Conference on Edge Computing (EDGE), pp. 80–86. IEEE (2018)

22. Qayyum, T., Malik, A.W., Khattak, M.A.K., Khalid, O., Khan, S.U.: Fognetsim++: a toolkit for modeling and simulation of distributed fog environment. IEEE Access **6**, 63 570–63 583 (2018)

23. Malik, A.W., Qayyum, T., Rahman, A.U., Khan, M.A., Khalid, O., Khan, S.U.: Xfogsim: a distributed fog resource management framework for sustainable IoT services. IEEE Trans. Sustain. Comput. **6**(4), 691–702 (2020)

24. Mayer, R., Graser, L., Gupta, H., Saurez, E., Ramachandran, U.: Emufog: extensible and scalable emulation of large-scale fog computing infrastructures. In: IEEE Fog World Congress (FWC), pp. 1–6. IEEE (2017)

25. Coutinho, A., Rodrigues, H., Prazeres, C., Greve, F.: Scalable fogbed for fog computing emulation. In: 2018 IEEE Symposium on Computers and Communications (ISCC), pp. 00 334–00 340. IEEE (2018)

26. Zeng, Y., Chao, M., Stoleru, R.: Emuedge: a hybrid emulator for reproducible and realistic edge computing experiments. In: 2019 IEEE International Conference on Fog Computing (ICFC), pp. 153–164. IEEE (2019)

27. Hasenburg, J., Grambow, M., Grünewald, E., Huk, S., Bermbach, D.: Mockfog: emulating fog computing infrastructure in the cloud, in 2019 IEEE International Conference on Fog Computing (ICFC), pp. 144–152. IEEE (2019)

28. Hasenburg, J., Grambow, M., Bermbach, D.: Mockfog 2.0: automated execution of fog application experiments in the cloud. IEEE Trans. Cloud Comput. (2021)

29. Andrés Ramiro, C., Fiandrino, C., Blanco Pizarro, A., Jiménez Mateo, P., Ludant, N., Widmer, J.: Openleon: an end-to-end emulator from the edge data center to the mobile users. In: Proceedings of the 12th International Workshop on Wireless Network Testbeds, Experimental Evaluation and Characterization, pp. 19–27 (2018)

30. Gomez-Miguelez, I., Garcia-Saavedra, A., Sutton, P.D., Serrano, P., Cano, C., Leith, D.J.: SRSLTE: an open-source platform for lte evolution and experimentation. In: Proceedings of the Tenth ACM International Workshop on Wireless Network Testbeds, Experimental Evaluation, and Characterization, pp. 25–32 (2016)

31. Symeonides, M., Georgiou, Z., Trihinas, D., Pallis, G., Dikaiakos, M.D.: Fogify: a fog computing emulation framework. In: IEEE/ACM Symposium on Edge Computing (SEC), pp. 42–54. IEEE (2020)

32. Ferreira, B.C., Dufour, G., Silvestre, G.: Mace: a mobile ad-hoc computing emulation framework. In: 2021 International Conference on Computer Communications and Networks (ICCCN), pp. 1–6. IEEE (2021)

33. Zhou, H., et al.: Cloudsstorm: a framework for seamlessly programming and controlling virtual infrastructure functions during the devops lifecycle of cloud applications. Software: Pract. Exper. **49**(10), 1421–1447 (2019)

34. Li, M., Su, J., Liu, H., Zhao, Z., Ouyang, X., Zhou, H.: The extreme counts: modeling the performance uncertainty of cloud resources with extreme value theory. In: Troya, J., Medjahed, B., Piattini, M., Yao, L., Fernandez, P., Ruiz-Cortes, A. (eds.) Service-oriented Computing: 20th International Conference, ICSOC: Seville, Spain, November 29-December 2, 2022. Proceedings, LNCS, pp. 498–512. Springer, Cham (2022). https://doi.org/10.1007/978-3-031-20984-0_35

35. Haklay, M., Weber, P.: Openstreetmap: user-generated street maps. IEEE Pervasive comput. **7**(4), 12–18 (2008)

36. Boeing, G.: Osmnx: new methods for acquiring, constructing, analyzing, and visualizing complex street networks. Comput. Environ. Urban Syst. **65**, 126–139 (2017)

37. Mystakidis, S.: Metaverse. Encyclopedia **2**(1), 486–497 (2022)

38. Grinberg, M.: Flask web development: developing web applications with python. O'Reilly Media, Inc. (2018)

39. Satyanarayan, A., Moritz, D., Wongsuphasawat, K., Heer, J.: Vega-lite: a grammar of interactive graphics. IEEE Trans. Visual. Comput. Graph. **23**(1), 341–350 (2016)

40. Pfaff, B., et al.: The design and implementation of open vswitch. In: 12th {USENIX} Symposium on Networked Systems Design and Implementation ({NSDI} 15), pp. 117–130 (2015)

41. Almesberger, W.: Linux traffic control-next generation. In: Proceedings of the 9th International Linux System Technology Conference (Linux-Kongress 2002), pp. 95–103 (2002)

42. Hemminger, S., et al.: Network emulation with netem. In Linux conf au, vol. 5. Citeseer, p. 2005 (2005)
43. Pedregosa, F., et al.: Scikit-learn: machine learning in python. J. Mach. Learn. Res. **12**, 2825–2830 (2011)
44. Haversine formula. https://en.wikipedia.org/wiki/Haversine_formula
45. Log-distance path loss model. https://en.wikipedia.org/wiki/Log-distance_path_loss_model
46. Crickard III, P.: Leaflet.js essentials. Packt Publishing Ltd (2014)
47. Malinen, J.: mac80211_hwsim: Software simulator of 802.11 radio (s) for mac80211, Online. Accessed, vol. 19 (2017)

Multifunctional Sitting Posture Detector Based on Face Tracking

Zhaoning Jin[1], Jiahan Wei[2], Zhiyan Yu[3], and Yang Zhou[2(✉)]

[1] School of Information and Electronic Engineering, Zhejiang University of Science and Technology, Hangzhou, China
[2] Chinese-German Institute for Applied Engineering, Zhejiang University of Science and Technology Hangzhou, Hangzhou, China
zybuaa@163.com
[3] School of Automation and Electrical Engineering, Zhejiang University of Science and Technology Hangzhou, Hangzhou, China

Abstract. To reduce the vision problems caused by improper sitting posture, the research group used Raspberry Pi as the main controller for a multifunctional sitting posture detector with functions such as sitting posture detection, face positioning, cloud monitoring, etc. UUsing technologies or algorithms such as machine vision and convolutional neural networks, our design can realize the user's sitting posture error detection, such as left, right, low head position, or forward body position with alarming, so that the user can maintain the appropriate sitting posture.

Keywords: sitting posture detection · face tracking · Raspberry Pi · machine vision · convolutional neural network

1 Introduction

Currently, myopia has become a very common eye disease in society. The number of myopia cases in our country has reached one-third of the national population, and the myopia rate of junior high school students has exceeded more than 70 of the important reasons. The incorrect sitting posture of students may make the distance between the eyes and the book too close, resulting in refractive errors of the eyes [1,2]. Some students write homework or read in a comfortable posture (such as lying on their back, lying on their side, etc.), and some students recline on the sofa at will, which is a major threat to physical development. [3] These incorrect sitting postures not only cause vision loss but may also cause problems such as cervical spine problems or scoliosis.

At present, various types of sitting posture detectors have been developed at home and abroad, but there are defects such as insufficient comfort, large environmental restrictions, low detection accuracy, low fault tolerance, short battery life, and easy dependence on users.

This paper uses the OpenCV function database, convolutional neural network architecture, face recognition and other technologies to design a multifunctional

Z. Yu et al. (Eds.): ICPCSEE 2023, CCIS 1880, pp. 116–129, 2023.
https://doi.org/10.1007/978-981-99-5971-6_8

sitting posture instrument for detecting sitting posture with an alarm function. The innovations are as follows:

Accurate Measurement: The infrared detection module and the visual detection module are used to jointly detect the sitting posture of the human body while collecting and analysing the head posture information and giving feedback after processing by the computer.

Face Tracking: When the user enters the detection range, the system will automatically turn on the camera to analyse the facial movement of the person and control the stepping motor to drive the device to lock the target to locate and track the device's face.

Cloud Storage: Based on the Alibaba Cloud platform, software and hardware are combined to build a complete "cloud" database, which is convenient for users to query their recent sitting posture data and provide future sitting posture suggestions.

Remote Monitoring: Users can view the real-time dynamics of users through mobile devices.

Electronic Product Usage Record: The device will automatically identify and record the time duration when the user uses the electronic product and deliver it to the user's monitor's mobile terminal.

2   Overall Scheme Design

2.1   Overall Design

The overall design of the sitting posture instrument is shown in Fig. 1 below.

The main controller performs data reading, processing, and output operations. The main controller is connected to data input devices such as cameras, sensors, and motors. The main controller uploads the data to the Alibaba Cloud through the Internet of Things technology and then uploads the data to the Alibaba Cloud. The data are further processed on the cloud, and finally transmitted to the APP and PC for processing.

2.2   Function Introduction

Human Sitting Posture Detection: This function can realize the detection of sitting posture, which includes two types of detection methods: infrared ranging detection and visual detection based on a convolutional neural network and OpenCV computer database. Accurate determination of sitting posture is achieved through two different detection methods.

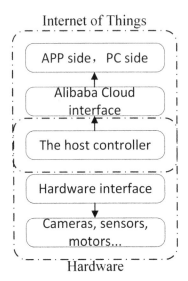

Fig. 1. Overall design of the sitting posture instrument.

Face Tracking: Because of the limitation of the desktop environment where the sitting posture instrument is placed, there is often a direction deviation in the measurement. Therefore, a face tracking module is added to the sitting posture instrument to realize the device. The angle of the device is actively adjusted following the face within a certain angle so that the infrared ranging sensor always faces the user and achieves the best measurement mode.

Gesture Recognition: This function can be applied to scenarios such as device function wake-up, device hibernation and shutdown. Compared with the preset gesture, some physical buttons can be replaced by gestures.

Remote Monitoring: Using the Internet of Things technology, monitors of users can use mobile devices to view the real-time dynamics of users anytime and anywhere and can also monitor their children's learning status at any time.

Electronic Product Usage Records: When a user uses an electronic product, such as a mobile phone or iPad, the device automatically identifies and records the usage time period and delivers it to the user's parent's mobile terminal.

3   Hardware Design

3.1   General Hardware Design

In view of the function of the sitting posture instrument, the project team conceived the hardware design scheme of the sitting posture instrument and passed the simulation verification. The main controller adopts Raspberry Pi 4B, with a sensor expansion board and motor driver board, connecting cameras, sensors, motors, etc., which achieve face recognition tracking and detection of human sitting posture. The overall hardware design scheme is shown in Fig. 2:

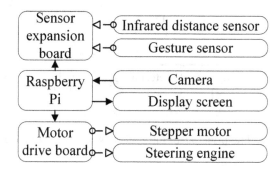

Fig. 2. General hardware design scheme.

3.2   Hardware Function Module Design

The hardware function module is shown in Fig. 3. The sitting posture intervention has functional modules such as a sitting posture detection module, a face positioning module, a gesture recognition module, a motor reset module, and an extended function module.

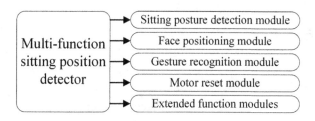

Fig. 3. Functional Module Design Scheme.

Design of Infrared Sitting Posture Detection Module: To make the measurement more accurate, two infrared ranging sensors are used in this module, the two sensors are placed side by side, and the upwards lifting angle is adjustable. Sensor A is responsible for measuring the distance from the device to the human chest, and sensor B is responsible for measuring the distance from the device to the human eye. The sitting posture is judged according to the two sets of

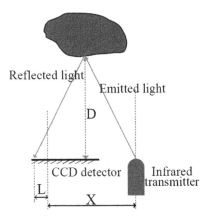

Fig. 4. The principle of infrared ranging sensor ranging.

data and the difference between the angles of the two sensors. Figure 4 shows the principle of infrared ranging sensor ranging:

The following figure shows the change in the analog output voltage of the sensor with distance (Fig. 5):

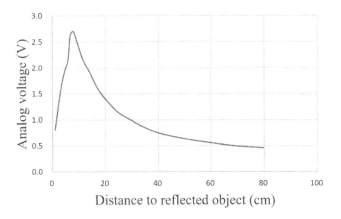

Fig. 5. General design of the Internet of Things.

The main controller analyses the data captured by the camera. After sending a signal to the motor drive board, it adjusts the raising angle of the infrared sensor through servos A and B so that sensor A faces the body to measure the distance between the device and the user; sensor B faces the eye. to measure the distance between the device and the human eye.

Design of Visual Sitting Posture Detection Module: The traditional sitting posture detector only has a single detection system, which is also the reason for the low accuracy. To this end, the project team added a visual distance detection module using a Raspberry camera and a visual sitting posture detection module based on the infrared sensor.

Design of Face Positioning Module: Because the user is limited by the desktop environment when placing the sitting posture instrument, there is often a deviation in the measurement direction of the device, which causes a large deviation in the measurement data and cannot be measured correctly. Therefore, a face positioning module is added to the sitting posture instrument to realize the device. The angle of the device is actively adjusted following the face within a certain angle so that the infrared ranging sensor always faces the user and achieves the best measurement mode.

The face positioning module uses the raspberry camera to capture the picture and uses the OpenCV computer database to write a program to actively identify and detect the face, record the face position, size and eye position, and provide the necessary data for the infrared ranging sensor B. and are the lifting angles of infrared sensors A and B. According to the data measured by infrared sensor A and the face position detected by this module, calculate the angle that sensor B needs to lift facing the human eye, and send and to the motor drive board, will be used as the rotation angle of the stepping motor, will be used as the rotation angle of the steering gear B after data processing, and the device will be rotated to the angle facing the face, which can make the measurement data of the infrared ranging sensor B in the later stage more accurate.

Design of Gesture Recognition Module: The gesture recognition sensor module used in this module has strong compatibility. The sensor integrates the optical module of ALS, infrared LED and proximity detector and the ambient brightness sensor for ambient brightness sensing, which can greatly reduce the demand for other modules. Additionally, the workload of camera recognition can be reduced, and the preset gestures can be directly converted into mathematical analog signals, which are sent to the sensor module for further processing.

Design of Motor Reset Module: To avoid the position deviation caused by the stepper motor and the external force, the motor reset operation after each boot is performed (or perform a single reset within the settings). This module uses the mechanical travel limit for the reset, and the stepper motor is set at a 60o clockwise position, and the steering gear is set at a 30o downwards angle of the infrared ranging sensor. Reset action is started after touching the switch.

4   IOT Design

For the usage of the sitting posture instrument and the sitting posture detection data, the sitting posture instrument uses the Internet of Things technology to access the Internet and divides the desktop interaction module, the sitting posture information summary module, and the cloud monitoring module. The overall design of the Internet of Things is shown in the following figure (Fig. 6):

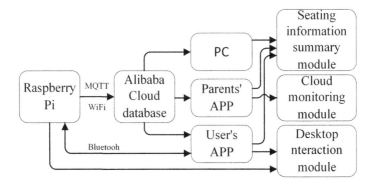

Fig. 6. The principle of infrared ranging sensor ranging.

4.1   Design of Desktop Interactive Module

Applications such as video calls, clocks, music, and settings are added to the desktop of the system.

Video Call: used for video communication between monitors and users, which can be initiated in both directions.

Clock: Built-in alarm clock, timer, countdown and other widgets for daily timing.

Music: It can be played through the speaker using a Bluetooth-connected device or play music stored on an SD card.

Settings: used for system settings, such as face information input, Bluetooth WiFi connection settings, monitor mobile device binding and unbinding, system device reset, factory reset, etc.

4.2   Design of Sitting Posture Information Summary Module

The system is based on the Alibaba Cloud platform and relies on the Internet of Things technology to transmit the detection data to the server, complete the data processing on the server side, and finally output it to the smartphone APP or PC client.

The app is designed with daily usage time, daily sitting status, sitting posture data comparison, sitting posture suggestions and other content, such as sitting posture reporting and viewing sections, as well as the remote setting section of the sitting posture meter.

Daily Usage Time: The user's usage time of the day is recorded in detail through the human body induction sensor system, and the histogram of the marked data is displayed. Press and hold the mouse to view the user's sitting posture data during this period.

Daily Sitting Posture: It records and stores the data detected by the multivariate sitting posture detection.

Comparison of Sitting Posture Data: Big data technology is used to compare the user's data with the average of all user's data to encourage users to maintain a good sitting posture.

Sitting Posture Suggestion: According to the user's sitting posture, the level is divided, and the user is given some good suggestions on the sitting posture.

4.3 Design of Cloud Monitoring Module

The user's monitors can use the mobile device bound to the sitting posture instrument, remotely observe the user's sitting posture, and can also send information to the user through this module. All the related information on the display screen and prompt relevant content.

5 Sitting Posture Instrument System and Program Design

5.1 The Underlying System Configuration of the Sitting Posture Instrument

The Underlying System Configuration of the Sitting Posture Instrument: Raspberry Pi OS is an ARM-based Linux operating system. The OpenCV computer database and other programs used by this device can run efficiently on the Linux system. Its built-in language environment, such as Python 3.9, can simplify the workload of the work concept, its desktop environment components and Bluetooth support are helpful for early hardware development, and its network application tools can help for the later network development of the sitting posture instrument.

ROS Is Configured in the Control System: ROS (Robot Operating System) is an open source secondary operating system for robots and is often used in the development of machine vision. It has functions such as bottom-level driver management, message transmission between programs, and execution of shared functions. Its design aims to improve the code reuse rate in the field of robot research and development, [4] which has a large number of built-in tool software, library codes and convention protocols, and help to reduce the difficulty and complexity of making this sitting posture instrument. This sitting posture meter uses accessories such as cameras, infrared ranging, stepper motors, steering gear, etc. Relying on ROS can greatly reduce the workload, and it is used for the underlying driver of sensors and motors.

OpenCV Computer Database Is Used for Visual Positioning: OpenCV is a computer vision library that can be written in C or C++ and run on systems such as Linux. OpenCV also provides interfaces in Python and other languages to facilitate program writing. [5] It has as many as 500 built-in functions, covering multiple application fields of computer vision. [6] For example, the detect Multi-Scale function used in this article can recognize multiple faces at the same time and save the coordinates and sizes of each face. The intelligent sitting posture instrument relies on OpenCV to complete face recognition positioning, tracking, and recognition for visually measuring the distance between the device and the human body.

Sensor Expansion Board Configuration: Since there is no analog to digital converter (ADC) in the Raspberry Pi, the analog value of the sensor cannot be directly measured by the GPIO of the Raspberry Pi, so it is necessary to import library files such as wiringPi.h and wiringPiI2C.h and use wiring PiI2CSetup (0x04) to open the I2C device, wiringPiI2CWrite write transfer information and wiringPiI2CReadReg16 read transfer information. The 10-bit ADC can be read with the help of the MCU (micro control unit) in this expansion board, which means that the analog voltage can be input to the 10-bit analog-todigital converter using an analog sensor on the Raspberry Pi. After the analog-to-digital converter converts the analog data into digital data, the digital data are input into the Raspberry Pi through an I2C terminal. This configuration file is written in C language.

5.2   Program Design of the Sitting Positioner

Design of Test Program for Infrared Ranging Sensor: In the test of this module, Arduino is used to process the signal first, the signal transmission frequency is 7 4880 Hz, and the program is written in C language. After several sets of data tests and analysis of the output curve, the approximate conversion formula of the analog digital signal "x" read from the GPIO serial port and the actual distance is as follows:

$$distance = 25/(2x) - 2(20 < distance < 150) \tag{1}$$

To reduce the error of the sitting posture judgment caused by the fluctuation of the signal, the data are sampled 10,000 times per second, and the average value is determined during the acquisition process. The final signal fluctuation range is less than 2%, which greatly reduces the error and increases the recognition performance. The measuring distance of this sensor is 20 cm–150 cm. The actual distance during the detection process is usually more than 20 cm, so the situation in which the measuring distance is less than 20 cm can be ignored.

Design of Visual Sitting Posture Detection

Visual Distance Detection Module: This module performs the face recognition detection captured by the camera, records the upper and lower width of the face recognition frame, and compares it with the user's initial setting data (the distance between the device and the human body and the face detection data before using the face size parameter), uses machine vision to estimate the distance between the device and the human body, and uses the detection data of the infrared sensor as the detection data of the sitting posture and to judge the sitting posture.

Visual Sitting Posture Detection Module: The convolutional neural network is used to train the human body posture and analyse the human head posture and the degree of inclination of the human sitting posture through machine vision to obtain the sitting posture state of the human body.

Program Design of Face Recognition Tracking Module: Considering that face recognition is complex, simpler and more convenient Python is used for development in the program of this module. In this program, OpenCV face is used for face recognition. After the face is detected, the four corners and eye coordinate data of the face detection frame are recorded. The procedure flow of face information acquisition is shown in Fig. 7:

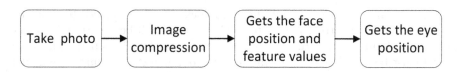

Fig. 7. Process flow of face information acquisition program.

After obtaining the face data, an instruction was sent to the stepping motor controller, and the face was automated in the center of the camera frame. Combined with the data of infrared ranging sensor A, calculate the required lift angle when infrared ranging sensor B faces the eye. The operation flow chart is shown in Fig. 8:

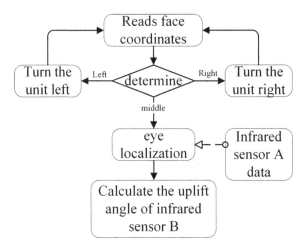

Fig. 8. Flow chart of face tracking.

Design of Electronic Product Usage Record Module: This module is based on OpenCV and uses machine vision to determine whether users use electronic products. The convolutional neural network algorithm is used for deep learning, and recognition training is performed on thousands of electronic product models (pictures such as mobile phones and tablets). The training process is shown in Fig. 9:

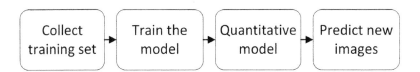

Fig. 9. Electronic product recognition training process.

First, training images are collected, and docker images are used to train the model and find the correct parameters. This module will be used in microcomputers with limited computing power and memory, such as Raspberry Pi, so the model must be quantified and finally used to complete the training. The model makes predictions on new images.

Programming of Gesture Recognition: The program of this module is written in C language. After configuring the library file, the program workload for gesture recognition is greatly reduced. It can recognize various preset gestures and convert them into simple analog signals, which are transmitted to the main control for judgment.

6   Appearance Design

The overall profile of the sitting posture meter is initially designed as a regular hexagonal prism shape with a side length of 8 mm and 2 mm. The whole set is divided into upper and lower sections, the height of the upper section is 150 mm, and the height of the lower section is 50 mm. The interior is driven by a stepping motor to rotate the whole. The front is equipped with a camera, a gesture recognition sensor, a display screen, and a pair of infrared ranging sensors on the top. The angle of the sensor can be adjusted through the rear steering gear link. The appearance design is shown in Fig. 10 and Fig. 11:

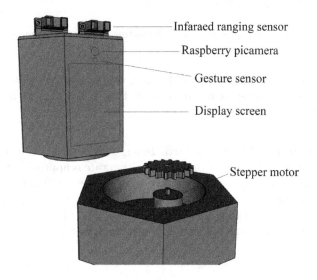

Fig. 10. Appearance design of the sitting posture instrument 1.

7   System Test

System debugging is divided into the following four parts: Sitting posture detection test: The device is placed between 60–100 cm in front of the user's body, which can detect and alarm the user's sitting posture errors such as left and right, low of head, and forward of body.

Face Recognition Tracking Test: The device is placed 60 cm in front of the user's body, and within a range of ±30°, the detection device can always acquire the human body and can always follow the movement of the human face. during the movement of the device or human body.

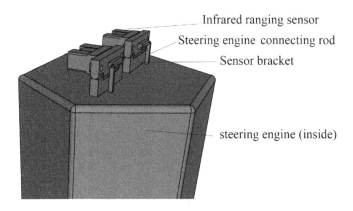

Fig. 11. Appearance design of the sitting posture instrument 2.

Gesture Recognition Test: By performing simple gesture operations in front of the gesture sensor, the system can recognize most gestures, and the recognition time is slightly longer.

Electronic Product Usage Record Test: When using electronic products within the detection range, the frontal recognition rate is approximately 79%, the side recognition rate is approximately 63%, and the recognition time is approximately 4 s. Through the above four tests, the functions of sitting posture detection, face recognition tracking, gesture recognition, and electronic product usage recording of the multifunctional sitting posture detector have been realized and verified.

8   Summary and Outlook

This paper describes the design of a face tracking-based sitting posture instruction. It uses ROS, OpenCV, convolutional neural networks, the Internet of Things and other technologies. After testing, the tracking of the face and the accurate measurement of the sitting posture can be realized. It has various recognition methods, such as gesture recognition and speech recognition. At the same time, it can store the sitting posture data in the Alibaba Cloud database and view the sitting posture data on the mobile device. It can remind the user in time when most of the sitting posture is incorrect. Use allows users to develop a habit of maintaining a good sitting posture.

Acknowledgements. Thanks for the funding of the 2021 Zhejiang University of Science and Technology College Students Extracurricular Science and Technology Innovation and Practice Project (Chunmeng Plan). Thank you for the funding of the 2022 Zhejiang University of Science and Technology New Engineering Training Program.

References

1. Sican, Y.: Maintaining health and preventing myopia with healthy sitting posture. Med. Diet Health **4**, 152–153 (2018)
2. Guofang, J.: A review on the exploration and prevention of myopia in adolescents. In: China Glasses Science-Technology Magazine, pp. 125–127 (2016)
3. Kun, C.: Talking about how to cultivate children's correct body posture. Value Eng. **29**, 252 (2010)
4. Meng, S.H.S., Liang, Y.: Motion planning of six degrees of freedom manipulator arm based on ros platform. China Equip. Eng. **50**, 94–97 (2009)
5. Bradski, A.K.G.: Learning Opencv (Chinese Version). Tsinghua University Press, Beijing (2009)
6. Qin, Q.W.X., Wen, Z.: Image processing based on opencv. In: Electronic Test, pp. 39–41 (2011)

Real-Time Analysis and Prediction System for Rail Transit Passenger Flow Based on Deep Learning

Xujun Che[1] (ID), Gang Cen[2] (ID), Shuhui Wu[1](✉) (ID), Jiaming Gu[2] (ID), and Keying Zhu[2] (ID)

[1] School of Science, Zhejiang University of Science and Technology, Hangzhou 310000, China
s.wu@zust.edu.cn
[2] School of Information and Electronic Engineering, Zhejiang University of Science and Technology, Hangzhou 310000, China

Abstract. With the rapid development of urban rail transit, rail transit plays an important role in alleviating city congestion. In recent years, with increasing passenger flow, there has been huge pressure on passenger flow management. To address this problem, we propose a novel system to provide real-time statistics and predictions of passenger flow based on big data technology and deep learning technology. Moreover, the passenger flow is visualized efficiently in this system. It can provide refined passenger flow information so that people can make more rational decisions in terms of operation and planning, deploy contingency plans to avoid emergency situations, and integrate passenger flow analysis with train production, scheduling and operation to achieve cost reduction and efficiency enhancement.

Keywords: Rail Transit · Passenger Flow · Deep Learning · Big Data

1 Introduction

With the rapid development of technological revolution and industrial transformation, China's traditional urban rail transit construction pattern has undergone tremendous change. On March 12, 2020, the China Association of Metros (CAMET) published and officially implemented the Development Outline of Smart Urban Rail in China's Urban Rail Transit to promote the development and evolution of smart urban rail with the mission of "transportation power and urban rail responsibility" [1]. Just one year later, the Hangzhou Comprehensive Transportation Special Plan (2021–2035) was issued by the Hangzhou Municipal People's Government in September 2021. It shows that Hangzhou has included all the stations that have not been constructed in the third phase of rail transit and the stations in the last phase of the fourth phase into the scope of TOD integrated development and vigorously advanced the TOD integrated development and construction process. Thus, in the future, the job coverage rate of the population within 800 m of the rail station will reach 45% [2], which means that the regional orbital density and the passenger flow density in Hangzhou will rise dramatically in the future. As a result of the iterative update of the city, the high volume and high density in the piece area will face a huge challenge in rail management.

Z. Yu et al. (Eds.): ICPCSEE 2023, CCIS 1880, pp. 130–138, 2023.
https://doi.org/10.1007/978-981-99-5971-6_9

Due to the advantages of large transportation volume, high punctuality rate and fast speed, the number of passengers in the metro continues to increase. In the meantime, the travel data of passengers becomes increasingly larger. For metro management, when decision makers face the difficulties of passenger flow management, they may not fully grasp the actual information of passenger flow distribution attributable to the unreasonable presentation of information or fail to make adjustments in advance due to emergency fluctuations of the passenger flow. Metro stations are usually public places with dense crowds. Once the peak passenger flow exceeds the saturation number without in-time adjustment, it could cause a crowded stampede accident, which likely causes casualties [3]. To overcome this issue, an effective visual analysis and prediction system for passenger flow is necessary and important.

As the scale of metro travel data constantly increases, traditional calculation methods can no longer meet the needs of today's large-scale passenger flow calculations. To process and analyse such huge data efficiently and precisely, big data processing technologies came into being. In fact, it has reached a relatively mature stage, with the emergence of many well-known big data parallel processing platforms, such as Dryad [4], Hadoop [5], and Spark [6]. These platforms not only solve the problem of insufficient single-machine resources but also have high computing efficiency.

For passenger flow prediction, many different models have been proposed, including the early Kalman filter, exponential smoothing, autoregressive integrated moving average model (ARIMA), random forest, and commonly used deep learning, such as the convolutional neural network (CNN) [7], long short-term memory (LSTM) [8], and gated recurrent unit (GRU) [9]. The accuracy of the models has reached a high level.

In this paper, we propose a novel system for multidomain decision-making groups with all-round analysis and highly user-friendly operation logic. It is implemented by big data technology and deep learning technology, which can provide users with real-time passenger flow statistics and predictions in the form of visualization. This could be a good solution for traffic managers to monitor the trend of passenger flow to prevent accidents.

2   System Design

The main functions of the real-time analysis and prediction system for rail transit passenger flow are gathering statistics and predicting future passenger flows. The system first receives real-time passenger card swiping data from the Automatic Fare Collection (AFC) system of the metro, second uses big data technology to calculate real-time passenger flow, and third displays it to users in the form of data visualization charts. At the same time, the system will save the passenger flow calculated in real time for subsequent analysis of historical passenger flow. The statistics and predictions in this system include four types of passenger flow: total flow, line flow, station flow and section flow. The overall design of the system is shown in Fig. 1.

Among them, the total flow, the line flow and the station flow can be calculated directly through the passengers' inbound and outbound information. However, the section flow is different. The section flow represents the passenger flow through the section between two stations. Its calculation requires the complete path of the passenger

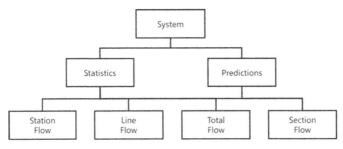

Fig. 1. Overall design of the system

travel process, not just inbound and outbound information [10]. Since the metro card swiping data only include the records of passengers entering and exiting the station, but not the stations that passengers pass through before leaving the station, the complete route of the passenger travel process cannot be determined. This makes it difficult to calculate the precise value of section flow.

The system uses the Bellman–Ford algorithm [11] to obtain the shortest path for each origin-destination, which is these passengers' virtual path. In other words, it is assumed that each passenger uses the shortest path to reach their destination. The path information obtained by the above algorithm will be used to calculate the estimated value of section flow.

Through the area chart of station flow, line flow, and total flow provided by this system, it is possible to accurately grasp the passenger flow trend of a station, a line, or the entire city. At the same time, the system displays the passenger flow of every station with the sizes of the bubble and the passenger flow of every section with the thicknesses of the lines on the metro map. Thus, the spatial distribution of passenger flow can be visualized directly among all stations and among all sections. The user interface design of the system is shown in Fig. 2.

3   Prediction Models

This system provides a passenger flow prediction function. According to the different features of different types of passenger flow, the system adopts three different deep learning-based models to predict them.

3.1   Total Flow and Line Flow

The system uses the GRU-based Seq2Seq [12] model with an attention mechanism [13] to predict total flow and line flow. The Seq2Seq model is a network composed of an encoder and a decoder. Its input is a sequence, and its output is also a sequence, which meets the needs of the system to predict multiple time steps. Our model adopts two GRUs as the encoder and decoder. Afterwards, an attention mechanism is introduced to improve the performance of the model. Such a structure can avoid the problem of error accumulation and thus achieve better performance under multistep prediction.

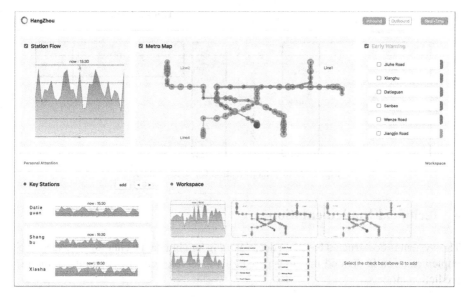

Fig. 2. User interface design

3.2 Section Flow

Due to the huge number of metro sections, a large number of parameters will be generated if the model is directly established. If so, the calculation efficiency will be quite low. Hence, our system adopts low-rank matrix factorization to solve this problem [14]. As shown in Fig. 3, for the section flow matrix $\mathbf{A} \in \mathbb{R}^{n \times T}$ with n sections and T time steps, it can be factorized into the form of multiplying two matrices: $\mathbf{A} \approx \mathbf{ST}^T$, where $\mathbf{S} \in \mathbb{R}^{n \times k}$ is the latent sectional embedding matrix and $\mathbf{T} \in \mathbb{R}^{T \times k}$ is the latent temporal embedding matrix. To obtain the optimal $\mathbf{S}$ and $\mathbf{T}$, we can solve:

$$\underset{\mathbf{S}, \mathbf{T}}{\mathrm{argmin}} \sum_{i=1}^{n} \sum_{t=1}^{T} \left(\mathbf{A}_{it} - \mathbf{S}_i \mathbf{T}_t^T\right)^2 + \lambda_{\mathbf{S}} \|\mathbf{S}\|_F^2 + \lambda_{\mathbf{T}} \|\mathbf{T}\|_F^2 \tag{1}$$

This optimization problem can be solved by alternating least squares or gradient descent. Then, the system needs to predict the temporal embedding matrix in the same way as the total flow and line flow. Hence, the section flow prediction result can be obtained as $\mathbf{A}_{new} = \mathbf{ST}_{new}^T$.

3.3 Station Flow

To make full use of the links and relative location relationships between stations, the system uses the T-GCN model [15] to predict station flow. This model combines the advantages of both GCN and GRU, using GCN to capture the spatial dependency of passenger flow between different stations and then GRU to capture the temporal dependency of passenger flow, resulting in the final prediction result, which achieves the comprehensive use of spatiotemporal information and improves the accuracy of the prediction result.

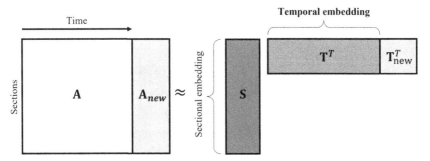

Fig. 3. Matrix factorization for section flow

4 Technical Implementation

This system is constructed based on deep learning and big data technology. Its technical implementation is divided into four parts: client, server, network communication and prediction models.

4.1 Client

The client is programmed by TypeScript. Based on JavaScript, TypeScript adds a series of type checking functions and provides generic programming, which reduces the possibility of errors in the development process and greatly improves the reliability of the system. The interface is rendered by React. Using React's declarative paradigm, applications can be easily described. The reusability brought by its componentized development reduces the development workload and improves code quality, and its virtual DOM feature minimizes page repainting, providing support for frequent real-time rendering of the system.

The client uses the open-source visualization chart library ECharts to draw diversified data visualization charts according to the characteristics of each statistical dimension. Echarts has rich chart types and high flexibility and extensibility. It can easily create all kinds of charts needed in this system and customize the styles so that we can present suitable and beautiful charts for users.

4.2 Server

The system uses Kafka as a message queue, the Hadoop Distributed File System (HDFS) as data storage, Spark as a big data computing engine, Redis as a cache, and Django as a service.

- Kafka is a distributed, high-throughput message queue system [16] that enables real-time messaging between different parts of the system. Kafka realizes decoupling between different applications or between different parts of the same application, reduces unnecessary dependencies, and plays the role of peak shaving, avoiding the adverse impact of instantaneous traffic on the availability of the system.

- HDFS is a highly fault-tolerant distributed file system that provides efficient data management and access. The system uses it for the storage of historical passenger flow.
- Spark is adopted as the big data computing engine of the system. For easier and more efficient use, we use Spark SQL [17], a module of Spark that provides a better API called Data Frame. Structured Streaming is a high-level streaming API built on Spark SQL [18], and we use it to calculate real-time passenger flow. In such a framework, we can achieve high performance.
- Redis is an in-memory nonrelational database with very fast reading and writing speed that is suitable for use as a cache. The system uses it to store real-time passenger flow over a period of time. These data will be used for prediction and deleted when not needed. In addition, by storing the queried historical traffic in Redis, the results can be retrieved directly from the cache when querying again without repeated calculation, which greatly improves the query speed, reduces the pressure on the server, and improves the operating efficiency of the system.
- Django is a web framework developed by Python that can integrate with Redis, GraphQL, WebSocket and other components well.

4.3 Communication

During historical analysis, the client communicates with the server through the GraphQL API via HTTP requests. GraphQL is a novel API query language that allows clients to obtain exactly the data they need, reducing data redundancy and request redundancy, increasing communication efficiency, reducing the need for major changes, and improving development efficiency.

During real-time analysis, the client communicates with the server via WebSocket, a protocol that supports persistent connections between the client and the server, which enables the server to actively push real-time passenger flow to the client.

4.4 Prediction Models

This system uses PyTorch to build the prediction models described in Sect. 3. PyTorch is a very popular machine learning framework among researchers due to its ease of use. It allows us to build models more easily.

5 System Operation Process

As shown in Fig. 4, the system obtains the passenger swiping card data constantly from the data source through Kafka, using Spark Structured Streaming to aggregate the data, using the deep learning models to predict future passenger flow, and communicating with the client via WebSocket. Finally, the statistical results and prediction results are presented to the user. At the same time, the system always saves the calculated real-time passenger flow to HDFS as the data of historical passenger flow for future calculation of historical statistics. For all the passenger card swiping data generated before the system is put into use, the system uses Spark to calculate the passenger flow every 10 min, stores it in the distributed file system, and then uses Spark again to calculate the passenger flow in units of days through these data. This ensures data consistency in HDFS.

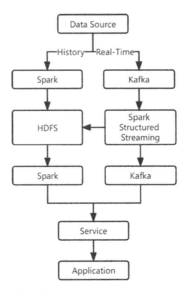

Fig. 4. System operation process

6 Features and Advantages

6.1 Efficiency

The system uses Spark, HDFS, Kafka and other technologies for distributed computing and data storage, with high computing efficiency, and various types of passenger flows can be quickly calculated.

6.2 Real-Time

Relying on high efficiency, the system can calculate the real-time passenger flow in time so that users can see the real-time passenger flow statistical charts, analyse them, and make decisions accordingly.

6.3 Accuracy

According to the different features of different types of passenger flows, the system adopts different models to predict them, which has higher accuracy and provides users with a reliable decision-making basis.

6.4 Intuitiveness

The system provides area charts of station flow, line flow, and total flow, enabling users to understand the corresponding passenger flow trends. Furthermore, the system displays station flow and section flow in the form of bubble size and line thickness in the metro map so that users can intuitively see the spatial distribution of passenger flow.

7 Conclusion

On the basis of big data technology and deep learning technology, we proposed a system to make statistics and real-time predictions of metro passenger flow based on real-time swipe card data. The system adopted an intuitive and efficient visualization to provide a reliable data guarantee in metro management as well as reduce the occurrence of accidents. Furthermore, it could also provide guidance for precautions of metro operation risk, optimal personnel allocation, engineering investment and business layout.

Acknowledgements. This work is supported in part by grants of Zhejiang Xinmiao Talents Program under No. 2021R415025 and the Innovation and Entrepreneurship Training Program for Chinese College Students under No. 202111057017.

References

1. China Association of Metros: Development outline of smart urban rail in China's urban rail. China Metros. **8**, 8–23 (2020)
2. Nanjing Institute of City & Transport Planning Company Limited and Hangzhou City Planning and Design Academy: Hangzhou comprehensive transportation special planning (2021)
3. Liu, Y., Wang, T., Ding, H., Wu, Z.: Research and application on risk assessment DEA model of crowd crushing and trampling accidents in subway stations. In: Procedia Engineering, pp. 494–498. Elsevier Ltd. (2012). https://doi.org/10.1016/j.proeng.2012.08.085
4. Isard, M., Budiu, M., Yu, Y., Birrell, A., Fetterly, D.: Dryad: distributed data-parallel programs from sequential building blocks. In: Proceedings of the 2nd ACM SIGOPS/EuroSys European Conference on Computer Systems 2007, pp. 59–72 (2007)
5. Ghazi, M.R., Gangodkar, D.: Hadoop, mapreduce and HDFS: A developers perspective. In: Procedia Computer Science, pp. 45–50. Elsevier B.V (2015). https://doi.org/10.1016/j.procs.2015.04.108
6. Zaharia, M., Chowdhury, M., Franklin, M.J., Shenker, S., Stoica, I.: others: spark: cluster computing with working sets. HotCloud. **10**, 95 (2010)
7. LeCun, Y., et al.: Backpropagation applied to handwritten zip code recognition. Neural Comput. **1**, 541–551 (1989)
8. Graves, A., Graves, A.: Long short-term memory. Supervised sequence labelling with recurrent neural networks, pp. 37–45 (2012)
9. Cho, K., Van Merriënboer, B., Bahdanau, D., Bengio, Y.: On the properties of neural machine translation: encoder-decoder approaches. arXiv preprint arXiv:1409.1259 (2014)
10. Chen, C., Liu, G.: Research on the calculation method of section passenger flow volume of urban rail transit. Technol. Econ. Areas Commun. **17**, 43–46 (2015)
11. Bellman, R.: On a routing problem. Q. Appl. Math. **16**, 87–90 (1958)
12. Sutskever, I., Vinyals, O., Le, Q.V.: Sequence to sequence learning with neural networks. Adv. Neural Inf. Process. Syst. **27** (2014)
13. Vaswani, A., et al.: Attention is all you need. Adv. Neural Inf. Process. Syst. **30** (2017)
14. Yu, H.-F., Rao, N., Dhillon, I.S.: Temporal regularized matrix factorization for high-dimensional time series prediction. Adv. Neural Inf. Process. Syst. **29** (2016)
15. Zhao, L., et al.: T-GCN: a temporal graph convolutional network for traffic prediction. IEEE Trans. Intell. Transp. Syst. **21**, 3848–3858 (2019)

16. Kreps, J., Narkhede, N., Rao, J., others: Kafka: a distributed messaging system for log processing. In: Proceedings of the NetDB, pp. 1–7 (2011)
17. Armbrust, M., et al.: Spark SQL: relational data processing in spark. In: Proceedings of the 2015 ACM Sigmod International Conference on Management of Data, pp. 1383–1394 (2015)
18. Armbrust, M., et al.: Structured streaming: a declarative API for real-time applications in apache spark. In: Proceedings of the 2018 International Conference on Management of Data, pp. 601–613 (2018)

Research on Driver Monitoring Systems Based on Vital Signs and Behavior Detection

Man Niu, Yanqing Wang[✉], Xinya Shu, and Xiaofeng Gao

Nanjing Xiaozhuang University, Nanjing Jiangsu 211171, China
wyq0325@126.com

Abstract. Aiming at drivers' dangerous driving behavior monitoring and health monitoring, this paper designs an intelligent steering wheel that can monitor dangerous driving behavior and a steering wheel sleeve that can monitor physical health. The MTCNN model is primarily used to obtain a driver's face image in real time. The PFLD algorithm was used to obtain the facial model positioning feature points, and the degree of driver fatigue was determined by combining the relevant parameters. The fatigue algorithm proposed in this paper can improve the effectiveness and accuracy of monitoring. Then, according to the LSTM network model, 11 groups of key point information of the human body are obtained, and the human motion track is identified and then combined with the facial information to complete the judgment of driving behavior such as drinking water, smoking and walking. Through the PPG and ECG fusion algorithm based on LSTM, the reliability of the system to collect vital signs such as body temperature, blood pressure, heart rate and blood oxygen of the driver is improved. It was determined that the system could monitor a driver's driving behavior in real time and consider its health management.

Keywords: Autonomous Driving · Artificial Intelligence · Face Recognition · Behavior Monitoring · Fatigue Detection

1 Introduction

Currently, accidents involving vehicles on the road have become inevitable. The World Health Organization (WHO) has revealed that more than 3400 people die in road traffic accidents every day, and billions of people suffer from nonfatal injuries and disabilities due to traffic accidents. A study by the Traffic Safety Foundation found that 37% of drivers have drowsiness while driving. This may be due to sleepiness or microsleep. Accidents can cause fatal injuries and increase mortality. According to statistics, from 2007 to 2010, 65% of accidents were caused by human negligence and human error. Tiredness is often referred to as fatigue or drowsiness, resulting in falling asleep at inappropriate times or situations. Sleepiness can cause drivers to lose control of their speed or become unaware of obstacles on the road, which can lead to fatal accidents.

The DMS (Driver Monitor System) is a driver monitoring system. The aim is to prevent drivers from dangerous behavior. Due to the advancement of computer vision

Z. Yu et al. (Eds.): ICPCSEE 2023, CCIS 1880, pp. 139–152, 2023.
https://doi.org/10.1007/978-981-99-5971-6_10

technology, the DMS [1] system has become a complete vehicle visualization scheme that can help drivers drive safely.

At present, human–machine codriving is the best choice for industry and the market [2]. In the stage of human–machine codriving [3], more attention is given to driver condition monitoring. DMS technology is an important part of intelligent driving. With the development of autonomous driving, DMSs have become necessary. President of the People's Republic of China pointed out in the forum of experts' representatives in the fields of education, culture, health, and sports: "People's health is the foundation of social civilization and progress, an important symbol of national prosperity and national prosperity, and is also the common pursuit of the broad masses of the people." At present, the detection equipment in the market has a single performance and cannot meet various health needs. Most of the equipment can only measure heartbeat, blood pressure, body temperature, etc., for deeper and more critical data. However, the accuracy and real-time performance of the data cannot be guaranteed.

Simultaneously, national car ownership continues to increase, and the demand for traffic safety is growing. According to the China Report Hall Network, in 2021, the number of traffic accidents in a city was 61703, the number of injuries was 250723, and the number of motor vehicle traffic accidents was 211074. Owing to the poor physical condition of drivers, such as physical discomfort, fatigue, heart attack, and arrhythmia, the proportion of traffic accidents is as high as 37%, which is a serious safety hazard.

In this regard, this paper proposes a driver's vital signs and driving behavior monitoring system.

2 System Process

The overall process of the driver's vital signs and the driving behavior monitoring system proposed in this study are shown in Fig. 1.

The system is mainly composed of two parts: monitoring vital signs and monitoring driving behavior. Driving behavior monitoring is mainly completed by face recognition and behavior recognition [4]. Using the MTCNN face detection model [5], the driver face data information in the camera head can be obtained in real time, and the key points are detected by the PFLD algorithm to obtain the feature points of the face model. The degree of fatigue was accurately determined based on the parameters of these feature points. Using the LSTM network model, we can extract useful information from 11 sets of key points, store joint point motion data of the human body, and combine facial features to achieve accurate recognition of driver behavior. The PPG and ECG fusion algorithm based on LSTM [6] is mainly used to monitor vital signs such as blood pressure, heart rate, and blood oxygen.

3 Fatigue Algorithm

The driving fatigue-state detection process is illustrated in Fig. 2. Using the MTCNN face detection model, the face image of the driver in the camera head can be immediately obtained. The PFLD algorithm, known for its high efficiency and accuracy, is used to carry out the key test and obtain facial feature points such as the mouth and eyes.

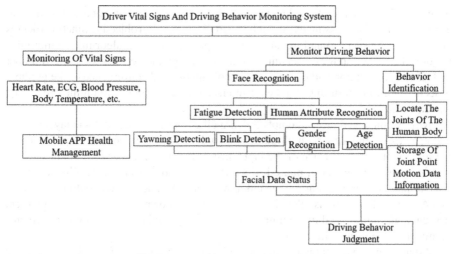

Fig. 1. The total process of the system

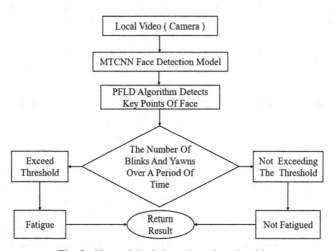

Fig. 2. Flow of the fatigue detection algorithm

The algorithm then calculates the frequency of blinking and yawning in a specified period of time to determine the driver's fatigue state.

3.1   MTCNN Face Detection Model

The MTCNN is a specific target detector with deep learning ideas. It combines cascading and coarse-to-fine concepts. The detection effect implemented on traditional hardware is very good, particularly in face detection tasks. Its detection level is superior, and there is still much room for optimization.

We built a neural network model with inputs from different databases. Neural networks consist of convolutional or fully connected layers whose coefficients (weights and deviation terms) are initially unknown. Each layer also applies component-style nonlinear activation before passing its output forward. We defined an objective function (details are given in the section) and iterated with the gradient descent algorithm to minimize the objective function for ownership weight. We design the network structure (number and size of convolution and dense layers) to obtain optimal convergence of the gradient descent algorithm, resulting in a better intersection than joint score on the predefined test data set.

A fully connected layer attempts to learn all connections between the layer's input and its output using a matrix multiplied by an unknown weight. For Ntrain × Nfeatures, the input moment A and size of the input size feature (the number of training samples multiplied by the number of numerical values representing each sample characteristic) are Ntrain × Nfeatures features of an unknown matrix W, and the layer output is $\phi(A) = W \cdot A + b^T$, where b is also the bias vector of the unknown weight. Then, the gradient descent algorithm is used in the training process to approximate the unknown elements relative to the objective function.

After designing the network (number of layers, type and size of each layer, etc.). We use the gradient descent algorithm to determine the ownership weight. We define a loss (target) function that gradient descent attempts to minimize with respect to ownership weight. If the network is designed correctly and the problem geometry is learned, the gradient descent algorithm will converge. Therefore, after a certain number of iterations (called epochs), we expect that a set of discovered weights with new data (measurements not used for training) as input will produce an output that predicts segmentation with high precision.

The output of the network is a probability matrix of size M × M. Let us call it a. Each coordinate of a is the probability that that coordinate is in omega s. Thus, ϑ is the probability that the pixel is inside the scatterer. We convert ϑ to a binary image using threshold values to obtain A and check accuracy. After training our segmentation model, we apply it to the test set of sensor measurements. Each of these inferences produces a predictive binary segmentation map of the face. The binary image is the equivalent numerical interpretation of Ω, so the scheme restores the medium for solving the PDE. We divided the data set into two subsets: a training set containing samples and a test set containing samples (Ntrain + Ntest = Nsamples). To evaluate performance, we use models of test subsets and compare each prediction to a known one, where $1 \leq q \leq$ Ntest Ntest. We have several methods of measuring error: mean square error:

$$\frac{1}{N^2 \cdot Ntest} \sum_{q=1}^{Ntest} \left\| \vartheta q - \tilde{\vartheta} q \right\|^2, where : \left\| \vartheta q - \tilde{\vartheta} q \right\|$$
$$= \sqrt{\sum_{i=1}^{M} \sum_{j=1}^{M} |(\vartheta q)ij - (\tilde{\vartheta} q)ij|^2} \tag{1}$$

What we are thinking of is a simple neural network with two perfectly connected layers followed by two convolutional layers. The purpose of this paper is to transform multiple timing data (network input) into a two-dimensional (network output), transforming from a complete connection to a convolution layer. The impetus for this structure comes from the networks that are commonly used to classify images. In a database, the input is a column-stacked record, and the output is a possible division of barriers. In this example, the input data are recorded in each sampled row of data stack values, resulting in an [n-digit] matrix. To reduce the loss of NLL and soft IOU, we use ADAM's preset parameters to train the network. The resulting network structure is shown in Fig. 3.

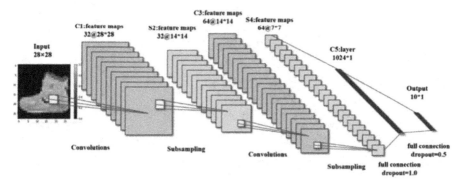

Fig. 3. CNN network architecture

3.2 Model Positioning Key Points

Facial landmark detection [8] faces many risks in practical applications: 1. Facial expression changes are complicated, the real environmental light changes greatly, and most of the face parts are obscured. 2. The face is 3D, pose changes significantly, and photographic equipment and the surrounding environment affect the image quality. 3. The types of training data are unbalanced, and the computing speed and model specifications need to be fully considered on machine (such as mobile phone) terminals with limited computing.

To improve the accuracy of face measurement, the PFLD algorithm is an effective face key point detection method that helps to ensure high accuracy in complex situations. In addition, to better capture the global change trend, the PFLD also uses auxiliary networks to evaluate the comprehensive situation of face samples to better achieve the goal of face detection.

First, we used the PFLD algorithm to examine the facial features of the driver. This method detects the driver's contours, eyebrows, eyes, nose, and mouth. In this system, we generally use the mouth and eyes [9] focus for detection (Fig. 4).

3.3 Eye Algorithm Optimization

Using the PFLD algorithm to detect the driver's face to detect the driver's drowsiness is the core of this article. This is done using a unique mathematical model and ratio called the eye aspect ratio. The driver's drowsiness is determined based on a feature called the blink rate. Through the eye aspect ratio (EAR) formula given by Eq. (2), a scalar value is obtained to quantitatively define sleepiness. For example, if the driver's eyes are closed or blink frequently within a certain period of time, it indicates the degree of drowsiness

Fig. 4. Faces 106 key points

that needs to be detected [10]. Therefore, to draw conclusions about the driver's state, calculating the blink rate becomes relevant. The flags detected in each video frame are applied to the EAR formula to determine the open state of the eyes.

These markers are usually scalar coordinates between the length and width of the eyes from each video frame of the driver. As shown in Eq. (2), the coordinates of the eyes are defined by scalar coordinates from 33 to 72. Coordinates 34, 36, 73 and 71 help to calculate the width of the eye, as shown below.

The eye aspect ratio is a fixed value when the eye is open, but when the eye is closed, the eye aspect ratio quickly drops to nearly 0. The ratio of width to height of the eyes open and close is shown in the following formula. When the eyes are closed, the eye aspect ratio is closer to 0, and when the eyes are open, the EAR value can be an integer value greater than 0.

$$\text{EAR} = \frac{|y_{67} - y_{73}| + |y_{68} - y_{72}||y_{69} - y_{71}|}{3|x_{70} - x_{66}|} + \frac{|y_{76} - y_{82}| + |y_{77} - y_{81}| + |y_{78} - y_{80}|}{3|x_{79} - x_{75}|} \quad (2)$$

Among them x_{66}, x_{70}, x_{75}, x_{79} x_{66}, x_{70}, x_{75}, x_{79} x_{66}, x_{70}, x_{75}, x_{79} are the abscissa of the 66th, 70th, 75th and 79th key points. y_{67}, y_{68}, y_{71}, y_{72}, y_{73}, y_{76}, y_{77}, y_{78}, y_{80}, y_{81}, y_{82} are the ordinates of the 67th, 68th, 69th, 71st, 72nd, 73rd, 76th, 77th, 78th, 80th, 81st, and 82nd key points (Fig. 5).

By opening the mouth, a threshold point can be set to determine whether the patient is in the yawning stage. When mouth opening reaches the prescribed level, the threshold continues to decline and will continue for a period of time (Fig. 6).

To determine the yawning condition, we used the average MAR of the outer contour fatigue threshold MAR_1 and the inner contour fatigue threshold MAR_2 to estimate the number of yawnings per unit time.

$$\text{MAR}_1 = \frac{|y_{85} - y_{95}| + |y_{86} - y_{94}| + |y_{87} - y_{93}| + |y_{88} - y_{92}| + |y_{89} - y_{91}|}{5|x_{84} - x_{90}|} \quad (3)$$

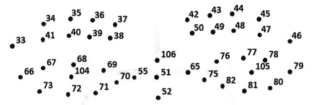

Fig. 5. Distribution of key points in the face mouth area

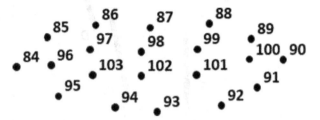

Fig. 6. The distribution of key points in the face mouth area

$$MAR_2 = \frac{|y_{97} - y_{103}| + |y_{98} - y_{102}| + |y_{99} - y_{101}|}{3||x_{96} - x_{100}||} \tag{4}$$

$$MAR = \frac{MAR_1 + MAR_2}{2} \tag{5}$$

4 Behavior Recognition Algorithm

Driving dangerous actions include calling, smoking, playing on mobile phones, and drinking water. The behavior recognition module is a recurrent neural network with the LSTM [11] (long short-term memory) structure using multifeature ECG signal fusion in TensorFlow. First, the keypoint information of the human body predicted by the network was extracted, and 11 groups of keypoint information shown in the following figure were obtained in the ideal state. Facial information is combined to determine the action according to the angle, direction, and other information design conditions. Because the driver generally only has a large body movement when entering and leaving the cab, only the neck, shoulders, and arms are identified to complete the action recognition required by the system (Fig. 7).

A long short-term network (LSTM) is a model that only retains relevant information for prediction and forgets irrelevant data. This applies to the long time series data in this paper. The LSTM neural network model is a unique RNN model [12]. LSTM has three gates, and the forgetting gate determines the information memory and forgetting at each moment. The input gate determines how much new information is added to the cell, the forgetting gate determines whether the information is forgotten at any given time, and the output gate determines whether the information is output at any given time (Fig. 8).

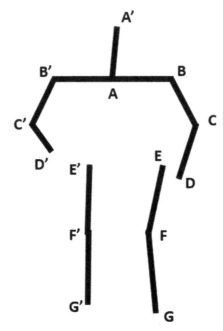

Fig. 7. Keypoint information map of the human body

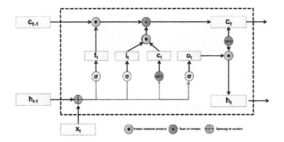

Fig. 8. LSTM schematic

In the picture, i, f, o, c, h represent the input gate, the output gate, the memory cell, the hidden layer and the output gate, respectively.

$$\mathbf{f_t} = \sigma\left(\mathbf{W_f} \cdot \left[\mathbf{h_{t-1}, x_t}\right] + \mathbf{b_f}\right) \tag{6}$$

The next step is to decide what new information to add to the cell state. First, use h_{t-1} and x_t to determine what information to update through the operation of the input gate, then use h_{t-1} and x_t to obtain $\tilde{c}_t$ through a tanh layer, and then update the old c_{t-1} to c_t

$$i_t = \sigma\left(W_i \cdot \left[h_{t-1}, x_t\right] + b_i\right) \tag{7}$$

$$\tilde{c}_t = \tanh(W_c \cdot [h_t, x_t] + b_c) \tag{8}$$

$$c_t = f_t * c_{t-1} + i_t * \tilde{c}_t \qquad (9)$$

Finally, a sigmoid test was performed to determine which states of cells needed to be output.

$$o_t = \sigma\left(W_i \cdot \left[h_{t-1}, x_t\right] + b_o\right) \qquad (10)$$

$$h_t = o_t * \tanh(c_t) \qquad (11)$$

The camera is mounted on the steering wheel, which can capture the main joint point activity information of the body and store it from a computer perspective to identify the body trajectory more accurately. Compared to traditional human pose recognition, human pose recognition based on motion capture technology is more accurate, can capture the details of the action well and is not limited by object color or occlusion.

5 Monitoring Vital Signs

An induction chip device was installed on a spoke of the steering wheel, and the driver could fit his wrist as needed to obtain information such as body temperature, heart rate, blood pressure, and electrocardiogram. This information was presented on the LCD screen of the steering wheel. Simultaneously, this information can also be connected to the charging-back cover through the line. Finally, the physical information is transmitted to the mobile app through the Bluetooth device.

Human body sign signal acquisition technology that integrates PPG photoelectric + ECG measurement technology and body temperature measurement technology can monitor the driver's vital signs in real time and analyse them through an intelligent system to take effective measures in time to ensure driver safety.

Because the ECG mark may be lost during driving, this may lead to serious deviations in the vital sign parameters. Therefore, lost information must be supplemented to ensure safe driving to recover the real signal as accurately as possible. In this study, the PPG and ECG fusion algorithms based on LSTM solved the above problems well. First, information was extracted from the ECG and PPG data, and then the data were normalized to generate the LSTM input time series.

Next, through modelling exercises and storage, a loss function waveform is produced, as shown in the following Fig. 9.

Using the same spectrum and duration, ECG and PPG data results were normalized to a 0–1 interval and converted into a time sequence for training, using a certain value mean difference function and Adam optimizer to improve the accuracy and reliability of modelling. Owing to the use of LSTM prediction technology, we can calibrate the original data to measure the time-domain and frequency-parameter changes of HRV signals in different time periods.

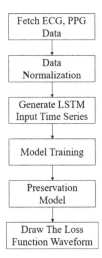

Fig. 9. Model training flow-chart

6   Experiment

After in-depth discussion of the experimental scene and the order of the test flow chart, we need to design the detection circuit and connect the peripheral devices required to successfully complete the task. Our circuit is based on a PIC18f4580 microcontroller connected to a 20 MHz quartz crystal. This electronic card is connected to the vehicle by 5 V power generated and adapted. State of the voltage generated on the LED dashboard. The printed circuit board is designed by Altium Designer software.

6.1   Driver Information Class

The system can identify driver identity information, including name, gender, and age (Fig. 10).

6.2   Driver Behavior Analysis Class

Leaving Behavior and Entering Behavior. To save space, when the driver left the driving position for 1 s, identity recognition and behavior prediction were no longer performed (Fig. 11).

"Fatigue Driving" Series of Behaviors. In this experiment, the blink threshold standard (PERCLOS) and yawn duration were used as two important indicators to determine fatigue status. P70 refers to the 'closed eye' duration, which accounts for more than 70% of the total duration, to determine whether the driver is in a state of fatigue. When the test ends, to ensure the maximum value of the yawning duration, the system extends the test time limit until there is no ongoing action. In this process, the following parameter standards were used:

Fig. 10. Driver facial information

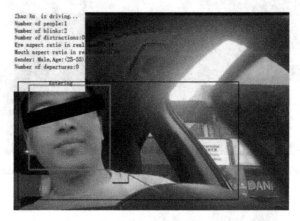

Fig. 11. Driver entering the car

(1) According to the P70 criterion, when the duration of eye closure exceeded 0.7, it was judged as fatigue.
(2) When the proportion of eye closure time is less than 0.5, it is regarded as eye closure; this threshold is called eye closure EAR.
(3) When the yawning time exceeded the MAR threshold of 0.40, it was regarded as yawning behavior and recorded in the database.
(4) When the yaw time exceeded 50, it indicated a fatigue state. The number of yawning frames was set to 50 (Figs. 12 and 13).

"Driving Distraction" Series of Behaviors. At present, the system also improves the accuracy of behavior recognition by using object recognition; however, its requirements for computing power are also higher, and there are higher requirements and challenges for real-time recognition and real-time reminders (Figs. 14, 15, 16 and 17).

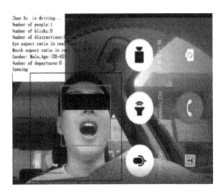

Fig. 12. Yawning

Fig. 13. Continuous blinking

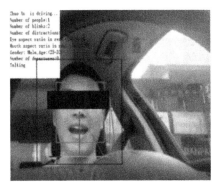

Fig. 14. Talking

Fig. 15. Drinking

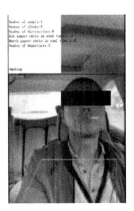

Fig. 16. Smoking

Fig. 17. Not seeing the road in 3 s

6.3 Driver Health Management

The mobile phone app connects to the steering wheel handle sleeve via BlueTooth. It displays the driver's body temperature, heart rate, blood pressure, ECG and other information to facilitate the driver's health management (Fig. 18).

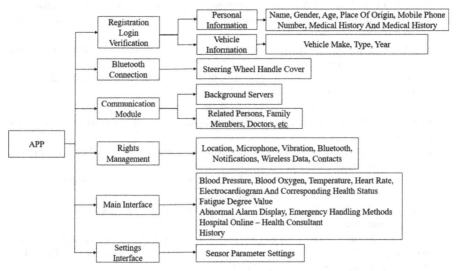

Fig. 18. APP page

7 Concluding Remarks

This paper proposes a more accurate and advanced driver vital signs and driving behavior monitoring system. The system mainly uses the MTCNN model to obtain face images in real time, uses the PFLD algorithm to locate facial keypoints, and combines relevant parameters to judge fatigue. Then, according to the LSTM network model, the key point information of the human body is obtained, the human motion track is identified, and the driving behavior, such as drinking water, smoking, and mind wandering, is determined by combining the facial information. The fusion algorithm based on LSTM can collect vital sign data and heart rate variability parameters more accurately, thereby further improving the reliability of the information system. This creates a safer and more comfortable driving environment for drivers. The experimental results show that the system can monitor the driver's health and driving state well during the driving process, and the accuracy of driving distraction action recognition is greatly improved. At present, the system also improves the accuracy of behavior recognition by using object recognition, but its requirements for computing power are also higher, and there are higher requirements and challenges for real-time recognition and real-time reminders. In the future, we will try new health monitoring methods and real-time monitoring of driving status, constantly optimize the system and provide more accurate services.

References

1. Jin, W.: Design and Implementation of After sales DMS System for Automobile Dealers. Shanghai Jiao Tong University, Shanghai (2018)
2. Zhang, X., Xu, Q., Gong, X., Li, X., Huang, J.: Integrated control strategy for path tracking of human machine codriving vehicle. J. Autom. Safe. Energy **41**(09), 130–139 (2022)
3. Zeng, J., Li, J., You, Z.: A review of progress on drivers' talk-over ability in automated driving human–machine codriving environment. Technol. Econ. Areas Commun. **24**(05), 30–37 (2022)
4. Xie, Z., Zhou, Y., Wu, K., Zhang, S.: Activity recognition based on spatial-temporal attention LSTM. Chin. J. Comput. **44**(02), 261–274 (2022)
5. Jia, X., Zeng, S., Pan, B., Zhou, Y.: Fast detection of target based on the inproved MTCNN network. Comput. Eng. Sci. **42**(7), 1262–1266 (2022)
6. Liu, W.: Thesis Submitted to Nanjing University of Posts and Telecommunications for the Degree of Master of Engineering. Jiangsu: Nanjing University of Posts and Telecommunications (2021)
7. Li, X., Bai, C.: Research on driver fatigue driving detection method based on deep learning. J. China Railway Soc. **23**(3), 123–126 (2022)
8. Xu, B., Cen, K., Huang, J., Shen, H., Cheng, X.: A survey on graph convolutional neural network. Chin. J. Comput. **43**(05), 755–780 (2020)
9. Pan, Z., Liu, R., Zhang, M.: Research on fatigue driving detection algorithm based on fuzzy comprehensive evaluation. J. Softw. **30**(10), 2954–2963 (2019)
10. Wang, Y., Sheng, R., Xu, Z.: Dynamic imitation learning algorithm based on surrounding vision. In: 2021 International Conference on Culture-oriented Science & Technology (ICCST), Beijing, China, pp. 47–51 (2021)
11. Yang, S., Yang, J., Li, Z., Wang, J., Li, D.: Human action recognition based on LSTM neural network. J. Graph. **42**(2), 174–181 (2021)
12. Ma, Z.: Research on Driver Hand Detection and Driving Behavior Analysis. University of Jinan, Shandong (2021)
13. Huang, Z., Ren, F., Hu, M., Liu, J.: A real-time expression mimicking method for humanoid robot based on dual LSTM fusion. Robot **41**(02), 137–146 (2019)

People Flow Monitoring Based on Deep Learning

Xinran Wang, Yanqing Wang$^{(\boxtimes)}$, Yiqing Xu, and Tianxin Wang

Nanjing Xiaozhuang University, Nanjing, China
wyq0325@126.com

Abstract. According to the application scenarios of the size of the human flow in different consumption places, to solve the problem of crowd detection, distance estimation between crowds and the inability to monitor and calculate the human flow in real time, this paper designs a real-time crowd detection scheme for application scenarios where consumers pay attention to the size of the human flow in consumption places. The main use of the YOLO algorithm with the Darknet53 network as the main network is to separate pedestrians from the background. Pedestrians' central two-dimensional coordinates are converted into three-dimensional coordinates, realizing crowd detection and apart from the distance estimation of crowds, real-time monitoring of current regional traffic and flow density, and solving the problem of being unable to monitor and calculate people in real time. It can be applied to many aspects, such as shop rating, traffic control and flow control of scenic spots. Existing monitors are affected by different lights and cannot provide accurate data. In addition, the processing algorithm of this scheme is stable and accurate, and preprocessing is performed before judging the human flow and the position of the human body to reduce the interference of light. This scheme has the performance of real-time monitoring and calculation through experimental verification.

Keywords: Target detection · Perspective transformation · Distance detection · Population density

1 Introduction

Currently, with the continuous improvement of the economy and people's level, an increasing number of consumers have begun to turn to cultural and spiritual consumption. However, with the growth of the flow of people, there are frequent crowds in places such as street attractions, and so-and-so shops queue up to buy overnight. Especially in the season of high incidence of infectious diseases, dense crowds are very likely to increase the risk of disease transmission, so a product is urgently needed to control the flow of people in related places.

In recent years, with the emergence of artificial intelligence, pedestrian flow monitoring has made some progress at home and abroad. For example, Viola proposed pedestrian flow monitoring based on motion characteristics [4]. Antonini et al. [5] proposed

Z. Yu et al. (Eds.): ICPCSEE 2023, CCIS 1880, pp. 153–167, 2023.
https://doi.org/10.1007/978-981-99-5971-6_11

a pedestrian flow estimation method based on track. At the same time, Kong et al. [6] proposed and designed a pedestrian flow monitoring method based on a neural network. The edge histogram and gray histogram are extracted as features to establish the classification model of personnel density by a feedforward neural network. Ali [7] provides a new idea for pedestrian flow monitoring by using particle dynamics for reference. This method can detect and separate foreground targets in an optical flow field and study the dynamics of group targets. Zhou et al. [8] proposed pedestrian flow monitoring based on feature regression and detection. Wu et al. [9] proposed a SURF algorithm based on linear interpolation angle correction for large-scale personnel statistical monitoring. The above performance will be weakened in the case of overlapping objects and dim light. In addition, the use of cameras will also bring privacy problems. In addition, Li et al. [10] proposed pedestrian flow monitoring based on smartphones. Mizutani et al. [11] proposed a method for monitoring pedestrian flow through equipment. These require people to carry equipment, so in some cases, they may not.

At present, traffic monitoring can be divided into two categories: image-based and nonimage-based. Image-based images usually need to be processed to extract head and shoulder features and then realize personnel statistics and flow monitoring. Object detection is one of the core problems in the field of computer vision. It is widely used in intelligent image monitoring. Target detection algorithms can be divided into algorithms based on feature extraction [1] and target detection algorithms based on convolution neural networks [2]. Target detection algorithms use feature extraction operators such as SIFT [3], LBP [4], HOG [5] or Haar [6] to extract features from candidate regions of targets and classify targets by classifiers such as SVM. Felzenszwalb et al. [7] combined HOG with SVM and proposed a deformable component model DPM, which stands out in the target detection algorithm. Although the algorithm has made some achievements, it has the disadvantages of high time complexity and many redundant windows. The robustness of manual design features is low, the detection accuracy is low and the generalization is poor. With the rapid development of deep learning, using a convolution neural network to detect objects by introducing deep semantic features shows great advantages compared with algorithms, among which the YOLO algorithm proposed by Joseph Redmon et al. [8] is one of the mainstream target detection algorithms at present.

In this paper, a real-time crowd monitoring system based on the YOLO algorithm is proposed. It is of great significance to calculate the crowd gathering situation and social safety distance in public places by detecting pedestrians and calculating the distance between pedestrians through the YOLO algorithm.

2   Brief Introduction of the System

There are also some problems based on images, such as blind spots, ease of being affected by light, and difficulty in identifying crowded scenes. Nonimages vary according to the perceptual technology adopted. The real-time crowd monitoring system proposed in this paper first captures and locates the pedestrians on each image by camera and uses the YOLO target detection algorithm and simultaneously generates the coordinates of each pedestrian. Considering the perspective effect, OpenCV is used for perspective transformation to realize the transformation of two-dimensional coordinates and three-dimensional coordinates of the image. Then, a circle with a set radius is generated at the foot of pedestrians to measure the actual distance between pedestrians and judge the degree of crowd gathering and staying in a period of time.

2.1 Safety Distance Detection of People Flow

With the continuous improvement of the target detection algorithm model based on deep learning in science and technology, great progress has been made in network accuracy and computing speed. However, in some specific scenes, only depending on the image information obtained by camera and the two-dimensional target detection algorithm cannot be applied in practice. It is impossible to make an accurate judgment on the positioning of pedestrians in shops with large traffic.

Since the end of 2019, the COVID-19 epidemic has raged all over the world, causing serious losses to all countries. Academician Zhong Nanshan pointed out that we need all people to consciously prevent epidemics if we want to implement the normalization prevention and control of epidemic situations. Because of the strong transmissibility of SARS-COV-2, there must be a certain safe distance between people.

Taking the maintenance of customer queuing orders as an example, this paper proposes a crowd real-time monitoring system. First, the camera randomly captures and collects the current team image, and then the image is transmitted to the detection system. The system measures the distance between the adjacent customers of the current team. If the actual distance between customers is less than the safe distance (generally 1 m), the system will give an alarm to remind the public places more accurately and quickly. Maintaining the order of public places more accurately and quickly The workflow is shown in Fig. 1.

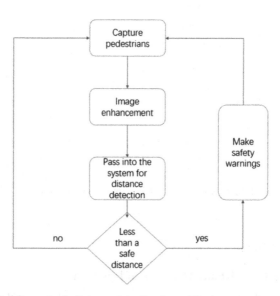

Fig. 1. Workflow of safe distance detection for epidemic prevention and control

2.2 Flow Control of Scenic Spots

In recent years, with the improvement of the income level, the tourism industry in China has greatly increased, and the number of tourists has increased exponentially. The surge in the number of tourists has brought considerable income to scenic spots, and at the same time, it has also caused a series of safety problems. For example, scenic spots do not control the flow of people, which leads to accidents such as stampedes. Similar news emerges one after another. To provide a more comfortable play experience for tourists, this system can be used to control the flow of people in scenic spots. The large flow of people in public places, high density of people and difficulty in evacuation bring many problems to fire safety management.

First, sampling points are randomly set in the scenic spot to collect the real-time tour situation of the sampling points through the camera. After the collected photos are transmitted to the system, the system judges the actual distance between each tourist and its adjacent tourists and compares each tourist's distance Wi ($1 \leq i \leq n$) to its nearest tourist. Formula (1) is used to obtain the congestion index S of this scenic spot:

$$S = \frac{\sum_{i=1}^{n} W_i}{n} \tag{1}$$

When S is not in the safety range of the scenic spot, the system will warn the staff to control the flow of people. The specific workflow is shown in Fig. 2.

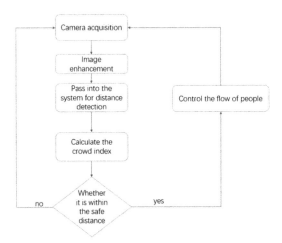

Fig. 2. Workflow of flow control in scenic spots

3 Key Technology

3.1 Target Detection Distance Detection of People Flow

The main steps of the target detection algorithm are as follows: the first step is to preprocess the input image, that is, to remove noise, detect the edge and grayscale the image; the second step is to slide small windows of multiple scales on the target image to construct candidate areas; the third step is to extract the features of candidate regions and select representative feature vectors; the fourth step is to classify the features; and in the fifth step, the redundant detection boxes are filtered out by nonmaximum suppression (NMS) (weighted box clustering (WBC) and so on).

The YOLOv3 algorithm proposed by Joseph Redmon and others is one of the mainstream target detection algorithms. It uses a convolution neural network to improve the accuracy and achieve multiscale prediction. DarkNet 53 has 53 layers, including 52 convolution layers and 1 fully connected layer. The residual structure ensures the super expression of features and avoids the gradient problem caused by a network that is too deep. DarkNet 53 is the backbone network commonly used by the YOLO algorithm. The DarkNet53 network model is shown in Fig. 3, where Conv represents convolution, Avgpool represents average pooling, BN represents batch normalization, LeakyReLU is the activation function, and we set it to 0.1. Its mathematical formula is:

$$\begin{cases} y = \max(0, x) + leak * \min(0, x) \\ leak = 0.1 \end{cases} \tag{2}$$

The target detection algorithm can be divided into a two-stage algorithm represented by the R-CNN series algorithm and a one-stage algorithm represented by the YOLO series algorithm. The two-stage target detection algorithm is divided into two steps: first, the candidate regions are generated, then the candidate frames are classified, and their positions are corrected by YOLO. This phase target detection algorithm transforms the detection task to regression problem processing, truly realizing that end-to-end detection can be used in video surveillance and other real-time detection tasks. The network structure of YOLO is relatively simple. First, the convolution layer and pooling layer extract image features, and then the fully connected layer is used to predict the object class probability and regress the bounding box position.

3.2 Distance Detection Addition

In this paper, the center coordinates of pedestrians in images are detected, and the distance between pedestrians is calculated according to the distance between the center coordinates of pedestrians. The key problem faced by distance detection is the conversion between two-dimensional pixel coordinates and three-dimensional world coordinates. In this paper, the center point of a two-dimensional image is taken as the origin to establish a coordinate system, and a pixel value is taken as a basic unit. The distance between the bottom center coordinates of the image is used to represent the actual distance of pedestrians in the street. The relationship between the two is related to the height, position, viewing angle range and inherent parameters of the camera. In this paper, the pinhole camera model [9] using the direct linear transformation method maps the two-dimensional pixel coordinates (x′,y′) to the corresponding three-dimensional world coordinates (x, y, z). The transformation relationship between the coordinates is shown in Fig. 4.

Where M is a three-dimensional coordinate point and M is a projection point. The mathematical formula of the direct linear transformation method for the pinhole camera

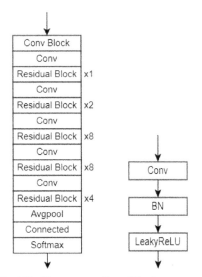

DarkNet53 structure Conv Block structure

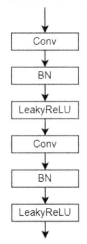

Residual block structure

Fig. 3. DarkNet 53 Network Structure Diagram

model is as follows:

$$s\begin{bmatrix} x' \\ y' \\ 1 \end{bmatrix} = KRT * \begin{bmatrix} x \\ y \\ z \\ 1 \end{bmatrix} \tag{3}$$

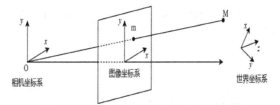

Fig. 4. Transformation relationship between coordinates

where S is the unknown scale factor and K is the internal parameter of the camera, as shown in the following matrix:

$$K = \begin{bmatrix} f_x & \gamma & c_x & 0 \\ 0 & f_y & c_y & 0 \\ 0 & 0 & 1 & 0 \end{bmatrix}, \begin{cases} f_x = \frac{f}{dx} \\ f_y = \frac{f}{dy} \end{cases} \tag{4}$$

RT is the camera external parameter, and R is the rotation matrix:

$$R = \begin{bmatrix} 1 & 0 & 0 & 0 \\ 0 & \cos\theta & -\sin\theta & 0 \\ 0 & \sin\theta & \cos\theta & 0 \\ 0 & 0 & 0 & 1 \end{bmatrix} \tag{5}$$

T is the translation vector:

$$T = \begin{bmatrix} 1 & 0 & 0 & 0 \\ 0 & 1 & 0 & 0 \\ 0 & 0 & 1 & -\frac{h}{\sin\theta} \\ 0 & 0 & 0 & 1 \end{bmatrix} \tag{6}$$

where f is the focal length, dx and dy are the physical dimensions of the horizontal axis and the vertical axis of the image, respectively, and γ is the tilt factor representing the nonperpendicular factor of the horizontal axis and the vertical axis. γ of the camera is usually 0. (C_x, C_y) is the optical center, that is, the intersection point between the optical axis of the camera and the image plane h is the height of the camera.

Robots have the basic characteristics of perception, decision-making and execution. Most of the street images in the data set are monocular vision, that is, a single camera is used, which leads to a perspective effect. That is, the projection of a three-dimensional world scene to a two-dimensional perspective image plane will lead to a difference between the pixel distance and the actual distance between objects. Under the perspective effect, it is difficult to perceive the uniform distribution of distance in the whole image. To solve this problem, this paper uses OpenCV to carry out perspective transformation to solve the difference between two-dimensional and three-dimensional coordinates.

Perspective transformation projects a plane to a designated plane through a projection matrix by using the condition that the three points of the perspective center image point target point are collinear (x, y, z) assuming that a three-dimensional coordinate is a two-dimensional coordinate of aqi perspective transformation is (x', y'), the relationship between (x, y, z) and (x', y') is:

$$\begin{cases} x' = \frac{x}{z} \\ y' = \frac{y}{z} \end{cases} \tag{7}$$

According to the principle formula of perspective transformation:

$$\begin{bmatrix} x \ y \ z \end{bmatrix} = \begin{bmatrix} u \ v \ z' \end{bmatrix} \begin{bmatrix} a_{11} \ a_{12} \ a_{13} \\ a_{21} \ a_{22} \ a_{23} \\ a_{31} \ a_{32} \ a_{33} \end{bmatrix} \tag{8}$$

where u and v represent that the original image coordinate z' is 1, and then the relationship between (x, y, z) and (x', y') is:

$$\begin{cases} x' = \frac{x}{z} = \frac{a_{11}u + a_{21}v + a_{31}}{a_{13}u + a_{23}v + a_{33}} \\ y' = \frac{y}{z} = \frac{a_{12}u + a_{22}v + a_{32}}{a_{13}u + a_{23}v + a_{33}} \end{cases} \tag{9}$$

3.3 Stereo Matching Algorithm Based on Deep Learning

The stereo matching network based on deep learning requires high real-time performance. The running speed of BGNet is 39 fps, which meets the requirement of real-time in automatic driving scene BGNet designs a cost upsampling module based on two-sided meshes, which uses two-sided meshes and slicing operations to upsample low-resolution cost quantities with high-resolution cost quantities. The module uses $3 \times 3 \times 3$ 3D convolution to convert the low resolution cost feature map into bilateral mesh, where the input of 3D convolution is the cost of dimension (X, Y, D, C) and the output is the bilateral mesh feature map of dimension (X, Y, D, G). D represents the disparity map, C represents the number of channels, and G represents the guiding feature. The high-dimensional feature map can be transformed into the leading feature G by two successive 1×1 convolutions. Finally, the slice operation based on linear interpolation is used for upsampling. The cost quantity upsampling module based on two-sided meshes greatly reduces the computation and is easy to transplant and can also be used in other stereo matching networks (Fig. 5).

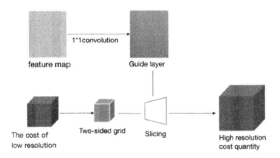

Fig. 5. Cost Quantity Upsampling Module (CUBG) based on bilateral grid

Single target detection algorithms or density map regression algorithms are mainly used in the crowd statistics scheme. In view of the advantages and disadvantages of the population statistics algorithm for object detection and density map regression described in the introduction, this paper designs a population statistics algorithm suitable for various population density scenes. By setting the threshold, the improved SSD object detection algorithm MNSSD (MobileNet-SSD) is selected to obtain a more accurate number of people when there are few people. It can accurately locate pedestrians and can be used to detect other crowd behaviors such as running, gathering and fighting. The algorithm switches to the population estimation algorithm based on the density map regression MCNN population density map and estimated population flow in the scene with large population flow. At the same time, it can also know the population distribution according to the population density and warn of abnormal gathering events in advance. In this solution, the algorithm switching part can be set by threshold or adaptively according to the computing power of the edge.

The SSD feature algorithm formula is as follows [x]:
$$T_n = S_n(T_{n-1}) = S_n(S_{n-1}(\dots S_1(I)))$$
$$R = D(d_n(T_n),\dots d_{n-k}(T_{n-k})), n > k > 0$$

where (T_n) is the characteristic diagram of the nth layer, (S_n) is the nth layer feature map created from the (n-1) layer map, and (S_1 (I)) is a first layer feature map created from the image input (I) (d_n (.)) is the detection result for the nth layer (D (.)) is the sum of all median values [7]. The system is designed to be deployed on a lightweight embedded platform for real-time flow detection. The influence of brightness and running speed of the model is very important. If VGG and other models are used for training, the model is very large [8]. The performance of VGG-SSD can be improved by using the MobileNet network for feature extraction. In addition, the smaller and deeper features of the detected object (head) are not important, so the algorithm deletes the deep 1×1 feature map of VGG-SSD for computational complexity.

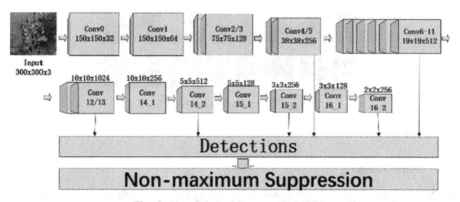

Fig. 6. Network Architecture of MNSSD

The core of MobileNet is that deep separable convolution can be divided into deep convolution and point-by-point convolution. The MobileNet principle is that each single input channel is filtered using 3 * 3 * 1 convolution and then output using 1 × 1 convolution combined with all depth convolution. This decomposition greatly increases the amount of calculation.

4   Experiments and Results

4.1   Experiment

The experiment in this paper uses the Windows 10 operating system. The hardware platform processor is an Intel Core i9-9900XRAM 64 GB graphics card model, the NVDIA GeForce RTX 2080TI × 2 video memory 22 GB language is Python 3.7, and the neural network is built based on the PyTorch deep learning framework.

4.2   Experiments and Results

The data set used in this paper is the public data set MOT15, which contains 22 videos, including 11 videos as training sets and 11 videos as verification sets. All videos are divided into frames. Images are uniformly named as 6 digits in jpeg format, such as 000001. Some data samples in the MOT15 data set are shown in Fig. 6, for example (Fig. 7).

Fig. 7. Sample data in MOT15

4.3   Model Training

CNN model training uses weak supervised learning [11]. According to cross-validation, 10% of the training data are randomly selected to determine the model parameters and select the optimal model. Every 2 iterations of H5PY are used to store data similar to arrays and directory containers. It can be seen from the figure that the overall trend of the loss curve with the total number of trainings is declining.

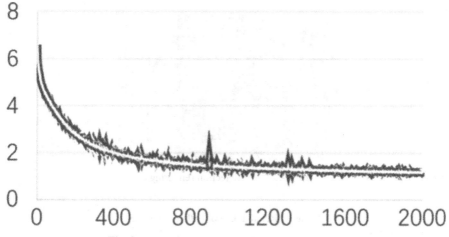

Fig. 8. Some data samples in CNN model training

Through data enhancement and random scrambling data enhancement, including image mixing random pixel transformation random clipping random horizontal flipping and other enhancement model robustness to avoid overfitting, we use the DarkNet 53 network model as the backbone network, use the YOLOv3 framework, input image size is 512 × 512 batches are set to 35 iteration rounds are set to 1000 rounds, select momentum to optimize the initial learning rate is set to 0.00025, the learning rate is decay rate is 0.1, and pretrain the model on Microsoft COCO data set. Weight model training time The accuracy rate of the final experiment is 84.3%, and the recall rate is 87.5%. The calculation formula of the accuracy rate and recall rate is as follows:

$$\begin{cases} precision = \frac{TP}{TP+FP} \\ Recall = \frac{TP}{TP+FN} \end{cases} \tag{10}$$

where TP represents the number of true positives, FP represents the number of false positives, and FN represents the number of false negatives. The target detection effect is shown in Fig. 6, in which the red mark is the gathering crowd with the pixel value as the unit, and the data such as coordinates and distance are shown in Table 1. At the same time, Fig. 6 shows the pedestrian distance detection in a two-dimensional image, that is, Euclidean distance, which is obviously different from the pedestrian distance in the three-dimensional real world. We will explain how to convert two-dimensional coordinates into three-dimensional coordinates and calculate the distance below. The number of pedestrians was derived from the test video (Fig. 9).

Since we cannot know the camera height, angle range, camera inherent parameters and other information in the data set, it is difficult to convert the pixel distance into the actual distance, so this paper directly sets 100 basic units as 1 m. The perspective transformation matrix obtained by calculation is:

$$\begin{bmatrix} 0.8092 & -0.2960 & 11.0 \\ 0.0131 & 0.0910 & 30.0 \\ 0.0001 & -0.0052 & 1.0 \end{bmatrix}$$

Fig. 9. Target detection renderings

Table 1. Related data of gathered people

Pedestrian number	IoU	Top left frame coordinates	Bottom right frame coordinates	Frame center coordinates	Distance
2	0.9908	(466, −1)	(487,40)	(476.5,19.5)	37.6430
9	0.8768	(425, −2)	(438,19)	(431.5, 8.5)	
8	0.9076	(484, −1)	(505,41)	(494.5,20.0)	18.3371
10	0.7556	(504, 2)	(521,45)	(512.5,23.5)	

Through perspective transformation, we obtain the three-dimensional coordinate (x, y, z) of the center of each pedestrian and choose to set z as 0 so that the center of each pedestrian detection box is converted to the bottom center of each pedestrian. The calculation of the distance of the three-dimensional coordinate system is converted to the calculation of the two-dimensional coordinate system, the height difference between pedestrians is eliminated, and the center of each pedestrian is detected as the center of the circle. If the radius is 0.5 m, that is, 50 basic units, the bottom circle is connected with other people's circles. If the circle is connected, both sides will be marked. The specific detection effect is shown in Fig. 8, where Figure (a) is the heatmap of the distance between pedestrians on the street, Figure (b) is the effect of image background removal to realize image enhancement, and Figure (c) is the effect map of crowd gathering detection in the real street scene. For distance detection, each pedestrian is numbered, and the distance between pedestrians is calculated using the L2 normal form. Each pedestrian is regarded as a class, and its distance from other classes is calculated using the clustering method. The two classes that are too close together are merged into a new class (Fig. 10).

$$\|D\|_2 = \sqrt{\sum_{i=1}^{3} (q_i - p_i)^2} \tag{11}$$

(a) **Heatmap of crowd size**

(b) **Street detection effect picture**

(c) **Street detection effect picture**

Fig. 10. Test effect diagram

5   Concluding Remarks

This paper mainly introduces the real-time crowd detection system, which uses the YOLO algorithm to detect pedestrians and realizes the estimation of three-dimensional realistic distance between pedestrians in two-dimensional images through perspective transformation technology. Through experiments, this system has achieved good results in crowd detection and crowd distance estimation. In the future, we will try new target detection algorithms and the technology of converting two-dimensional coordinates into three-dimensional coordinates to continuously optimize the system and provide more accurate services.

References

1. Jingcheng, Z., Xinru, F.U., Zongkai, Y., et al.: UAV detection and identification in the Internet of Things. In: 2019 15th International Wireless Communications & Mobile Computing Conference (IWCMC), pp. 1499–1503. IEEE (2019)
2. Wenxuan, D., Hongtao, L., Guozhu, L., et al.: A review of deep convolution algorithms for object detection [J/OL]. Comput. Sci. Explor. 1–20 [2022–02–07]
3. Yang, Z., Wang, H., Liu, C.: Traffic monitoring method based on wireless perception and integration of learning. Comput. Eng. Des. Lancet (11), 3243–3249 (2022). https://doi.org/10.16208/j.issn1000-7024.2022.11.032
4. Cai,J., Wang, S., Gan, X., Wang, L.: Traffic monitoring system design and implementation based on LabVIEW. J. Tonghua Normal Univ. J. lancet (10), 67-722022). https://doi.org/10.13877/j.carolcarrollnkicn22-1284.2022.10.011
5. Liu, X., et al.: Development of intelligent analysis system for human flow Monitoring and query based on photoelectric sensing. Phys. Eng. 32(04), 218–223 (2022). (in Chinese)
6. Chen, J., Yantao, Q., Feng, Y., Fu, S., Wang, L., Fan, C.: Human flow monitoring method with millimeter wave Radar based on dual time point detection. Telecommun. Technol. **202, 62**(11), 1593–1599
7. king inside. Traffic monitoring method based on the CSI research. Nanchang aviation university (2021). https://doi.org/10.27233/dcnki.GNCHC.2021.000293
8. Liao, J., Zong, F.: Traffic monitoring system based on deep learning study. J. Electron. (9), 54–55 + 81 (2021). https://doi.org/10.16589/j.carolcarrollnkicn11-3571/tn.2021.09.017
9. Wang, C., Shi, G., Chen, T., Liu, D., Wang, Z., Zhou, S.: Traffic monitoring system based on OpenCV. Comput. Knowl. Technol. (7), 235–236+241 (2021). https://doi.org/10.14004/j.carolcarrollnkiCKT.2021.0802
10. Wang, C., Yang, Y., Ren, X., Zhang, X.: Indoor traffic monitoring system design . Autom. Instr. 33(09) 6, 49–51 (2018). https://doi.org/10.16086/j.carolcarrollnki. issn1000–0380.2018020013
11. Shi, Y., Chang, C., Liu, X., et al.: Advances in calibration methods of internal and external parameters of planar array cameras. Laser Optoelectron. Progress, **58, 707**(24), 9–29 (2021)
12. Jia Shi-na. The small target detection algorithm based on improved YOLOv5 research. Nanchang university (2022). 10.27232/,dc nki. Gnchu. 2022.004449
13. Huang. Target detection based on deep learning applied research. Nanjing university of posts and telecommunications (2022). https://doi.org/10.27251/dcnki.GNJDC.2022.001250
14. Zhu, J., Feng, F., Shen, B.: People counting and pedestrian flow statistics based on convolutional neural network and recurrent neural network. In: 2018 33rd Youth Academic Annual Conference of Chinese Association of Automation (YAC) (Nanjing, China, 18–20 May 2018), pp. 993–998 (2018)
15. Wu, D., Wang, J., Li, B., Guo, T.: Large crowd count based on improved surf algorithm. J. Xi'an Univ. Sci. Technol. 27, 650–655 (2015)
16. Mizutani, M., Uchiyama, A., Murakami, T., Abeysekera, H., Higashino, T.: Towards people counting using Wi-Fi CSI of mobile devices. In: 2020 IEEE International Conference on Pervasive Computing and Communications Workshops (PerCom Workshops) (23–27 March 2020), pp. 1–6 (2020)

17. He, C.T., Zhu, M.: No-reference face image quality assessment based on deep learning for mobile devices. J. Chin. Mini-Micro Comput. Syst. **40**(02), 407–412 (2019)
18. Long, J., Shelhamer, E., Darrell, T.: Fully convolutional networks for semantic segmentation. In: Proceedings of the IEEE Conference on Computer Vision and Pattern Recognition, pp. 3431–3440 (2015)

Research on the Influencing Factors
of Passenger Traffic at Sanya Airport Based
on Gray Correlation Theory

Yuanhui Li[✉], Haiyun Han[✉], Zhipeng Ou, and Wen Zhao

Sanya Aviation and Tourism College, Sanya 572000, Hainan, China
576735855@qq.com, 103238991@qq.com

Abstract. Many factors have an effect on the passenger traffic of Sanya Airport. In this paper, seven factors that have a great influence on passenger demand, namely accommodation and catering turnover, number of overnight visitors received, total tourism revenue, average room opening rate of tourist hotels, flight movements, airport passenger flow and railway station passenger flow, are selected to conduct gray correlation analysis on these factors to provide a research basis for passenger flow prediction.

Keywords: Passenger traffic of Sanya airport · Influencing factors · Gray correlation analysis

1 Introduction

The degree of development of air transport as a modern mode of transport reflects the level of economic development of a region. Air transport not only meets the working needs of the population, but is also an indispensable tool for the development of tourism and the promotion of foreign contact [1].

The analysis of civil aviation passenger transportation systems and their related factors usually has two types of research: qualitative and quantitative. Qualitative research, mainly based on the analysis of historical data and the study of future conditions, relies on subjective experience and logical reasoning ability to make speculative judgments on its influencing factors, but because of the lack of the necessary research on the magnitude of the influence of the factors on their degree, it is not possible to clearly understand the priority of the influencing factors. In quantitative research, regression analysis, variance analysis, principal component analysis and other methods are generally used to calculate and analyse the correlation, determine the degree of the factors and their correlation, and predict the future trend according to the main factors and the relevant regression equation. However, it requires the system to have a large number of complete statistics, requires the samples to obey the typical probability distribution, and is computationally intensive. However, civil aviation passenger transport started late and its development process is mostly influenced by domestic politics and policies; on the other hand, various social and economic data including civil aviation are not complete [2].

The gray prediction model can make more accurate predictions in the case of incomplete information. The single-data differential model has better fitting and extrapolation characteristics and has the characteristics of requiring less sample data and simplicity of operation [3]. In view of this objective reality and the characteristics of gray system theory, this paper uses gray correlation theory to analyse the correlation of many factors affecting the civil aviation passenger transportation system and determine the order of the degree of influence of the influencing factors to lay the foundation for the subsequent study of the civil aviation passenger transportation system.

1.1 Factors Affecting Passenger Traffic at Sanya Airport

According to Sanya's location and industrial characteristics and summarizing the opinions of experts and scholars in related fields, the factors that may affect the passenger volume of the local airport are summarized according to the principle of easy availability and quantifiability, which are as follows:

Accommodation and catering turnover (ten thousand yuan), the number of overnight tourists (10,000 people), total tourism revenue (100 million yuan), the average occupancy rate of tourist hotels (%), flight take-off and landing flights, airport passenger flow (ten thousand person-times), railway station passenger flow (ten thousand person-times).

Because of the complicated influence of road passengers (such as self-driven travel) and water transport passengers (such as cruise ships) on the passenger volume of Sanya Airport, they are not included in this study and will be studied separately. Considering the special island location of Hainan, tourists from outside the province come to Haikou and then turn their car to Sanya; or fly to Sanya and return to Haikou. The railway system can also largely provide transfer services for airport passengers, describing the relationship as a positive correlation.

2 The Gray Correlation Theory

2.1 Gray System and Gray Correlation Principle

A system with partially clear and partially unclear information is called a gray system. Gray system theory is characterized by the study of "small sample" and "information-poor" uncertainty, so the full exploitation of the information possessed to explore the inherent laws of the system itself is the basic criterion of gray system theory research. In a developing and changing system, there are many factors, various factors are not clearly related, and the influence is unclear. Generalizing the gray system theory and measuring the similarity or different degrees of development between the factors is called the correlation analysis method. The measure of the magnitude of the correlation between two systems or two factors is called the degree of correlation. The degree of correlation describes the relative changes between factors during the development of the system, that is, the relativity of the magnitude, direction and speed of change, etc. If the relative changes are basically the same between the two in the development process, the two are considered to be highly correlated; conversely, they are less correlated [4].

He Ying (2012) used the gray correlation degree analysis method to conduct empirical research to explore the internal system factors affecting the development of international and domestic tourism economies in Xinjiang and summarized and analysed the main factors affecting the development of tourism in Xinjiang [5]. Zhou Junpei (2021) used the improved gray correlation analysis method to perform a quantitative analysis on the importance of many factors affecting the development of Gansu civil aviation. The results show that the industrial structure, population size, income and expenditure level indicators have a significant impact on the development of Gansu civil aviation [6]. Liu Xiaoyan (2013) analysed the relationship between Hainan's science and technology investment and economic growth by gray correlation theory, and the results show that the investment in science and technology, personnel and internal expenditure have a positive effect on Hainan's economic growth [7]. Yin Fu (2013) used gray correlation analysis to build an evaluation model of aviation equipment technical support capability, providing a quantitative analysis model [8] for the evaluation of aviation equipment technical support capability. Zhang Liyun (2013) used the gray correlation analysis method to make decisions on the recruitment of universities and combined qualitative and quantitative methods, which can exclude the subjectivity and arbitrity of decision makers to some extent [9].

2.2 Gray Correlation Degree Analysis

Let the comprehensive evaluation problem contain n evaluation objects and m indicators, and the corresponding index observation value is

$$a_{ij} \ (i = 1, 2, \cdots, n; j = 1, 2, \cdots, m).$$

The specific steps of the gray correlation analysis are as follows:

(1) The evaluation indicators are pretreated
    That is, the index consistency and dimensionlessness, and the evaluation matrix are constructed $B = (b_{ij})_{n \times m.}$
(2) Determine the comparison series (evaluation objects) and the reference series (evaluation criteria)
    The comparison series are

$$b_i = \{b_{ij} | i = 1, 2, \cdots, n; j = 1, 2, \cdots, m\}.$$

b_i is the standardized index vector value of the ith evaluation object.

$$b_0 = \{b_{0j} | j = 1, 2, \cdots, m\} b_{oj} = \max_{1 \le i \le n} b_{ij}, j = 1, 2, \cdots, m.$$

The reference sequence is equivalent to each index value of a virtual best evaluation object.

(3) Calculate the gray correlation coefficient

$$\xi_{ij} = \frac{\min\limits_{1 \le s \le n} \min\limits_{1 \le k \le m} |b_{0k} - b_{sk}| + \rho \max\limits_{1 \le s \le n} \max\limits_{1 \le k \le m} |b_{0k} - b_{sk}|}{|b_{0j} - b_{ij}| + \rho \max\limits_{1 \le s \le n} \max\limits_{1 \le k \le m} |b_{0k} - b_{sk}|},$$

$$i = 1, 2, \cdots, n; j = 1, 2, \cdots, m.$$

$\rho \in [0, 1]$. Where is the resolution coefficient. Generally, the larger the resolution coefficient ρ is, the greater the resolution; the smaller ρ is, the smaller the resolution.

(4) Calculate the gray correlation degree

The formula for calculating the gray correlation degree is given as follows:

$$r_i = \sum_{j=1}^{m} w_j \xi_{ij}, i = 1, 2, \cdots, n.$$

r_i is the gray correlation degree of the relatively ideal object of the ith evaluation object. Where is the weight of the j th indicator variable. If the weight is not determined, each index variable should be the same weight, $w_j = \frac{1}{m}$.

(5) Evaluation and analysis

According to the magnitude of the gray correlation, the correlation of each evaluation object can be established by ranking the objects, and the larger the correlation is the better the evaluation results [4].

3　Gray Correlation Analysis of Factors Related to Passenger Traffic at Sanya Airport

3.1　Collect Data on Related Factors

Table 1. Data on passenger volume and related factors of Sanya Airport

Year	2015	2016	2017	2018	2019
Accommodation and catering turnover	119.87	129.71	143.84	152.46	165.89
Number of overnight visitors	1495.73	1651.58	1830.97	2150.52	2396.33
Total tourism revenue	302.31	322.4	406.34	492.43	633.19
The average occupancy rate of tourist hotels is (%)	64.5	66.02	69.57	71.46	71.81
Flight take-off and landing sorties	106869	112771	119788	121490	123292
Airport passenger flow	1619.19	1736.13	1938.99	2003.90	2016.37
Railway Station Passenger flow	862.09	1055.62	1096.94	1162.20	1380.72

According to the public data of the Sanya Municipal Bureau of Tourism, Culture, Radio, Film, Television and Sports, the statistical data collected for the five years from 2015 to 2019 are shown in Table 1. Due to the impact of the epidemic from 2020 to 2022, the data of all civil aviation passenger transport fluctuated abnormally, so the data of the past three years are not used.

3.2 The Gray Association is Calculated

The turnover of accommodation and catering is A1, the number of overnight tourists is A2, the total tourism revenue is A3, the average occupancy rate of tourist hotels is A4, flight take-off and landing flights is A5, airport passenger flow is A6, and railway station passenger flow is A7. The vector normalization method is adopted to standardize each index, and the evaluation matrix can be obtained.

R =	0.0011	0.0011	0.0012	0.0013	0.0013
	0.0140	0.0146	0.0153	0.0177	0.0194
	0.0028	0.0029	0.0034	0.0041	0.0051
	0.0006	0.0006	0.0006	0.0006	0.0006
	0.9998	0.9997	0.9997	0.9997	0.9996
	0.0151	0.0154	0.0162	0.0165	0.0163
	0.0081	0.0094	0.0092	0.0096	0.0112

Table 2. Gray correlation coefficient and correlation degree calculation data

Factors	2015	2016	2017	2018	2019	Correlation degree
A1	0.3334	0.3335	0.3335	0.3335	0.3335	0.3335
A2	0.3363	0.3365	0.3366	0.3372	0.3376	0.3369
A3	0.3338	0.3338	0.3340	0.3341	0.3344	0.3340
A4	0.3333	0.3333	0.3333	0.3334	0.3334	0.3333
A5	1.0000	1.0000	1.0000	1.0000	1.0000	1.0000
A6	0.3366	0.3367	0.3368	0.3369	0.3369	0.3368
A7	0.3350	0.3353	0.3353	0.3354	0.3357	0.3353

Taking the resolution coefficient $\rho = 0.5$, each index variable takes the same weight, that is $w_j = 0.2$. The gray correlation coefficient table and the gray correlation degree of each influencing factor are calculated, as shown in Table 2.

According to the gray correlation degree, the importance of various factors affecting Sanya Airport and passenger volume is ranked as follows:

$$A5 > A2 > A6 > A7 > A3 > A1 > A4$$

3.3 Analysis of the Calculation Results

According to the above calculation results, among the factors affecting Sanya Airport, the flight takeoff and landing number is most directly related to the round-trip passenger flow, so the correlation degree is 1. The difference in the correlation degree of the other factors is relatively small.

In the qualitative analysis and prediction of the influencing factors of airport passenger volume in the past, past passenger flow data are usually considered the main factor, and then a A RIMA model or GM (1,1) is established to predict the passenger flow. However, in this paper, through the gray correlation calculation, the number of flights taking off and landing and the number of overnight tourists have a greater impact on the passenger flow. The reason is that flights are relatively stable and less volatile; most overnight visitors come from outside the island, and the fast nature of civil aviation makes airplanes their first choice, so these two factors have a greater impact on airport passenger traffic than simply looking at the previous passenger traffic.

4 Conclusions

In summary, this research result can provide a reference basis for the passenger traffic forecasting study of Sanya Airport.

Gray correlation theory can better express the relative comparative advantage of each factor series on the degree of influence of the characteristic series when dealing with the degree of influence of many factor series on civil aviation passenger traffic and can quantify this degree of influence. It has been proven that gray correlation theory is not a good method for dealing with this kind of problem.

Acknowledgments. Project supported by the Education Department of Hainan Province, project number: Hnky2022ZD-25.

References

1. Li, C.P.: Analyzing the influential factors in China's civil aviation passenger. Sci. Technol. Ind **11**, 59–61 (2011)
2. Xiong, C.: Study on the Influencing Factors of Civil Aviation Passenger Transport Based on the Gray Relation Theory. J. Stat. Decis. **01**, 52–53 (2006)
3. Li, Y.: The application research of gray forecasting model. J. Math. Pract. Theory. **43**(11), 90–95 (2013)
4. Si, S.: Python Mathematical Experiment and Modelling. Science Press, Beijing (2020)
5. He, Y.: A research of impacting xinjiang tourism economic factors based on gray correlative analysis. J. Ecol. Econ. **01**, 160–162+170 (2012)
6. Zhou, J.: Analysis of influencing factors of gansu civil aviation development based on improved gray correlation degree. J. Civ. Aviat. **5**(02), 15–18+61 (2021)
7. Liu, X.: Empirical research on science and technology investment and economic growth in Hainan based on gray correlation analysis. J. Changchun Univ. Sci. Technol. (Soc. Sci. Ed). **26**(01), 102–103+120 (2013)
8. Yin, F.: The evaluation model of the technical support capability for aviation equipment based on gray relations analysis. J. Math. Pract. Theory. **43**(08), 104–109 (2013)
9. Zhang, L.: Application of gray association analysis in college teacher recruitment. J. Border Econ. Cult. **117**(09), 173–175 (2013)

Qinghai Embroidery Classification System and Intelligent Classification Research

Xiaofei Lin[1,2,3,4], Zhonglin Ye[1,2,3,4(✉)], and Haixing Zhao[1,2,3,4]

[1] College of Computer, Qinghai Normal University, Xining 810008, China
zhonglin_ye@foxmail.com
[2] The State Key Laboratory of Tibetan Intelligent Information Processing and Application,
Xining 810008, China
[3] Tibetan Information Processing and Machine Translation Key Laboratory of Qinghai
Province, Xining 810008, China
[4] Key Laboratory of Tibetan Information Processing, Ministry of Education, Xining 810008,
China

Abstract. Qinghai embroidery is an artistic treasure of folk embroidery in Qinghai Province. Classifying them to understand the differences between them is an important task. However, currently, there is a lack of a systematic classification method for Qinghai embroidery. First, by studying the history of Qinghai embroidery and a large number of Qinghai embroidery patterns, this paper divides Qinghai embroidery patterns into three categories: animals and plants, auspicious meanings, and geometric decoration. This method breaks through the regional classification system for the Qinghai embroidery. Second, this article utilizes four CNN models to classify the Qinghai embroidery datasets, exploring the differences in the classification of Qinghai embroidery by different models and finding the optimal classification model. The results show that the GoogLeNet model performs the best in the classification of Qinghai embroidery images, achieving the highest accuracy rate. This is mainly due to the small size of the Qinghai embroidery datasets and the application of the Inception structure and batch normalization technology in the GoogLeNet model, enabling it to better extract features and classify Qinghai embroidery images. Through this research, we can provide a certain reference and assistance for the classification of Qinghai embroidery images and provide technical support for the protection and inheritance of cultural heritage.

Keywords: Qinghai Embroidery Culture · Convolutional Neural Network · Image Classification

1 Introduction

With the development of information technology, computer vision processing technology plays an important role in human production and life. As a branch of artificial intelligence, deep learning has achieved great success in computer vision processing. Image classification is a vital task, and it is based on convolutional neural networks and has become an important research task in deep learning [1].

© The Author(s), under exclusive license to Springer Nature Singapore Pte Ltd. 2023
Z. Yu et al. (Eds.): ICPCSEE 2023, CCIS 1880, pp. 174–190, 2023.
https://doi.org/10.1007/978-981-99-5971-6_13

Convolutional neural networks (CNNs) are mainly used for image classification in deep learning [2], and their core task is to assign labels to images from a given set of categories and output the category. First, image preprocessing is performed on the Qinghai embroidery datasets [3]; then, the image is input to the convolutional neural network, and the feature map is formed through image preprocessing and feature extraction. Finally, image classification is completed by using a classifier. Feature extraction [4] is the most critical step in CNN. Most existing classification techniques are supervised learning techniques that require large amounts of labelled data to train neural network models [5]. Zero-shot classification is a hot research area that has expanded from early applications in image classification tasks to other fields [6, 7]. Currently, CNNs have achieved numerous research results in image classification tasks [8–10].

"Qinghai embroidery" [11] is one of China's outstanding traditional national crafts and a shining pearl in the historical and cultural heritage of Qinghai. Its unique artistic value and cultural significance lies in the fact that it has become a cultural symbol symbolizing the exchange between various ethnic groups in Qinghai. However, there is currently no systematic or scientific computer classification mechanism or method for classifying Qinghai embroidery. This situation not only applies to Qinghai embroidery but also exists in many other relics and artworks, and the lack of effective classification methods makes it difficult for us to accurately describe and compare the differences between each sample. Traditional Qinghai embroidery is divided into five categories: Tu nationality pan embroidery, Mongolian embroidery, Huangzhong dui embroidery, Hainan Tibetan embroidery and Hehuang embroidery.

However, this classification system lacks overall consistency, and the content between different categories is complex, making it difficult to distinguish. Due to the close relationship between embroidery, different categories in the existing five classification systems have the same embroidery elements and patterns (see Fig. 1), which leads to the inability of the classification system to be applied to computerized intelligent Qinghai embroidery classification tasks.

Therefore, a new classification system is proposed, where Qinghai embroidery is divided into three categories: Animal and plant, Auspicious meanings, and Geometric decoration, which breaks through regional limitations with greater universality, while pattern-based classifications better match people's cognitive understanding of Qinghai embroidery, making it easier for acceptance. At the same time, this new method can retain the connections among all kinds of traditional embroidered works, which will benefit research on intelligent classifications.

In addition, few people have focused on the classification of Qinghai embroidery images using deep learning techniques in the field of computer vision. Therefore, this paper proposes a method for classifying Qinghai embroidery images based on the AlexNet, DenseNet, GoogLeNet, and ResNet50 models. This will contribute to the protection and promotion of the Qinghai embroidery culture and provide ideas and references for exploring new directions in related fields.

(a)Hainan Tibetan embroidery   (b) Tu nationality pan embroidery

(c) Hehuang embroidery   (d) Mongolian embroidery   (e) Huangzhong dui embroidery

Fig. 1. Flowers in the five categories of the Qinghai embroidery classification system

2   Related Work

In recent years, CNNs have achieved significant results in the field of image classification. CNN is a forward-propagating artificial neural network based on the design of an artificial neuron network structure, which performs well in image processing. By performing multilevel and highly nonlinear feature extraction on input data, it can effectively solve complex problems that traditional methods cannot solve5. Currently, many researchers have used CNN models to study classification tasks for various types of images. For example, Krizhevsky et al. [12]. Proposed a CNN model named "AlexNet" and won the imageNet competition in 2012; Karen Simonyan and Andrew Zisserman et al. [13]. Proposed VGG, which uses very deep hierarchical structures to improve image classification performance in 2014. The model used multiple convolutional layers and fully connected layers, with the deepest VGG-19 model having 19 layers achieving a top-1 accuracy rate of 74.4% and a top-5 accuracy rate of 92% on imageNet datasets. Subsequently, a series of improved models emerged, such as GoogLeNet [14], ResNet [15], DenseNet [16], and SENet [17]. GoogLeNet was proposed by He et al. and achieved excellent results at the ImageNet large-scale visual recognition competition in 2014. The model adopts the Inception structure and introduces a 1x1 convolution layer to reduce computational complexity. Many researchers have also proposed various variants based on this idea. Another is Residual Network (ResNet), which was introduced by Microsoft Asia Research Institute and won the championship at the imageNet competition in 2015. This model solves the deep network degradation problem by introducing residual blocks and makes it easy to add more layers. Hence, it is widely used in many image classification tasks. In addition, Liu [18] et al. proposed Swin-Transformer in 2021, which can effectively transfer to various visual tasks that require higher resolution images.

An important design of the Swin-Transformer is shifted windows, which can significantly reduce the computational complexity and linearly increase the computational complexity with the input image size. Unlike traditional sliding windows, the design of nonoverlapping windows is more hardware-friendly, which results in faster actual runtime speed.

In the field of image classification, many studies have also conducted experiments using GoogLeNet and ResNet. For example, Zhang et al. [19]. Proposed a method called "CGR-DCNN" for blue and white porcelain bowl classification in 2020. This method used Resnet50 and GoogLeNet models for feature extraction and integrated them into a new deep convolutional neural network. The experimental results show that CGR-DCNN achieves excellent classification performance on public datasets. Additionally, Li et al. proposed a method called the residual attention network for Qinghai embroidery image classification in 2019 [20], which combines residual blocks and attention mechanisms to build a more effective CNN model, and the experimental results show that the residual attention network is significantly better than other traditional algorithms and CNN models. Qian et al. [21]. Proposed a flower recognition and classification method based on deep neural network transfer learning in 2022, achieving the effect of completing model training with few data samples. The experimental results show that using the ResNet34 network trained by transfer learning has good recognition performance and robustness in the field of flower image classification recognition, which can meet practical application needs.

Based on the above methods, this paper performs the image classification task of Qinghai embroidery on GoogLeNet and ResNet50. We analyse and summarize the experimental results from different perspectives. We attempt to compare and analyse the performance of the two models on the Qinghai embroidery datasets.

3 Research on the classification system of Qinghai embroidery

The existing classification system divides Qinghai embroidery into five categories: Hainan Tibetan embroidery, Tu nationality pan embroidery, Mongolian embroidery, Huangzhong dui embroidery and Hehuang embroidery. In this article, based on the patterns of Qinghai embroidery and historical heritage and referring to research results from other classification systems, such as Suzhou embroidery, we classify Qinghai embroidery into three categories: animal and plant, suspicious meaning, and geometric decoration (see Fig. 2).

3.1 Tibetan Embroidery Motifs Constitute Elements

The most common elements of the patterns in Tibetan embroidery are geometric patterns, animal motifs, plant patterns, and textual symbols. Geometric patterns in Tibetan embroidery mainly include fretwork, crosshatching, spiral patterns and triangle designs. Combining these geometric shapes with squares and arcs in various ways creates polygonal designs [22].

3.2 Tu Nationality Pan Embroidery Patterns Constitute Elements

Plate embroidery has cleverly designed patterns with a strong ethnic style, including the Tai Chi diagram, five-petal plum blossom, divine immortal headwear, walnuts, cloud patterns, diamonds, sparrow heads and endless wealth styles. These patterns have extremely high ornamental and collection value. Different parts of Tu nationality clothing are decorated with different plate embroidery designs that represent different auspicious meanings. For example, an "endless wealth" pattern is used to decorate the collar, indicating wishes for prosperity and longevity without interruption [22]. In terms of pattern composition, repetition is often used to create a series of designs using characters such as "ten thousand".

3.3 Hehuang Embroidery Motifs Constitute Elements

The flowers mainly embroidered in Hehuang embroidery are peonies, pomegranates, plums, orchids, chrysanthemums, and bamboo. The embroidered products are gifted to parents-in-law, implying enduring hardships, integrity, and nobility. Embroidered pine and cypress trees symbolize longevity and can also be used to decorate birds and animals. The animals embroidered are mostly butterflies, magpies, mandarin ducks, cranes, fish, cats, lions, tigers, etc., vivid and lively, full of vitality [22].

3.4 Mongolian Embroidery Motifs Constitute Elements

The Mongolian embroidery patterns mainly include flower and plant motifs such as plum blossoms, apricot blossoms, peonies, and begonias; animal motifs such as butterflies, bats, deer, horses, sheep, cows, camel lions, tigers, and elephants; and auspicious patterns such as mountains, water, fire, and soon [22].

3.5 Huangzhong Dui Embroidery Motifs Constitute Elements

The pattern elements of Huangzhong dui embroidery mainly include flowers, among which flowers are one of the main design elements of Huangzhong dui embroidery, with common ones being peonies, chrysanthemums, roses, etc.; animals, in addition to flowers, animals are also an important part of the pattern in Huangzhong dui embroidery, such as mandarin ducks, peacocks, butterflies, etc.; geographical landscapes, this embroidery technique often uses natural scenery and geographical landscapes as one of the pattern elements, such as rivers, peaks, and grasslands such as landscape paintings.

3.6 Studies the Classification System of Qinghai Embroidery in this Paper

Based on the pattern elements of Tibetan embroidery, Tu nationality pan embroidery, Hehuang embroidery and Mongolian embroidery mentioned above, we can classify Qinghai embroidery patterns into three categories: Animal and plant, Auspicious meanings, and Geometric decoration, which can help people better understand and appreciate the embroidery culture of the ethnic groups in Qinghai Province. It also facilitates the promotion and publicity of Qinghai embroidery products.

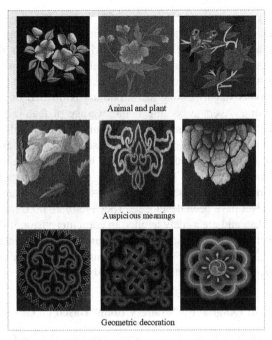

Fig. 2. Qinghai embroidery image classification standard

1. Animal and plant: This category includes all patterns designed based on flora and fauna. For example, peony, peach blossom, plum, orchid, chrysanthemum, bamboo, and begonia usually emphasize the curvature of plant lines, and the colors are full and natural to reproduce the original colors and their hierarchical changes as much as possible. The overall balance is beautiful; the entire pattern should present aesthetic characteristics such as neatness, symmetry and balance.

2. Auspicious meanings: This category includes all patterns that are considered auspicious or represent specific meanings in traditional culture. Examples include the Eight Auspicious Symbols, plate-long patterns, Ruyi cloud head patterns, meander patterns, and scroll vine patterns.

3. Geometric decoration: This category includes simple combination design elements that are used solely for decoration on clothing or other items and do not have a clear theme or purpose. It involves the repetition of simple geometric elements to form complex and exquisite patterns. For example, Flower of Life, Tai Chi diagram, endless wealth head, wave lines, dotted arrays, etc.

4   Research on Intelligent Classification of Qinghai Embroidery

CNN is a feedforward neural network that achieves better results in image and speech recognition compared to other deep learning architectures. The main purpose of CNN is to extract features from images, and its basic structure consists of an input layer, convolutional layer, pooling layer, fully connected layer, and output layer. In the process of

Qinghai embroidery image classification using CNN, first, the image is input into the convolutional layer, and then the feature map is formed by extracting features, and the pooling layer compresses the input feature map, which is aimed at reducing features on the one hand, resulting in a reduction of parameters, thereby simplifying the computational complexity of the convolutional neural network and keeping certain features unchanged on the other hand. Among them, pooling is divided into average pooling and maximum pooling. Average pooling is the average of feature points, and maximum pooling is the maximum value of feature points. Then, the fully connected layer enters the features into the Softmax classifier to classify the Qinghai embroidery pattern and finally obtains the classification result through learning and training. As networks become deeper, problems such as large parameter counts, overfitting, and vanishing gradients arise. In response to these issues, the classic convolutional neural networks GoogleNet and ResNet have been developed. They are often used for extracting image features from raw data and are widely applied in computer vision tasks such as image classification, object detection, image generation, and image segmentation.

Therefore, GoogLeNet and ResNet models are employed for intelligent classification research of Qinghai embroidery patterns in this study.

4.1 Classification of Qinghai Embroidery Image Based on GoogLeNet Model

GoogLeNet is a new deep learning architecture proposed by the Google team in the Large Scale Visual Recognition Challenge in 2014. Its main feature is to improve the utilization of computer internal resources and increase network depth (number of layers) and width (number of neurons) without increasing computational complexity. Compared with AlexNet, which was proposed by the champion team of the image classification contest in 2012, GoogLeNet has 12 times fewer parameters and higher accuracy. Although GoogLeNet has up to 22 layers, its parameter quantity is much less than that of previous network structures such as AlexNet and VGG. Therefore, using GoogLeNet for image classification is a good choice when computer resources are limited.

Model design: The main structure of GoogLeNet consists of an input layer, 5 sets of convolutional modules and an output layer, including a total of 22 parameter layers and 5 pooling layers. The input layer is an image with a size of $224 \times 224 \times 3$, the first and second sets of convolutional modules include convolutional layers and max-pooling layers, and the third, fourth, and fifth sets of convolutional modules mainly consist of Inception module structures and max-pooling layers. The output layer consists of a global average pooling layer, dropout, and a fully connected layer. The GoogLeNet structure (see Fig. 3).

To achieve better training results, AlexNet, VGG and other network models increase the depth or width of the network. However, increasing the depth of the network leads to a series of problems: (1) due to insufficient datasets, the model cannot estimate the distribution of the entire dataset during the training process, resulting in overfitting and high accuracy on the training set but low accuracy on the test set; (2) as network models grow larger, the number of parameters and computational complexity increase, leading to unnecessary resource waste; and (3) deeper networks are more prone to the gradient disappearance problem, which makes gradients almost zero and early layers' parameters no longer update, eventually leading to poor performance of the model that is difficult to optimize.

The introduction of the Inception structure improves training results from another perspective and solves the problems mentioned above. It can more efficiently utilize resources and extract more features with the same amount of computation, thereby improving training effectiveness. Inception module structure (see Fig. 4).

The Inception structure has four branches, which process input data in parallel and obtain four outputs. These outputs are then concatenated to obtain the final output. The first three branches use convolutional kernels with sizes of 1 × 1, 3 × 3, and 5 × 5. The fourth branch is a max-pooling downsampling operation with a pool size of 3 × 3. Additionally, 1x1 convolution is added to the second, third and fourth branches mainly for reducing model complexity and parameter quantity so that the network can become deeper and wider to better extract features.

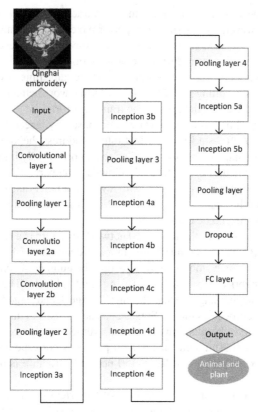

Fig. 3. GoogLeNet structure

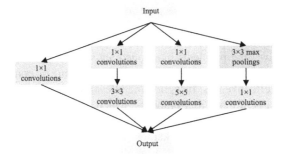

Fig. 4. Inception module structure

4.2   Classification of Qinghai Embroidery Image Based on ResNet Model

More layers of the network means that the model extracts richer levels of features in the neural network. However, its classification performance does not necessarily improve as the network deepens; in contrast, the network convergence speed is slow, and the gradient will gradually disappear during the propagation process so that the parameters of the previous network layer cannot be adjusted. At the same time, it will cause network degradation problems, resulting in shallow networks that work better than deep networks, and the more layers there are, the greater the training error.

The residual neural network (ResNet) was proposed by researchers, including Kai Ming He from Microsoft Asia Research Institute, to address the above issue. In 2015, it won the ImageNet image recognition challenge and profoundly influenced the design of deep convolutional neural networks that followed. Compared to GoogleNet network models, ResNet has deeper network layers with fewer parameters and better classification performance. Currently, there are various ResNet structures available, among which ResNet-50 is one of the classic ones.

Model design: The core of the ResNet50 network structure is a stack of overlapping convolution blocks (Conv block) and identity blocks (Identity block). The dimensionality of the network system can be modified by adjusting the number of convolutional blocks, while changes in network depth are achieved using identity blocks. The input to ResNet50 passes through five stages: stage 1, stage 2, stage 3, stage 4, and stage 5. Stage 1 serves as an input layer that preprocesses inputs. Stages 2–5 each consist of three bottleneck layers with 3, 4, and 6 bottleneck units. Finally, global average pooling is followed by a fully connected layer with 1000 dimensions, which produces corresponding classifications through the softmax classifier. The ResNet50 network structure is used in this article (see Fig. 5).

The deep residual network uses the idea of "shortcut connections" and residual functions to weaken gradient vanishing and effectively solve the problem of network degradation. The residual module structure (see Fig. 6) transforms the original regression target function $H(x)$ into $F(x) + x$, where $F(x) = H(x) - x$ is called a residual.

The residual formula is as follows:

$$F(x) = H(x) - x \qquad (1)$$

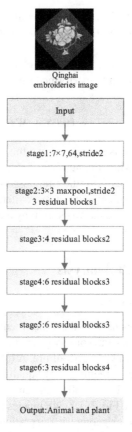

Fig. 5. ResNet50 structure

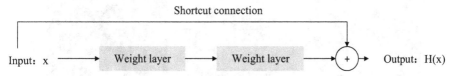

Fig. 6. Residual module structure

In the equation, x is the input of the residual unit, *H(x)* is the output of the residual unit, and *F(x)* is the target mapping.

Shortcut connection in two ways:

If the residual network and the constant mapping are of the same dimension, then *F(x)* and *x* are summed element by element:

$$y = F(x, \{Wi\}) + x \qquad (2)$$

where *x* and *y* represent the input and output of the layer respectively. The function *F(x, Wi)* represents the learned residual mapping. This approach passes the input *x* directly through shortcuts without introducing additional parameters or increasing the computational complexity of the module and therefore allows a fair comparison of the residual network with the normal network.

(2) If the dimensions of the two are different (changing the input/output channels), then a linear mapping Ws needs to be performed on x to match the dimensions of the two, and the equation becomes:

$$y = F(x, \{Wi\}) + Wsx \qquad (3)$$

The purpose of this approach is simply to keep the dimensions of x consistent with $F(x)$, so it is usually only used when the number of channels changes between adjacent residual blocks, and in most cases only the first approach is used. Therefore, ResNet models are easier to optimize and can extract more hierarchical features by increasing network depth. This speeds up neural network training and significantly improves model accuracy.

5 Experiment and Result Analysis

5.1 Dataset and Preprocessing

The Qinghai embroidery images used in the experiments were collected from the preserved physical objects of Qinghai embroidery by examining the local cultural museums and local inheritors of Qinghai embroidery culture. A total of 5000 Qinghai embroidery images were collected, and after screening, 1949 qualified images remained. The resolution of the images is 6240 pixels × 4160 pixels among them, there were 716 images for animal and plant patterns (see Fig. 7(a)), 548 images for suspicious meaning patterns (see Fig. 7(b)), and 649 images for geometric decoration patterns (see Fig. 7(c)).

(a) Animal and plant;   (b) Auspicious meanings;   (c) Geometric decoration

Fig. 7. Images of Qinghai embroidery with different patterns.

The quality of the datasets is closely related to the experimental results in image recognition. Due to the limited number of Qinghai embroidery images collected and the use of a deep network in this experiment, preprocessing was performed on the Qinghai embroidery datasets. First, complete and clear pictures containing full embroidery patterns were selected from original images, and some nontarget pattern areas were deleted.

Then, the obtained datasets were divided into a training set, a validation set, and a test set at a ratio of 7:2:1. Second, random cropping was applied to both the length and width of each picture to reduce interference from redundant background information. After that, random vertical flipping with a probability p = 0.5 was applied. Finally, data were converted into Tensor format before normalizing pixel values for Qinghai embroidery images between [−1,1], making it easier for models to converge during the training process.

5.2 Experimental Environment

The experiment builds models using the PyTorch deep learning framework. PyTorch is an open-source third-party machine learning library for Python that is widely used in fields such as computer vision and natural language processing. It is mainly used to construct deep learning network models, providing two core functionalities: (1) it supports GPU-accelerated tensor computing and provides dynamic changes in the network model function, that is, adding or removing layers from the model; and (2) it is an automatic differentiation mechanism for conveniently optimizing the model.

Therefore, this article can use the PyTorch deep learning framework to build the required neural network model. The software and hardware environment of the experiment are shown in Table 1.

Table 1. Experimental hardware and software environment.

Hardware environment	Software environment
CPU: Intel(R) Xeon(R) CPU E5–2603 v4 @ 1.7 1.70 GHz	Windows 10
Memoey: 32.0 GB	Python 3.8 + PyTorch1.13
GPU: NVIDIA Quadro K1200 video memory: 4.0GB	CUDA 7.5.9

5.3 Experimental Parameter Settings

First, the experimental datasets are preprocessed by cropping, vertical flipping, and angle adjustment. Then, the Qinghai embroidery datasets are divided into a training set, a validation set, and a test set at a ratio of 7:2:1. Afterwards, a convolutional neural network model is constructed using the PyTorch deep learning framework to extract features of the targets in Qinghai embroidery images.

In machine learning, the purpose of hyperparameters is to optimize the network model, making the training of the model more effective and faster. After experimental testing, the training result is more stable when the batch size is 54. The total number of images in the training set is 1364, the validation set has 390 images, and the test set has 195 images. To compare the classification results of the Qinghai embroidery datasets on the GoogLeNet and ResNet50 network models, the epoch is set to 100, the batch size is 32, the learning rate for AlexNet, GoogLeNet, and DenseNet is 0.0003, and the learning rate for ResNet50 is 0.0001. The hyperparameters are the best parameters for the model in this article. After each epoch of training, the average classification accuracy is immediately saved, and the training results and the best training model are saved. Finally, the best model is used to test the test set to obtain the test accuracy and classification results.

5.4 Results and Analysis of the Experiment

The main goal of this experiment was to extract features from Qinghai embroidery images, classify them using four CNN models with different structures and features, and evaluate the models based on their accuracy on the test set and the loss rate. As shown in Table 2, the classification results on the AlexNet, DenseNet, GoogLeNet, and ResNet50 models are 0.834, 0.830, 0.864, and 0.716, respectively. The final loss functions were 0.094, 0.370, 0.361, and 0.093, respectively. The average training times of the models were 1.12 s, 1.30 s, 1.10 s, and 1.78 s, respectively.

The experimental results showed that for the image classification task of the Qinghai embroidery datasets, the GoogLeNet model performed better than the other three models, with significant advantages in accuracy and model training time.

Table 2. Performance comparison of the four classification models.

Model	Loss	Accuracy	Training time
AlexNet	0.094	0.834	1.12
DenseNet	0.370	0.830	1.30
GoogLeNet	0.361	0.864	1.10
ResNet50	0.093	0.716	1.78

Reasons for the excellent performance of the GoogLeNet model on Qinghai embroidery datasets are as follows:

1. Inception structure: GoogLeNet uses an Inception structure, which can effectively utilize convolutional kernels of different sizes to extract richer image features. This design allows the network to simultaneously focus on information at different scales and abstraction levels, thereby better capturing the rich and diverse texture and shape features in images.

2. Auxiliary classifiers: GoogLeNet introduces auxiliary classifiers to help the network learn more robust and generalizable feature representations. These auxiliary classifiers are located inside the network and generate a set of outputs for each stage during training, which is then fused with the final classification result. This design can alleviate gradient vanishing problems and enhance model recognition capabilities for small objects or local areas.

3. Parameter sharing: GoogLeNet reduces model complexity by using parameter sharing technology, reducing computational requirements and storage space needs.

Specifically, in Inception structures, numerous identical convolutional kernels are used that are applied to different channels to extract multiple types of features while avoiding overfitting issues.

4. Data augmentation: During the training process, GoogLeNet utilizes a series of data augmentation techniques, such as random cropping or horizontal flipping operations, that effectively expand dataset size diversity while improving the robustness and generalization ability of networks.

Considering the GPU, CPU, and memory performance, the four models mentioned above were run 100 times on the same platform, and the comparison results of the validation accuracy and loss function for the four models are shown below.(see Figs. 8(a) and Fig. 8(b)).

According to the comparison chart of model accuracy and loss rate. For the green embroidery datasets, the GoogLeNet model is more suitable for the classification task of green embroidery images. The GoogLeNet model outperforms the other three CNN models in terms of both accuracy and training time for the following possible reasons:

The AlexNet model has a shallow network depth and fewer convolutional layers compared to the other three models, resulting in poorer feature extraction, which may not be suitable for complex image datasets such as green embroidery; hence, the accuracy performance is not as good as it could be.

Although the DenseNet model has a strong feature extraction ability, it is more computationally intensive, while it is prone to gradient disappearance and overfitting, which in turn reduces the classification accuracy of the model, so the accuracy and loss rate are not dominant.

The optimization method of the ResNet50 model uses a residual structure that effectively mitigates the gradient disappearance problem while allowing the depth of the network to increase, thus improving the capability and accuracy of feature extraction. However, the ResNet50 network model is deep in layers compared to the Qinghai embroidery datasets, and the existing green embroidery artifacts and the collected green embroidery images are limited, so the shortage of the Qinghai embroidery datasets leads to insufficient training and causes model overfitting during the training process, and the final test set is not as accurate as the GoogLeNet model.

The GoogLeNet model, on the other hand, is suitable for the classification task of the Qinghai embroidery datasets, and it achieves better feature extraction while maintaining fewer parameters by composing the Inception module with multiple convolution and pooling methods. Backwards links (auxiliary classifiers) are also employed to improve the robustness of the model. Together, these features allow the GoogLeNet model to perform well on the Qinghai embroidery datasets and have a significant advantage in training time.

6   Conclusion

This paper proposes a new Qinghai embroidery classification system and achieves intelligent classification of Qinghai embroidery patterns through the CNN method. Traditionally, Qinghai embroidery is classified into five categories based on geographical regions, but this classification method has many limitations. Therefore, in this

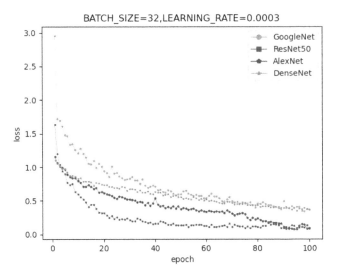

(a). Comparison chart of loss values

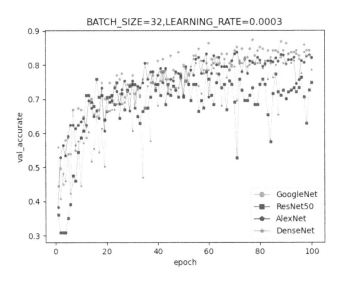

(b). Comparison chart of accuracy

Fig. 8. Comparison of model accuracy and loss rate

paper, pattern classification is simpler and clearer, breaking through geographical limitations. At the same time, this study uses four different deep learning models - AlexNet, DenseNet, GoogLeNet, and ResNet50 - to classify Qinghai embroidery image datasets and compares their classification performance. The experimental results show that the GoogLeNet network model performs better in recognizing and classifying Qinghai embroidery images, and under the same conditions, it has a higher accuracy rate than the other three models. This is because of the Inception structure, auxiliary classifiers, parameter sharing, and data augmentation introduced in GoogLeNet. In conclusion, in this paper, we proposed a new classification system and successfully used deep learning technology to solve the problem of Qinghai embroidery image classification based on CNN. The research results lay a foundation for Qinghai embroidery research and provide technical solutions.

References

1. Zeng, X.Q., Liu, R., Yang, X.: Analysis of convolutional neural network based research in gestural digital recognition. Internet Things Technol. **11**(06), 10–13+1 (2021)
2. Zhang, X.Q.: CNN based image classification. Electron. Technol. Softw. Eng. (**07**), 182–185 (2022)
3. Wei, Y.H., He, X.M., Xu, X., Wei, Y., Zhong, W.J., Wang, Y.: A research and application of an adaptive image pre-processing method. Modern Electron. **45**(07), 53–57 (2022)
4. Nguye, H., Maclagan, S.J., Nguten, T.D., et al.: Animal recognition and identification with deep convolutional neural networks for automated wild life monitoring. In: Proceedings of the2017 IEEE International Conference on Data Science andAdvanced Analytics, pp. 40–49. Piscataway: IEEE, China (2017)
5. LeCun, Y., Bengio, Y., Hinton, G.: Deep Learn. Nature **521**, 436–444 (2015)
6. Ji, Z., Yu, Y.L., Pang, Y.W., et al.: Manifold regularized cross-modal embedding for zero-shot learning. Inf. Sci. **378**, 48–58 (2017)
7. Xu, B.H., Fu, Y.W., Jiang, Y.G., et al.: Heterogeneous knowledge transfer in video emotion recognition, attribution and summarization. IEEE Trans. Affect. Comput. **9**, 255–270 (2018)
8. Zhao, X.Q., He, H.L., Yang, D.D.: Application of improved convolutional neural network based on image classification. High Tech Commun. **28**(11), 930–936 (2018)
9. Yan, H., Wang, P., Dong, Y.Y.: Improved convolutional neural network for image classification and recognition. Comput. Appl. Softw. **35**(12), 193–198 (2018)
10. Chang, L., Deng, X.M., Zhou, M.Q.: Convolutional neural networks in image understanding. J. Autom. 42(9), 1300–1312 (2016)
11. E, C.R., Li, W.Q, Zhang, L.: Research on the conservation and industrial development of Qinghai embroidery culture. Qinghai-Tibetan panau Altar **7**(04), 1–5 (2019)
12. Krizhevsky, A., Sutskever, I., Hinton, G.E.: Imagenet classification with deep convolutional neural networks. Commun. ACM **60**(6), 84–90 (2017)
13. Simonyan, K., Zisserman, A.: Very deep convolutional networks for large-scale image recognition. arXiv preprint arXiv1409, 1556 (2014)
14. Szegedy, C., Liu, W., Jia, Y.: Going deeper with convolutions. In: IEEE Conference on Computer Vision and Pattern Recognition, pp. 1–9. IEEE Xplore, United States (2015)
15. He, K., Zhang, X., Ren, S.: Deep residual learning for image recognition. In: IEEE Conference on Computer Vision and Pattern Recognition, pp. 770–778. IEEE Xplore, United States (2016)
16. Huang, G., Liu, Z., Van Der Maaten, L., et al.: Densely connected convolutional networks. In: Proceedings of the IEEE Conference on Computer Vision and Pattern Recognition, pp. 4700–4708. IEEE Xplore, United States (2017)
17. Hu, J., Shen, L., Sun, G.: Squeeze-and-excitation networks. In: Proceedings of the IEEE Conference on Computer Vision and Pattern Recognition, pp. 7132–7141. IEEE Xplore, United States (2018)

18. Liu, Z., Hu, H., Lin, Y.: Swin transformer v2: Scaling up capacity and resolution. In: Proceedings of the IEEE/CVF Conference on Computer Vision and Pattern Recognition, pp. 12009–12019(2022)
19. Hhang, Z., Liang, L.: CGR-DCNN: A novel deep convolutional neural network for Qinghua ceramics classification. Neural Comput. Appl. (**32**), 12153–12163 (2020)
20. Li, J., Wang, G.: Residual attention network for image classification. In: IEEE/CVF Conference on Computer Vision and Pattern Recognition Workshops, pp. 3156–3164. IEEE Xplore, United States (2017)
21. Qian, R.X.: Deep neural network based flower image recognition and classification. Inform. Technol. Inform. (**03**), 210–213 (2022)
22. Wang, X.M.: Qinghai embroidery modern design applications. Qinghai People's Publishing House, Xining (2022)

A Machine Learning-Based Botnet Malicious Domain Detection Technique for New Business

Aohan Mei[1], Zekun Chen[1], Jing Zhao[2], and Dequan Yang[3(✉)]

[1] School of Cyberspace Science and Technology, Beijing Institute of Technology,
Beijing 100081, China
[2] School of Continuing Education, Beijing Institute of Technology, Beijing 100081, China
[3] Network Information Technology Center, Beijing Institute of Technology, Beijing 100081,
China
yangdequan@bit.edu.cn

Abstract. In the new network business, the danger of botnets should not be underestimated. Botnets often generate malicious domain names through DGAs to enable communication with command and control servers (C&C) and then receive commands from the botmaster, carrying out further attack activities. Therefore, a system based on machine learning to dichotomize DNS domain access is designed, which can instantly detect DGA domain names and thus quickly dispose of infected computers to avoid spreading the virus and further damage. In the comparison, the bidirectional LSTM model slightly outperformed the unidirectional LSTM network and achieved 99% accuracy in the open dataset classification task.

Keywords: Botnet · Machine Learning · LSTM · Domain Generation Algorithm Detection

1 Introduction

1.1 Botnet

Recently, with the continuous development of information technology, botnets have gradually become a new type of network hazard, posing a major challenge to cybersecurity. A botnet [1] is a network consisting of a group of malware-infected computers (called "bot hosts"). These infected computers can be remotely controlled by malware (called "Bots") and coordinate their actions through commands and control centers (C&C servers) to accomplish various malicious activities such as DDoS attacks, phishing, and mining. The botnet structure is shown in the figure below.

The life cycle of a botnet consists of three main phases: infection, stealth, and attack [2]. In the infection phase, the C&C server walks the malware by various means, and the infected computers then become bots and start to be controlled by the botmaster. In the stealth phase, the bots avoid the security software for search activities and transmit the information to the botmaster while receiving software updates. In the attack phase, the bots will stop hibernating and start to be under the control of the botmaster simultaneously carrying out attack activities, such as DDoS and spreading malicious code.

Supported by Hainan Provincial National Science Foundation of China, 621MS0789.

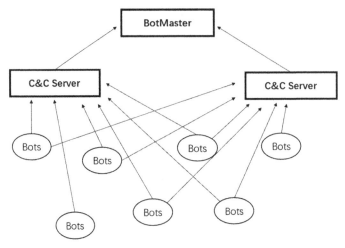

Fig. 1. Typical architecture of a botnet

1.2 Domain Generation Algorithm (DGA)

As shown in Fig. 1, communication between bots and C&C servers is crucial in a botnet. Therefore, bots will need the IP address of the C&C server. However, if a hard-coded IP address is used, once the malware is reversed, the IP will be directly banned; thus, the controlled machine can no longer communicate with the bot host. Therefore, modern botnet systems use the domain name generation algorithm (DGA) to hide the real IP of the host and then obtain the IP by DNS querying the generated domain name.

The DGA [3] generates a set of domains containing hundreds of domains called algorithmically generated domains (AGDs) through a random seed shared by the bots and the botmaster. The malware will use local DNS servers to resolve each of these domains until it finds the ones that are actually registered. It will then communicate with the C&C server through these domains to receive instructions.

A DGA domain table is huge, and an attacker picking a small number of registrations would enable communication with the infected hosts, while theoretically, as a defender, it would need to block all DGA domains to block communication between the malicious program and the bot hosts. Therefore, the prevention means through blacklist bans failed, and we had to adopt a new method.

1.3 Malicious Domain Detection

Bots communicate with the C&C server through DNS query AGDs, so DNS traffic information contains the key to botnets. There are many DNS-based malicious domain detection methods [4], mainly divided into the configuration of a honeynet and the intrusion detection system (IDS); the latter can be further divided into signature-based IDSs and anomaly-based IDSs.

Intrusion Detection System

Signature-Based Detection. Signature-based detection is mainly based on the characteristics of the type of attack. By building up a knowledge base of characteristics of known types of attacks, if the network traffic found in the detection of the information in the knowledge base matches the situation, the system is considered to be under attack. This method is very dependent on the knowledge base, and if the data are not updated in a timely manner, the detection system may not be able to detect new means of attack.

Anomaly-Based Detection. Anomaly-based detection is based on the current traffic behavior and the degree of deviation from the normal traffic behavior. If the detection of the current behavior is clearly abnormal and the known normal behavior is significantly different, the system is considered to be under attack. The key to this method is the extraction of normal feature information, so machine learning methods can then be considered. Deep learning algorithms can avoid the tedious work of manual feature extraction and achieve twice the result with half the effort.

Honeynet

Honeynet is an intrusion decoy technique, a network architecture formed by combining multiple honeypot systems together. It is a security resource whose value lies in being probed, attacked, or compromised [5]. Honeynet is often deployed in network environments deliberately for attackers to gain network intelligence. In this way, we can analyse the traffic captured by the honeynet to obtain information about some unknown attack methods and the set of suspicious IPs. Then, we can deploy targeted defense efforts and improve the vulnerability of the real system based on the attack situation of the honeynet. It can also be used in conjunction with IDS to achieve better results.

1.4 Design Motivation

For the detection of malicious DGA domain names, manual feature extraction is both tedious and does not achieve very good detection results. Therefore, this paper established an anomaly-based detection system through a neural network model to detect whether the domain names in DNS queries are generated by DGAs. Among the neural network models, the first one applied to DGA detection with good results is the LSTM model. A variation of it, the bidirectional LSTM network model, is used in this paper to observe the performance improvement it shows compared to the unidirectional one.

2 Related Work

For the AI-based DGA detection problem, there are two main approaches: feature-based and featureless. In the early days, it was mainly based on manual extraction of features, such as string length, and vowel-consonant ratio. However, this approach is both tedious and easily circumvented by DGA algorithms and often fails to achieve good results.

The first featureless detection method was proposed by the authors of [6]. It used an HMM for training, which eventually proved to be less effective than it could be.

Pioneering work was done by Woodbridge et al. [7], who first implemented a well-performing DGA real-time detection classifier using LSTM, which was easy to deploy and could achieve a 90% detection rate and a 10-4 false positive rate on an open dataset.

Catania [8] et al. and Saxe [9], on the other hand, both used CNNs to implement DGA classifiers and achieved good results.

Yu et al. [10] implemented an inline real-time DGA detection system using real traffic data instead of synthetic data to test their deep learning model.

Tran D et al. [11] focused on the multiclass imbalance problem in DGA detection, and they designed an algorithm based on the LSTM framework to solve the problem.

Baruch et al. [12] studied the application of machine learning in DGA detection and found that unsupervised methods performed better than supervised methods.

3   Machine Learning Based Detection Techniques

3.1   Bidirectional LSTM Network

RNN is a deep neural network related to time series. Its characteristic is that it can capture the context information of input content, so it is often used in language processing. LSTM is a classic RNN improvement that utilizes special mechanisms such as unique cell states and forget gates to achieve long-term memory and solve the gradient explosion problem. The schematic diagram is as follows (Fig. 2):

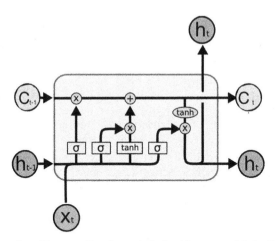

Fig. 2. LSTM network architecture (the picture is derived from an article by Zhihu, "Explanation of the Famous LSTM", https://zhuanlan.zhihu.com/p/518848475)

Each structural cell in this network accepts three inputs, cell state Ct-1 at the previous moment, hidden state ht-1 at the previous moment and current input Xt. The hidden layer acts on the cell state through the forget gate, update gate to decide whether the information is retained or forgotten, and finally decides the output of the cell based on the cell state and the current input, and the updated cell state and hidden state continue to pass forward. The three structures are as follows:

Forget Gate. The forget gate selectively forgets the last moment of the hidden state and the current input information by a sigmoid function.

$$f_t = \sigma\left(W_f \cdot [h_{t-1}, X_t] + b_f\right) \tag{1}$$

Update Gate. The update gate then has two parts, one part normalizes the hidden state with the input using the tanh function as the candidate cell state $\widetilde{C}_t$, and the other part is the same as the forgetting gate, using sigmoid to perform a filter on the input to obtain i_t. Then the two parts are combined and updated into the cell state.

$$i_t = \sigma\left(W_i \cdot [h_{t-1}, X_t] + b_i\right) \tag{2}$$

$$\widetilde{C}_t = tanh\left(W_C \cdot [h_{t-1}, X_t] + b_C\right) \tag{3}$$

Cell State. The cell state is the core of LSTM, and this network continuously preserves the information before updating the current time series by updating the cell state to achieve long-term memory.

$$C_t = f_t * C_{t-1} + i_t * \widetilde{C}_t \tag{4}$$

The two-way LSTM network is a further improvement on LSTM by combining two LSTM sequences together and inputting them forward and backwards at the same time, so that the content of the preceding and following texts can be combined and analysed at the same time to achieve better results (Fig. 3).

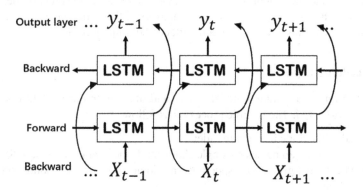

Fig. 3. Bidirectional LSTM network structure

4 System Design

Solving the DGA malicious domain name identification problem using bidirectional LSTM is essentially a binary classification problem.

4.1 DGA Domain Name Dataset Preprocessing

The process aimed to encode the domain name from a character perspective, which is vital to malicious domain name detection. At the same time, it is also the preprocessing of the word embedding. The specific operation is as follows:

Step 1. Create the data structure describing the domain name.
Step 2. Iterate through the black and white sample dataset in order.
Step 3. For each domain name, extract the required attributes from the data structure and store them uniformly. Then, we obtain the Vocab that records the domain name information (Table 1).

Table 1. The data structure

domain_name	domain_len	char_id	label
string	integer	list	bool

Char_id is the sequence obtained by mapping each character of the domain name to an integer. The length of the initially formed char_id is not consistent, so we unify it by setting a reasonable standard length. The label is used to mark whether a domain is malicious or not.

After the above processing, the dataset containing black and white samples is organized into data that can be applied for training. Of course, due to the order of file reading, the black and white samples are distributed in layers in the training data, so we need to shuffle the dataset. This operation directly affects the training results.

4.2 Embedding Layer

Word embedding is a common model in the field of NLP. In our model, it is the first layer of the Bi-LSTM. The specific process is as follows:

Step 1. Read in the domain name in order.
Step 2. Traverse domain names by single character. If the character is not in the Vocab, store it in Vocab, and give a corresponding char-id according to the storage order. If it is, we do nothing.
Step 3. Repeat step 1 and step 2 until the domain names in the dataset are coded. Due to the different lengths of the domain names, we use padding to uniform length.

The generated word vectors differ in each loading data due to different initial parameters (Table 2).

The white sample dataset is Alexa's top 1 million website domains in the world, and the black sample dataset is the DGAdrive dataset with 15,000 samples. After randomly disrupting them, 80% were used as training data and 20% as validation data. In the complete training process, the GPU is used to train the model to improve the training efficiency. Each epoch takes approximately 2 min.

Table 2. Parameters of the Model

Parameter	Value
Embedding_num	128
Hidden_size	100
Layers	2
Judge	True
Dense_size	100
Learing rate	1e-4
Batch size of train	2000
Batch size of valid	500
Epoch	15

Embedding num: the size of the word vector (Input)
Hidden_size: the size of the hidden layer
Layers: the number of hidden layer(s)
Judgement: True represents Bi-LSTM
Dense_size: the size of the fully connected layer

4.3 LSTM Layer

Compared with the one-way LSTM, the Bi-LSTM adds future information to the computational structure to improve the training effect. Therefore, LSTM layer is divided into a forward layer and a backwards layer.

The two layers are designed to compute and update the parameters. The former takes into account past information and the latter takes into account future information to adjust the current state from two perspectives. From the logical point of view, the two layers have the same function, and both update the state at the same moment. Form the implementation point of view, the forward layer is the forward calculation process. When the calculation of a batch of data is completed, the backwards calculation is performed from moment t to moment 1.

Both the forward and reverse calculations have the output of the implicit layer, making modifications to the above one-way formula:

$$h_t = f(w_1 x_t + w_2 h_{t-1}) \tag{5}$$

$$h'_t = f\left(w_3 x_t + w_5 h'_{t+1}\right) \tag{6}$$

$$O_t = w_4 h_t + w_6 h'_t \tag{7}$$

F is the LETM computational unit. h_t and h_t' record the past and future information, respectively, and O_t synthesizes the information. This is what makes Bi-LSTM unique.

4.4 Fully Connected Layer

From the LSTM layer, we can see that O_t is a vector. However, we want to obtain the binary result, i.e., true or false, which is a contradiction. The fully connected layer is the structure used to solve this contradiction and obtain the binary result.

The layer is divided into four parts: Fore-Linear, Tanh, Linear and Softmax.

Fore-Linear. Performing a linear transformation on the result of the previous layer is essentially a dimensionality reduction process.

$$Y_1 = O_t \cdot W_1 + b_1 \tag{8}$$

Tanh. After the linear transformation, the numbers in Y_1 are not regularized by the hyperbolic tangent function. The range is between -1 and 1.

$$Y_2 = \tan h(Y_1) \tag{9}$$

Linear. The linear transformation is performed again, directly changing the dimension from dense_size to 2. The value characterizes the probability of the categorical category, but at this point, only this relationship is presented, not the real probability.

$$Y_3 = Y_2 \cdot W_2 + b_2 \tag{10}$$

Softmax. The two-dimensional vector Y_3 is softmaxed, and the elements are scaled to the interval $[0,1]$, which sums to 1. This is the real true category probability.

$$Y_4 = Softmax(Y_3) = \frac{e^{\left(Y_3^i - shift\right)}}{\sum_{i=1}^{2} e^{\left(Y_3^i - shift\right)}} \tag{11}$$

$$shift = max\left(Y_3^i\right) \tag{12}$$

The result is obtained by selecting the category with the highest probability.

Loss Function. Cross-entropy is used as the loss function. The training objective is to minimize the cross-entropy. The input parameter of this function is the two-dimensional probability vector obtained from the fully connected layer.

$$Loss = -\sum_{i=1}^{2} p(x_i) log(q(x_i)) \tag{13}$$

Cross-entropy characterizes the distance between the predicted and actual values, with smaller values indicating that the prediction is closer to the actual value.

The above is the part about the structure of the model, now it's the training algorithm and learning rate.

Training Algorithm. The algorithm's goal is to make the loss function decrease to an acceptable level by adjusting different weight matrices. In the Bi-LSTM model, the network information is updated using back propagation. The Adam algorithm is adopted instead of the traditional stochastic gradient descent algorithm. The choice of this algorithm is based on three main considerations:

1. The word vector has a large dimensionality and a large number of samples, and Adam is more applicable to this case.
2. The Adam algorithm is excellent at handling sparse gradients and nonsmooth targets.
3. It can calculate different adaptive learning rates for different parameters.

Learning Rate. The learning rate update strategy in this model is similar to multistep decay, in which lr is designed to be large at the beginning of training to avoid falling into a local optimum and decreases as the training progresses to ensure the stability of the model.

5 Analysis of Results

The model effectiveness is judged by combining the training results and validation results. In the binary classification problem we used the following metrics for measurement (Table 3).

Table 3. Metrics for measurement

	Result is true	Result is false
Negative	TN	FN
Positive	TP	FP

Negative: the predicted result is right URLs.
Positive: the predicted result is false URLs.

The metrics for judging the model are shown below:

Accuracy. Accuracy is defined as the proportion of data that are correctly predicted in all training samples. It is calculated by the following equation:

$$Acc = \frac{TN+TP}{TN+TP+FN+FP} \tag{14}$$

This metric measures the model's ability to make judgements.

Precision. For the samples judged by the model to be Positive, the rate of correct judgement by the model. A high precision rate indicates that the vast majority of malicious domains detected by the model are indeed malicious, i.e., the probability of error is low. From the perspective of probability, it can be expressed as follows:

$$P = \frac{TP}{TP + FP} \propto P(truely\ malicious|detected\ as\ malicious) \tag{15}$$

Recall. For all positive samples in the dataset, the percentage of correct detections by the model. A high recall rate indicates that the model is strong in detecting malicious domains, i.e., strong in searching the malicious domain.

$$R = \frac{TP}{TP+FN} \tag{16}$$

F1-measure. F1-score is a combination of precision and recall, and generally speaking, The higher the F1-score, the more robust the classification model.

$$F1 = \frac{2 \cdot P \cdot R}{P+R} \qquad (17)$$

Of course, depending on the preference for the model, different weights can be assigned to recall and precision.

Comparing the LSTM network with the bidirectional LSTM network, the results are as follows (Table 4):

Table 4. Comparison of results

	ACC	precision	recall	F-score
LSTM	0.996	0.885	0.822	0.843
Bi-LSTM	0.996	0.915	0.826	0.847

The bidirectional LSTM network has a slight improvement over the unidirectional LSTM. However, in general, the bidirectional LSTM network performs better on the DGA binary classification problem, and can achieve 99% accuracy on the open dataset.

6 Discussion

We have shown that Bi-LSTM can achieve binary classification detection of DGA domain names. Domain names are strings, so it is necessary to combine LSTM with word embedding. If the network parameters are properly designed, the DGA domain names will be identified with extremely high accuracy. At the same time, the high performance of Bi-LSTM also proved that the domain name had strong contextual relevance. Follow-up research can provide more information on this aspect. With the guarantee of high accuracy, we can apply the model to the actual system for testing.

7 Conclusion

Machine learning has been well applied in DGA botnet detection, and deep learning algorithms have found an excellent solution to this problem. The bidirectional LSTM network is very effective in binary detection of malicious domain names in DGA, and the accuracy rate can be very close to 100%. By deploying the above model into a new business network management system, malicious domain name query information from DNS traffic can be detected in a timely manner to quickly find hosts that may be infected by botnets. However, this paper does not discuss the multiclassification problem of DGA families, and may not be able to identify some uncommon families of DGA domains, which is an area we will continue to improve in the future.

References

1. Kaur, N., Singh, M.: Botnet and botnet detection techniques in cyber realm. In: 2016 International Conference on Inventive Computation Technologies (ICICT), Coimbatore, India, pp. 1–7 (2016)
2. Tuan, T.A., Long, H.V., Taniar, D.: On detecting and classifying DGA botnets and their families. Comput. Secur. **113**, 102549 (2022)
3. Sood, A.K., Zeadally, S.: A taxonomy of domain-generation algorithms. IEEE Secur. Priv. **14**(4), 46–53 (2016). https://doi.org/10.1109/MSP.2016.76
4. Alieyan, K., Almomani, A., Manasrah, A., et al.: A survey of botnet detection based on DNS. Neural Comput. Appl. **28**, 1541–1558 (2017)
5. Spitzner, L.: The honeynet project: trapping the hackers. IEEE Secur. Privacy **1**(2), 15–23 (2003). https://doi.org/10.1109/MSECP.2003.1193207
6. Antonakakis, M., Perdisci, R., Nadji, Y., Vasiloglou, N., Abu-Nimeh, S., Lee, W., et al.: From throw-away traffic to bots: Detecting the rise of DGA-based malware. In: USENIX Security Symposium, vol. 12 (2012)
7. Woodbridge, J., Anderson, H.S., Ahuja, A., et al.: Predicting domain generation algorithms with long short-term memory networks. arXiv preprint arXiv:1611.00791 (2016)
8. Catania, C., García, S., Torres, P.: Deep convolutional neural networks for DGA detection. In: Pesado, P., Aciti, C. (eds.) CACIC 2018. CCIS, vol. 995, pp. 327–340. Springer, Cham (2019). https://doi.org/10.1007/978-3-030-20787-8_23
9. Saxe, J., Berlin, K.: eXpose: a character-level convolutional neural network with embeddings for detecting malicious URLs file paths and registry keys (2017)
10. Yu, B., Gray, D.L., Pan, J., Cock, M.D., Nascimento, A.C.A.: Inline DGA detection with deep networks. In: 2017 IEEE International Conference on Data Mining Workshops (ICDMW), New Orleans, LA, USA, pp. 683–692 (2017). https://doi.org/10.1109/ICDMW.2017.96
11. Tran, D., Mac, H., Tong, V., et al.: A LSTM based framework for handling multiclass imbalance in DGA botnet detection. Neurocomputing **275**, 2401–2413 (2018)
12. Baruch, M., David, G.: Domain generation algorithm detection using machine learning methods. In: Lehto, M., Neittaanmäki, P. (eds.) Cyber Security: Power and Technology. ISCASE, vol. 93, pp. 133–161. Springer, Cham (2018). https://doi.org/10.1007/978-3-319-75307-2_9

GPDCCL: Cross-Domain Named Entity Recognition with Span-Based Domain Confusion Contrastive Learning

Ye Wang[1], Chenxiao Shi[1], Lijie Li[1(✉)], and Manyuan Guo[2]

[1] College of Computer Science and Technology, Harbin Engineering University, Harbin, China
{wangye2020,shichenxiao,lilijie}@hrbeu.edu.cn
[2] Harbin Institute of Technology Software Engineering Co. Ltd., Harbin, China

Abstract. The goal of cross-domain named entity recognition is to transfer models learned from labelled source domain data to unlabelled or lightly labelled target domain datasets. This paper discusses how to adapt a cross-domain sentiment analysis model to the field of named entity recognition, as the sentiment analysis model is more relevant to the tasks and data characteristics of named entity recognition. Most previous classification methods were based on a token-wise approach, and this paper introduces entity boundary information to prevent the model from being affected by a large number of nonentity labels. Specifically, adversarial training is used to enable the model to learn domain-confusing knowledge, and contrastive learning is used to reduce domain shift problems. The entity boundary information is transformed into a global boundary matrix representing sentence-level target labels, enabling the model to learn explicit span boundary information. Experimental results demonstrate that this method achieves good performance compared to multiple cross-domain named entity recognition models on the SciTech dataset. Ablation experiments reveal that the method of introducing entity boundary information significantly improves KL divergence and contrastive learning.

Keywords: Transfer Learning · Named Entity Recognition · Domain Adaptation · Contrastive Learning · Adversarial Training

1 Introduction

Named entity recognition (NER) is a task in natural language processing that aims to identify entities such as people, locations, and organizations in text. The performance of NER models relies heavily on labelled data, and limited labelled data often impede effective cross-domain knowledge transfer. Transferring knowledge features from one domain to another, known as transfer learning, is a major challenge in NER.

Domain adaptation is a critical research direction in transfer learning, with the domain offset problem being a common challenge. This problem occurs when the model's performance declines while transferring from one domain to another. Numerous algorithms have been suggested to address the issue of domain shift, including domain adversarial

Z. Yu et al. (Eds.): ICPCSEE 2023, CCIS 1880, pp. 202–212, 2023.
https://doi.org/10.1007/978-981-99-5971-6_15

neural networks (DANN) [1] and distribution matching [2]. However, these algorithms have limitations. DANN suffers from instability in joint optimization, which requires an extensive amount of hyperparameter tuning [3–5]. Conversely, when attempting to achieve alignment at the instance level, the distribution matching algorithm may encounter difficulties in preserving the discriminative power for the target task [6, 7].

In recent years, contrastive learning (CL) has emerged as an effective method for self-supervised learning (SSL). By leveraging data surrogate tasks in primitive learning, CL aims to define instance-level representations [8, 9]. CL leverages data augmentation to generate positive pairs and treats instances as classes, which can lead to improved performance. In previous tasks, CL used the same label to mark cross-domain images as positive instance pairs in the contrastive loss [10, 11]. However, this method is not suitable for NLP tasks due to significant semantic and syntactic changes between cross-domain sentences. Previous work has focused on designing text transformations that preserve labels, such as back-translation, paraphrasing, omission, and their combinations [12, 13].

NER plays a vital role in natural language processing, where flat NER has been extensively studied as a sequence labelling problem. However, nested entities are of great significance in various real-world applications because of their multigranularity semantic meanings. Although several span-based methods have been proposed to address the nested NER problem, they only consider the combination of the head and tail of a span, disregarding the boundary information. This limitation hinders their effectiveness in addressing the nested NER problem.

This paper draws inspiration from Long et al.'s [14] DCCL algorithm and explores the effect of incorporating entity boundary information on cross-domain named entity recognition. The experimental results demonstrate that this approach yields promising results. The primary contributions of this paper are as follows:

- The utilization of Global Pointer improves the model's ability to learn boundary information.
- The named entity recognition task is accomplished using Long et al.'s cross-domain sentiment analysis model and utilizes both contrastive learning and KL divergence techniques at the span level to address domain shift problems.
- The model achieved excellent performance compared to multiple cross-domain named entity recognition models.

2 Related Work

Named entity recognition (NER) is an increasingly popular research topic in the field of natural language processing. Many studies have focused on this task, and deep learning has become a widely used and effective approach for NER, showing promising results in various applications.

Several methods have been proposed for NER. [15] proposed the Iterated Dilated Convolutional Neural Network (IDCNN) which uses dilation in a jumping pattern to capture long-range relationships between characters but may lose local information. Huang et al. [16] proposed a BiLSTM-CRF model, the first to use a bidirectional LSTM-CRF structure for sequence labelling tasks. Chiu et al. [17] proposed a method that effectively combines BiLSTM and CNN to detect word and character-level features. Zhang et al. [18] proposed the lattice-structured LSTM model for Chinese NER, which can incorporate word meanings to avoid the impact of Chinese segmentation errors.

NER often requires a considerable amount of labelled data, which can be costly and time-intensive to obtain. To mitigate this issue, transfer learning techniques can be applied, leveraging knowledge learned from a source domain to improve performance on a target domain. This approach reduces the need for large amounts of annotated data and can improve the efficiency and effectiveness of NER models.

Cao et al. [19] proposed a Chinese NER model that combines information from NER and Chinese word segmentation tasks with a self-attention mechanism. Zhou et al. [20] developed a dual adversarial transfer network model that addresses representation differences and data resource imbalance for cross-language and cross-domain NER transfer tasks. Chen et al. [21] approached the problem of reannotating data and retraining models caused by new entity types by adding new neurons to the target model's output layer and transferring parameters from the source model, followed by fine-tuning. Tong et al. [22] introduced a model MUCO to improve NER performance in small sample scenarios by automatically extracting undefined classes from other classes.

This paper proposes a method that uses a span-based domain confusion contrastive learning approach to encourage the model to learn similar representations of the original sentence and domain-confused counterpart. Additionally, it introduces entity boundary information to convert sequential labels into a global boundary matrix, enabling the model to learn explicit span boundary information at the sentence level.

3   Method

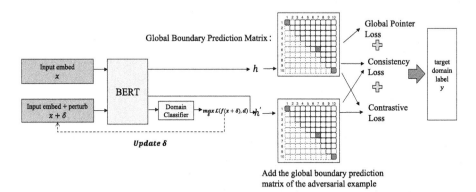

Fig. 1. GPDCCL framework

In our paper, we propose a model called GPDCCL, as illustrated in Fig. 1. First, the source domain is embedded into BERT to generate a global prediction matrix and calculate the loss of Global Pointer. Then, a domain-confused counterpart is created by synthesizing adversarial perturbations using the extreme direction that most perplexes the domain classifier. These adversarial samples, along with the source domain embeddings, are added together to generate a global prediction matrix, which is used to calculate the loss of KL divergence and contrastive learning. Finally, the model predicts the label y of the target domain.

3.1 Adversarial Training

For labelled tasks, the model aims to learn a function $f\left(x; \theta_f, \theta_y\right) : xC$:

$$\min_{\theta_f, \theta_y} \sum_{(x,y) \sim \mathcal{D}^S} \left[\mathcal{L}(f\left(x; \theta_f, \theta_y\right), y) \right] \tag{1}$$

Research has shown that adversarial training with perturbations significantly improves model performance. The common approach is to perturb word embeddings and minimize the adversarial loss, with the goal of maximizing internal loss and minimizing external loss in a single domain.

$$\min_{\theta_f, \theta_y} \sum_{(x,y) \sim \mathcal{D}} \left[\max_\delta \mathcal{L}(f\left(x + \delta; \theta_f, \theta_y\right), y) \right] \tag{2}$$

The weight α_{adv}, which determines the relative importance of the two types of losses, is commonly set to 1:

$$\min_{\theta_f, \theta_y} \sum_{(x,y) \sim \mathcal{D}} \left[\mathcal{L}(f\left(x; \theta_f, \theta_y\right), y) + \alpha_{adv} \right. \\ \left. \max_\delta \mathcal{L}(f\left(x + \delta; \theta_f, \theta_y\right), f\left(x; \theta_f, \theta_y\right)) \right] \tag{3}$$

Projected gradient descent (PGD) is a commonly used method for addressing internal maximization in adversarial training. PGD has the advantage of relying solely on the model and can generate samples with varying degrees of adversarial properties, improving the model's generalization ability on unseen data. To approximately generate adversarial perturbations, the following iterative steps are typically used:

$$\delta_{t+1} = \Pi_{\|\delta\|_F \le} \left(\delta_t + \eta \frac{g_y^{adv}(\delta_t)}{g_y^{adv}(\delta_t)_F} \right) \tag{4}$$

$$g_y^{adv}(\delta_t) = \nabla_\delta \mathcal{L}(f\left(x + \delta_t; \theta_f, \theta_y\right), y) \tag{5}$$

3.2 Crafting Domain Puzzles

One way to overcome the limitations of matching distributions between domains without target domain labels is to use a puzzle to match the distributions. This can improve transferability and reduce negative transfer and noise introduced by instance-based matching strategies. Another approach is to generate domain-invariant representations by identifying and masking domain-specific tags related to sentence topics and types. However, identifying the appropriate tags can be challenging, especially for complex sentence structures. To address this challenge, a method is proposed that searches for domain puzzles in the representation space and introduces adversarial perturbations to generate targeted puzzles. The goal is not to enhance robustness, but to improve domain invariance and this approach, which relies on the model to generate perturbations, can achieve better results than other methods.

To generate domain-confusion enhancement, adversarial attacks with perturbations are employed for domain classification, and domain-specific losses are learned for the domain classifier using adversarial attacks.

$$\mathcal{L}_{\text{domain}} = \mathcal{L}\big(f\big(x; \theta_f, \theta_d\big), d\big) + \\ \alpha_{adv}\mathcal{L}\big(f\big(x + \delta; \theta_f, \theta_d\big), f\big(x; \theta_f, \theta_d\big)\big) \tag{6}$$

$$\delta = \Pi_{\|\delta\|_F \le}\left(\delta_0 + \eta\frac{g_d^{adv}(\delta_0)}{g_d^{adv}(\delta_0)_F}\right) \tag{7}$$

δ_0 refers to the initial noise, while θ_d corresponds to the computation of domain classification, and d represents the domain label (Fig. 2).

3.3 Span Extraction

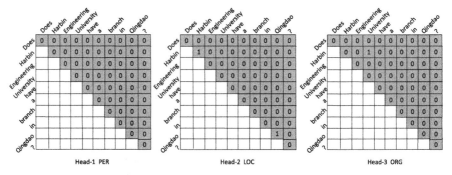

Head-1 PER   Head-2 LOC   Head-3 ORG

Fig. 2. Schematic diagram of multihead recognition nested entities

The purpose of span extraction is to generate all possible entity spans. Suppose $S = [s_1, s_2, \cdots, s_m]$ are possible spans in a sentence. A span s is denoted as $s[i : j]$, where i and j are the head and tail indices, respectively. The goal of named entity recognition is to identify all $s \in E$, where E is the set of entity types. Given a sentence $X = [x_1, x_2, \cdots x_n]$ with n tokens, first associate each token in X with its corresponding representation in the pretrained language model, resulting in a new matrix $H \in \mathbb{R}^{n \times v}$, where v is the number of dimensions:

$$h_1, h_2, ..., h_n = PLM\,(x_1, x_2, \ldots x_n) \tag{8}$$

To compute the span representation of H using the acquired sentences, two feedforward layers are used that depend on the start and end indices of the span.

$$q_{i,\alpha} = W_{q,\alpha}h_i + b_{q,\alpha} \tag{9}$$

$$k_{j,\alpha} = W_{k,\alpha} h_j + b_{k,\alpha} \tag{10}$$

where $q_{i,\alpha} \in \mathbb{R}^d$, $k_{j,\alpha} \in \mathbb{R}^d$ is the vector representation of the token used to identify the entity of typeα.Specifically, the representation of the start and end position is $q_{i,\alpha}$ and $k_{j,\alpha}$ for span s[$i: j$] of typeα. Then, the score of the span $s[i:j]$ to be an entity of type α is calculated as follows:

$$s_\alpha(i,j) = q_{i,\alpha}^\top k_{j,\alpha} \tag{11}$$

While absolute position encoding can provide positional information to word vectors, it cannot represent contextual information at a fixed position. To address this, Su et al. [23]proposed rotational position encoding (RoPE) which relies solely on relative positions. To leverage boundary information, RoPE is applied to entity representation, satisfying the condition $\mathcal{R}_i^\top \mathcal{R}_j = \mathcal{R}_{j-i}$. This improves the model's sensitivity to relative positions between entities and enhances entity recognition performance. Using this encoding, the scoring function can be calculated as follows:

$$\begin{aligned} s_\alpha(i,j) &= \left(\mathcal{R}_i q_{i,\alpha}\right)^\top \left(\mathcal{R}_j k_{j,\alpha}\right) \\ &= q_{i,\alpha}^\top \mathcal{R}_i^\top \mathcal{R}_j k_{j,\alpha} \\ &= q_{i,\alpha}^\top \mathcal{R}_{j-i} k_{j,\alpha} \end{aligned} \tag{12}$$

As a result, the score matrix for the category of each sentence can be obtained Similar to Su et al., a cross-entropy loss function is designed to encourage the model to learn the boundary information of each training support instance.

$$\mathcal{L}_{gp} = \log\left(1 + \sum\nolimits_{(q,k) \in P_\alpha} e^{-s_\alpha(q,k)}\right) + \log\left(1 + \sum\nolimits_{(q,k) \in Q_\alpha} e^{s_\alpha(q,k)}\right) \tag{13}$$

where q, k represent the start and tail indices of a span, P_α represents a collection of spans with entity type α, Q_α represents a collection of spans that are not entities or whose entity type is not α, and $s_\alpha(q, k)$ is the score that a span $s[q:k]$ is an entity of type α.

3.4 Learning Invariance with Domain Puzzles

In cross-domain learning, distribution matching and instance-based matching may have limitations, but contrastive learning (CL) can learn domain invariance without labelled data in the target domain. By considering both the similarity between instances and their importance from a global perspective, this method is effective in cross-domain learning tasks. The goal of contrastive loss is to learn a distance metric that maps similar samples to neighboring positions and dissimilar samples to distant positions. In the domain transfer task, positive samples are constructed from source and domain-perturbed samples, while negative samples are constructed from different span categories. This optimization approach improves the model's classification performance by minimizing the distance between positive samples and maximizing the distance between negative samples and the target sample.

The score matrix obtained in the previous step can be used to calculate the contrastive loss. The goal of contrastive learning is to bring similar spans closer to each other at the span level and push dissimilar spans farther away from each other. The formula for calculating the contrastive loss is as follows:

$$\mathcal{L}_{\text{constract}} = -\sum_{i=1}^{N} \sum_{j=1}^{N} \frac{e^{\cos(\text{span}(i,j),\text{span}(i,j)^+)}}{\sum^{M} e^{\cos(\text{span}(i,j),\text{span}(i,j)^-)}} \tag{14}$$

where N is the maximum length of the sentence, M is the number of negative examples, $\text{span}(i, j)$ is the span representation, and $\text{span}(i, j)^+$ is the positive example for the current sentence, i.e. the data augmentation for adversarial training on the source domain data, and $\text{span}(i, j)^-$ is the negative example for the current sentence, i.e. spans that differ from the current token label. The distance between the original sample and the positive and negative samples is calculated using cosine similarity.

Adversarial training generates perturbed data to improve model robustness. To ensure similarity and continuity, a latent variable controls the distance between samples. KL divergence is used to compare the generated samples to the model's predicted distribution. Lower KL divergence means higher similarity. This improves the quality of adversarial samples used to train the model and improves its robustness.

$$\mathcal{L}_{\text{consist}} = \sum_{n=1}^{N} \sum_{n=1}^{N} KL\big(\text{span}(i, j), \text{span}(i, j)^+\big) \tag{15}$$

This article trains the neural network end-to-end with four loss functions: source domain classification, domain classifier, contrastive learning, and KL divergence. The weighted loss is as follows:

$$\min_{\theta_f, \theta_y, \theta_d} \sum_{(x,y) \sim \mathcal{D}^S} \mathcal{L}_{\text{gp}} +$$
$$\sum_{(x,y) \sim \mathcal{D}^S, \mathcal{D}^T} [\alpha \mathcal{L}_{\text{domain}} + \lambda \mathcal{L}_{\text{contrast}} + \beta \mathcal{L}_{\text{consist}}] \tag{16}$$

where, α, and $\lambda\beta$ are the hyperparameters used to control the weight of each loss. By training end-to-end and optimizing this loss function, a robust model with good classification performance on original samples and high-quality adversarial samples can be obtained.

4   Experimental Analysis

4.1   Dataset

This article uses the CoNLL03 [24] English dataset in the news domain as the source domain and the SciTech [25] English dataset in the technology domain as the target domain. Both datasets contain four types of entities: PER (person), LOC(location), ORG(organization), and MISC (miscellaneous), with annotations in the BIO format. Table 1 displays the number of sentences and words in each dataset.

Table 1. Dataset statistics.

Dataset	Type	Train	Dev	Test
CoNLL03	#sentences	15.0k	3.5k	3.7k
	#tokens	203.6k	51.4k	46.4k
SciTech	#sentences	-	-	2.0k
	#tokens	-	-	4.1k

4.2 Experiment Setting

We use BERT-base-uncased with a hidden layer size of 768. The AdamW optimizer is used with the weight decay set to 1e-8, and the learning rate set to 1e-5. The L-infinity norm limits the magnitude of adversarial perturbation, the upper limit ε is set to 0.02, and the adversarial step size η is set to 0.05. α_{adv} is set to 1 for adversarial weighting. α=1, λ=6, and β=1 are used to weight the loss functions.

4.3 Analysis of Results

In this article, the target domain results are compared with several cross-domain NER models, including the following:

- BiLSTM + CRF [26]: This model utilizes character-level Bi-LSTM to capture morphological and spelling features and word-level Bi-LSTM to integrate sentence syntax features. Finally, the model stacks a CRF layer to predict labels.
- BERT + CRF: This model replaces traditional BiLSTM with the powerful pretraining language model BERT to obtain more informative and context-enhanced word representations and then incorporates a CRF layer for label prediction.
- JIA2019 [27]: This model combines language modelling with the NER task to build a multitask learning structure and then utilizes tensor decomposition to learn task embeddings for cross-domain NER prediction on these task embeddings.
- Multi-Cell [28]: This model uses a multicell composed LSTM and multitask learning for cross-domain NER.
- EAAT [29]: Proposing a model that leverages adversarial training, this approach addresses the domain shift problem by utilizing an entity-aware attention module to guide the training process, effectively reducing the disparities in entity features across diverse entities.

This paper presents an experimental study on cross-domain transfer learning using Conll2003 as the labelled source domain and SciTech as the unlabelled target domain. It can be observed from the table that our proposed model performs the best in terms of precision, and F1 score, with an F1 score of 77.40, significantly better than other models. The multicell model also performs well but with a slightly lower F1 score than our model. The results in Table 2 demonstrate that the GPDCCL model achieves a high level of performance, indicating its strong cross-domain transfer capabilities. The proposed contrastive learning method based on span-based domain confusion improves

Table 2. Model evaluation on the target domain dataset.

Model	SciTech		
	Pre	Rec	F1
BiLSTM + CRF	73.53	61.56	67.01
BERT + CRF	68.57	80.97	74.26
JIA2019	74.28	72.91	73.58
EAAT	68.96	79.16	73.71
Multi-Cell	77.01	73.03	75.01
Ours	77.40	77.41	77.40

the acquisition of domain-invariant features, with the KL divergence further enhancing the model's generalization ability.

To analyse the contribution of each loss function in the proposed method, ablation experiments are conducted on the target dataset using both the GPDCCL and baseline models, and the results are presented in Table 3.

Table 3. Ablation experiment of GPDCCL on the target domain dataset.

Method	SciTech		
	Pre	Rec	F1
GPDCCL	**77.40**	**77.41**	**77.40**
Base Model + GP	75.42 (-1.98)	75.13 (-2.28)	75.27 (-2.13)
GPDCCL w/o $\mathcal{L}_{constrast}$	76.56 (-0.84)	76.95 (-0.46)	76.75 (-0.65)
GPDCCL w/o $\mathcal{L}_{consist}$	75.96 (-1.44)	76.77 (-0.64)	76.36 (-1.04)
GPDCCL w/o $\mathcal{L}_{constrast} + \mathcal{L}_{consist}$	75.49 (-1.91)	76.00 (-1.41)	75.74 (-1.66)

Based on the data in the table, the GPDCCL model outperforms the Base Model + GP with significant improvements in accuracy, recall, and F1 scores. Ablation experiments show that each loss function has a positive impact on domain adaptation to varying degrees, and the contrastive learning at the span level has a great effect on learning the domain invariant characteristics of entity spans.

Comparing the first and last rows of the table, adding only domain confusion with adversarial perturbation decreases the F1 value by 1.66%. Additionally, changing domain puzzle training to simple data augmentation without contrastive learning only slightly improves model performance, suggesting that domain puzzle data enhancement is not very helpful for named entity recognition. The GPDCCL model's performance improvement is not due to data augmentation, and building resilience against adversarial attacks does not significantly impact domain adaptation.

In this paper, Global Pointer was used as the method of span extraction, and the method of calculating contrastive learning and KL divergence at the span level played an active role. Contrastive learning brought possible similar spans closer at the span level while pushing away dissimilar spans. KL divergence helped the model learn a more consistent distribution with the model predictions at the span level. Therefore, each module contributed to the model performance of cross-domain named entity recognition.

5 Conclusion

This paper proposes domain-confused contrastive learning based on span for named entity recognition using a model originally designed for sentiment analysis and adapting it to named entity recognition through parameter adjustments and contrastive loss function design. Experiments on the SciTech news domain dataset show that the proposed method outperforms various other cross-domain named entity recognition models in terms of accuracy and recall on the target domain unlabelled dataset.

Acknowledgements. This work was supported by the National Key R&D Program of China under Grant No. 2020YFB1710200.

References

1. Ganin, Y., Ustinova, E., Ajakan, H., et al.: Domain-adversarial training of neural networks. J. Machine Learn. Res. **17**(1), 2096–2030 (2016)
2. Zhuang, F., Cheng, X., Luo, P., et al.: Supervised representation learning: Transfer learning with deep autoencoders. In: Twenty-Fourth International Joint Conference n Artificial Intelligence (2015)
3. Shah, D.J., Lei, T., Moschitti, A., et al.: Adversarial domain adaptation for duplicate question detection. arXiv preprint arXiv:1809.02255 (2018)
4. Du, C., Sun, H., Wang, J., et al.: Adversarial and domain-aware BERT for cross-domain sentiment analysis. In: Proceedings of the 58th Annual Meeting of the Association for Computational Linguistics, pp. 4019–4028 (2020)
5. Karouzos, C., Paraskevopoulos, G., Potamianos, A.: UDALM: Unsupervised domain adaptation through language modeling. arXiv preprint arXiv:2104.07078. (2021)
6. Saito, K., Ushiku, Y., Harada, T.: Asymmetric tri-training for unsupervised domain adaptation. In: International Conference on Machine Learning. PMLR, pp. 2988–2997 (2017)
7. Lee, S., Kim, D., Kim, N., et al.: Drop to adapt: Learning discriminative features for unsupervised domain adaptation. In: Proceedings of the IEEE/CVF International Conference on Computer Vision, pp. 91–100 (2019)
8. Chen, T., Kornblith. S., Norouzi, M., et al.: A simple framework for contrastive learning of visual representations. In: International Conference on Machine Learning. PMLR, pp. 1597–1607 (2020)
9. Khosla, P., Teterwak, P., Wang, C., et al.: Supervised contrastive learning. Adv. Neural. Inf. Process. Syst. **33**, 18661–18673 (2020)

10. Rui, W., et al.: Crossdomain contrastive learning for unsupervised domain adaptation. ArXiv preprint, arXiv:2106.05528 (2021)
11. Park, C., Lee, J., Yoo, J., et al.: Joint contrastive learning for unsupervised domain adaptation. arXiv preprint arXiv:2006.10297 (2020)
12. Qu, Y., Shen, D., Shen, Y., et al.: Coda: Contrast-enhanced and diversity-promoting data augmentation for natural language understanding. arXiv preprint arXiv:2010.08670 (2020)
13. Gao, T., Yao, X., Chen, D.: Simcse: Simple contrastive learning of sentence embeddings. arXiv preprint arXiv:2104.08821 (2021)
14. Long, Q., Luo, T., Wang, W., Pan, S.J.: Domain Confused Contrastive Learning for Unsupervised Domain Adaptation. arXiv:2207.04564v1 [cs.CL] (2022)
15. Strubell, E., Verga, P., Belanger, D., et al.: Fast and accurate entity recognition with iterated dilated convolutions. arXiv preprint arXiv:1702.02098 (2017)
16. Huang, Z., Xu, W., Yu, K.: Bidirectional LSTM-CRF models for sequence tagging. arXiv preprint arXiv:1508.01991, (2015)
17. Chiu, J.P.C., Nichols, E.: Named entity recognition with bidirectional LSTM-CNNs. Trans. Assoc. Comput. Linguist. 4, 357–370 (2016)
18. Zhang, Y., Yang, J.: Chinese NER using lattice LSTM. arXiv preprint arXiv:1805.02023 (2018)
19. Cao, P., Chen, Y., Liu, K., et al.: Adversarial transfer learning for Chinese named entity recognition with self-attention mechanism. In: Proceedings of the 2018 Conference on Empirical Methods in Natural Language Processing, pp.182–192 (2018)
20. Zhou, J.T., Zhang, H., Jin, D, et al.: Dual adversarial neural transfer for low-resource named entity recognition. In: Proceedings of the 57th Annual Meeting of the Association for Computational Linguistics, pp. 3461–3471 (2019)
21. Chen, L., Moschitti, A.: Transfer learning for sequence labeling using source model and target data. In: Proceedings of the AAAI Conference on Artificial Intelligence, vol. 33(01), pp. 6260–6267 (2019)
22. Tong, M., Wang, S., Xu, B., et al.: Learning from miscellaneous other-class words for few-shot named entity recognition. arXiv preprint arXiv:2106.15167 (2021)
23. Su, J., Lu, Y., Pan, S, et al.: Roformer: Enhanced transformer with rotary position embedding. arXiv preprint arXiv:2104.09864 (2021)
24. Sang, E.F., De Meulder, F.: Introduction to the CoNLL-2003 shared task: Language-independent named entity recognition. arXiv preprint cs/0306050 (2003)
25. Jia, C., Liang, X., Zhang, Y.: Cross-domain NER using cross-domain language modeling. In: Proceedings of the 57th Annual Meeting of the Association for Computational Linguistics, pp. 2464–2474 (2019)
26. Madry, A., Makelov, A., Schmidt, L., et al.: Towards deep learning models resistant to adversarial attacks. arXiv preprint arXiv:1706.06083 (2017)
27. Lample, G., Ballesteros, M., Subramanian, S., et al.: Neural architectures for named entity recognition. arXiv preprint arXiv:1603.01360 (2016)
28. Jia, C., Zhang, Y.: Multi-cell compositional LSTM for NER domain adaptation. In: Proceedings of the 58th Annual Meeting of the Association for Computational Linguistics, pp. 5906–5917 (2020)
29. Peng, Q., Zheng, C., Cai, Y., et al.: Unsupervised cross-domain named entity recognition using entity-aware adversarial training. Neural Netw. 138, 68–77 (2021)

Data Analyses and Parallel Optimization of the Regional Marine Ecological Model

Yanqiang Wang[1,2] (ID), Jingjing Zheng[1], Tianyu Zhang[3,4](✉), Peng Liang[3], and Bo Lin[1]

[1] National Marine Environmental Forecasting Center, Beijing 100081, China

[2] College of Oceanic and Atmospheric Sciences, Ocean University of China, Qingdao 266100, China

[3] Key Laboratory of Climate, Resources and Environment in Continental Shelf Sea and Deep Sea of Department of Education of Guangdong Province, Guangdong Ocean University, Zhanjiang 54008, China
zhaangty@sina.com

[4] Southern Marine Science and Engineering Guangdong Laboratory (Zhuhai), Zhuhai 519000, China

Abstract. Under the joint influence of high-intensity human activities and climate change, the coastal ecological environment is deteriorating, and the ecological environment security and the sustainable development of the marine economy are seriously threatened. Therefore, it is of great significance to establish a high-resolution ecological environment operational forecasting system. To meet the run time requirements of the ecological operational forecasting system, a variety of parallel optimization methods were proposed to improve the operation efficiency of the model. First, based on the National Marine Environmental Forecasting Center's Lenovo cluster, the ROMS benchmark experiment was expanded to the 4000 Processes scale. A good speedup was obtained by the experiment. The ROMS model was analysed with strong scalability. Second, in the hydrodynamic-ecological simulation experiment of the Bohai Sea - Yellow Sea - East China Sea, by optimizing Vector, InfiniBand, and Parallel I/O, the performance of the model can be improved by 270% while maintaining the same computing resources. That computing resources were more reasonably used lay the foundation for the operational forecast.

Keywords: marine ecological model · ROMS · high-performance computing · parallel optimization

1 Introduction

Since the development of heavy industry, the marine ecological environment has faced severe challenges due to the discharge of industrial wastewater, domestic sewage and other human activities, as well as the impact of emergency accidents at sea. It is mainly manifested in the eutrophication of seawater, the change in nutrient structure, the expansion of hypoxic and anoxic regions, ocean acidification, frequent occurrence of red tide, green tide, jellyfish, coral bleaching, coral reef degradation and other environmental

Z. Yu et al. (Eds.): ICPCSEE 2023, CCIS 1880, pp. 213–224, 2023.
https://doi.org/10.1007/978-981-99-5971-6_16

degradation and frequent occurrence of biological disasters. The problem of the off-shore ecological environment has long received attention from domestic and foreign researchers. It is of great significance to study water dynamics and ecology and establish a regional ecological operational prediction system from marine environmental protection and governance.

There has been much research and application on the simulation and prediction of marine ecological environments at home and abroad. The study of a three-dimensional marine ecological dynamics model coupled with a physical model and biological model began in the mid-1980s. The biological model began with a simple NPZD model (nutrient - phytoplankton - zooplankton - debris) and gradually introduced more nutrient cycles, more functional groups, and even higher nutrient levels. The description of the ecosystem is becoming increasingly complex and close to reality. At present, many mature marine ecological dynamics models have been widely used, such as Fennel [1, 2], CoSiNE (Carbon, Silicate, Nitrogen Ecosystem Model) [3,4], NEMURO (North Pacific Ecosystem Model for Understanding Regional Oceanography) [5, 6], and ECOHAM (Ecosystem Model Hamburg) (Moll[7]). In addition to large-scale research projects, some regional marine ecological environment prediction models at home and abroad have begun to be operational or quasioperational. In recent years, the National Oceanic and Atmospheric Administration (NOAA) of the United States has established a prediction system for near-shore catastrophic events (red tide, hypoxia, and coral bleaching) (https://oceanservice.noaa.gov/ecoforecasting) The British Ocean Forecast Center (NCOF) has carried out a comprehensive study on the quasiprediction of the ecological environment by using the ERSEM ecological model and can provide product services such as ecosystem health, water quality auxiliary monitoring, eutrophication, and harmful algal bloom prediction in the continental shelf sea of Northwest Europe [8]. The China National Marine Environmental Forecasting Center has established a high-resolution operational numerical prediction system for China's offshore ecological environment, covering the Bohai Sea, the Yellow Sea, the East China Sea, the South China Sea, and the offshore demonstration areas, and has released a set of forecasting products for marine ecological environment elements such as nutrients, dissolved oxygen, chlorophyll, and pH.

With the continuous increase in ocean observation data and the continuous deepening of research on marine ecological models, marine ecological models are gradually developing towards higher resolution and more complex ecological dynamic processes, and their demand for computing resources is also increasing in geometric progression. According to the existing ocean numerical model, each time the resolution is doubled, the calculation amount will increase by 8 to 10 times. In addition, if the resolution is increased to 1 km, the corresponding calculation amount will increase by 100 to 1000 times [9]. Therefore, the rapid growth of marine ecological numerical models has put forward higher requirements for computing capacity, and it has become more necessary to make full use of and play to the performance of high-performance computers to improve the computing speed of marine ecological numerical models. The organic combination of marine numerical models and high-performance computing has become an important means of marine scientific research.

Due to the high resolution, many ecological elements, and complex calculation process, the ecological operational prediction system consumes a large amount of computing resources. Parallel optimization is an effective way to improve the performance of high-performance numerical simulations. Therefore, it is urgent to meet operational needs through parallel optimization methods. The regional ecological dynamics model established in this paper is based on the ROMS (Regional Ocean Modeling System) coupled Fennel model. This paper first analyses and tests the extensibility of the ROMS model according to its characteristics. Second, in the Yellow Sea and East Bohai Sea dynamic and ecological coupling simulation experiment, aiming at the low efficiency of the physical ROMS coupled ecological Fennel model in the initial stage, the model was optimized from the aspects of vectorization, high-speed switching network, and parallel I/O.

2 Model Configuration

2.1 Configuration of the Benchmark Experimental Model

The regional ocean model ROMS (Regional Ocean Modeling System) adopted in this paper is a primitive equation ocean model [10, 11] based on the static assumption, free surface, and terrain-following coordinate system using stretched sigma coordinates, which is mainly applied to the prediction of the estuary marine environment and the coastal waters. The dynamic core of ROMS is composed of four independent models, namely the nonlinear model (NLM), tangent linear model (TLM), representative tangent model (RPM), and adjoint model (ADM). These models have multiple drivers running at the same time [12]. The model has been favored by oceanographers at home and abroad since the 1990s. The ROMS model can be coupled with wave, sediment, ecological, and atmospheric models, and applied to ocean prediction [13]. The benchmark experiment is a performance test experiment of a standardized preset power module in ROMS mode. The power module of the experiment is set as follows:

The horizontal and vertical advection terms, Coriolis force terms, and second-order horizontal viscosity terms are opened in the seawater motion equation. The salinity term, solar radiation term, and second-order horizontal diffusion term are opened in the scalar advection-diffusion equation of seawater. LMD [14] mixed parameterization scheme is adopted for mixed parameterization, and the nonlinear state equation is adopted for the seawater state equation. The specific parameters of the mode in the experiment are as follows:

1) Simulation area: 20 latitudes (70° S ~ 50° S) in the north–south direction, 360 longitudes (0° ~ 180° ~ 0°) in the east–west direction, and the water depth of the grid point is 4000 m.
2) Model resolution: uniform grid in the horizontal direction; the resolution depends on the number of model grid points in the horizontal direction. It is divided into 30 layers in the vertical direction, which is a nonuniform grid, and the closer to the surface, the higher the resolution.
3) Sea surface wind field: The sea surface wind field adopts the analysis wind field, the north–south wind speed of the analysis wind field is 0, and the specific calculation method of the east–west wind speed is as follows:

$$Uwind = 15\exp\{-[0.2(60 + lat)]^2\} \tag{1}$$

where, Uwind is the zonal wind speed, in m/s, and lat is the current grid latitude (south latitude is negative).

4) Sea surface heat flux: The sea surface heat flux is calculated using the block formula. During the calculation, the atmospheric temperature is uniform, the size is 4 °C, the atmospheric pressure is 1025 Mb, the air humidity is 0.8, the cloud cover is 0.6, and the sea surface precipitation is 0.

5) Side boundary conditions: the east–west boundary adopts periodic boundary conditions, and the north–south boundary adopts closed boundary conditions.

The specific number of grids and time steps are shown in Table 1. The ROMS built-in benchmark test is benchmark 3. In the strong expansion test, the number of grids in the I and J directions is increased by 10 times, and the scalability is calculated by adjusting to the appropriate time steps.

Table 1. Benchmark and strong expansion experiment configuration

Configure	Benchmark3	Strong expansion experiment
Number of points in direction I	2048	20480
Number of points in direction J	256	2560
NTIMES	200	10
DT	150.0	10.0
NDTFAST	20	10

2.2 Actual Sea Area Experimental Model Configuration

The dynamic and ecological environment of the Yellow Sea and the East Bohai Sea are simulated in the actual sea area. The hydrodynamic model adopts the rectangular orthogonal Arakawa C grid in the horizontal direction, and the S coordinate varying with the terrain in the vertical direction. In terms of time, the algorithm adopts the calculation scheme of separation of internal and external models and solves the internal and external models separately to improve the calculation efficiency. The internal mode time step is 90 s, and the external mode time step is 9 s.

The ecological model is based on the Fennel model. This model is developed from the NPZD model of Fasham [15]. The model takes the nitrogen cycle as the core and contains 12 ecological variables, namely, nitrate (NO3), ammonium salt (NH4), chlorophyll (Chla), phytoplankton (Phy), zooplankton (Zoop), large detrital nitrogen (LDetN), small detrital nitrogen (SDetN), large detrital carbon (LDetC), small detrital carbon (SDetC), dissolved oxygen (DO), inorganic carbon (DIC), and total alkalinity (TALK). The concentration change of ecology-related state variables in the model is affected by the physical transport process, as well as the biological and chemical effects, which can be shown in Eq. (2).

$$\frac{\partial C}{\partial t} = -u\frac{\partial C}{\partial x} - v\frac{\partial C}{\partial y} - w\frac{\partial C}{\partial z} + \frac{\partial}{\partial x}\left(A_h\frac{\partial C}{\partial x}\right) + \frac{\partial}{\partial y}\left(A_h\frac{\partial C}{\partial y}\right) + \frac{\partial}{\partial z}\left(A_v\frac{\partial C}{\partial z}\right) + S + W_0 \qquad (2)$$

where, C is the relevant state variable of ecology; u, v and w are the current velocity components, A_h and A_v is the horizontal diffusion coefficient and vertical diffusion coefficient, respectively; S is the change in ecological variable concentration caused by biochemical processes; and W_0 is the impact of external inputs such as rivers on ecological variables.

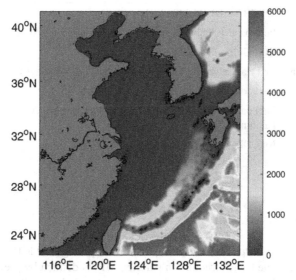

Fig. 1. The actual sea area of the marine ecological model

The actual sea area study area is 22° N-41° N, 114° E-133° E, which is shown in 错误!未找到引用源。, and the horizontal resolution is 1/30° × 1/30°. The total number of grids is 564 × 565, vertically divided into 30 layers. The water depth data of the model adopt a resolution of 0.5' × 0.5' GEBCO (General Bathymetric Chart of Oceans) data. The driving conditions of the model include initial conditions, open boundary conditions, and meteorological forcing fields. The initial temperature and salinity of the physical model are from the monthly average data of the GDEM (Generalized Digital Environment Model) climate state. The south, north and east of the model are open boundaries, and the west is a closed boundary. The data of temperature, salinity, current and water level on the boundary are derived from the monthly average data of the SODA (Simple Ocean Data Association) climatic state, and the tidal drive of the boundary is derived from the data of TPX07 [16]. The nutrients, total inorganic carbon, dissolved oxygen and total alkalinity of the initial and boundary fields of the ecological model are all from the monthly average data of the WOA09 climatic state. The forced field data come from NCEP-CFSRv2 (National Centers for Environmental Prediction Climate Forecast System Analysis Version 2) daily reanalysis data.

Based on the process of the coupled model system, considering the number and performance of the test platform's processors, the system is deployed on the Lenovo cluster of the National Marine Environmental Forecasting Center. The specific configuration is shown in Table 2.

Table 2. HPC facilities used for the tests

System processor	Intel(R) Xeon(R) CPU E5–2680 v4 @ 2.40 GHz
Cores/Node	14Core*2
Mem./Node	128GB TruDDR4 2400 MHz
MPI network	Infiniband 100Gb/s
Filesystem	IBM GPFS (General Parallel File System)
Operating System	Red Hat Enterprise Linux Server release 6.8 (Santiago)
Complier	Intel compilers_and_libraries_2017.3.191
Analysis Software	Intel parallel_studio_xe_2017.3.053; Intel oneAPI 2021.3; Paramon 8.4.0; Paratune 7.2.0
Other libraries	NetCDF 4.4.4 etc.

3   Parallel Optimization

ROMS supports serial, OpenMP and MPI parallel computing. The MPI and OpenMP of ROMS share the same basic structure in parallel. MPI is defined by precompiling in C language. The parallel code of the model is consistent with the result of the serial code. Therefore, the scalability of the model can be tested based on a single serial CPU core.

3.1   Baseline Test

The research takes four CPU cores as the benchmark. The test method is to use Paramon to record the benchmark running status of the program, increase the computing resources to record again, and predict the acceleration that can be obtained by further increasing the computing resources by comparing the acceleration ratio and other numerical changes. Based on the ROMS benchmark experiment benchmark3, the number of grids in the I and J directions can be expanded to 3920 cores (140 computing nodes), with an acceleration ratio of 533.8 times, as shown in Fig. 2.

3.2   IB Optimization

Use the Trace Analyzer tool in the Intel Parallel Studio and Intel oneAPI Base&HPC toolkit to further analyse the communication and computing share of the model, as

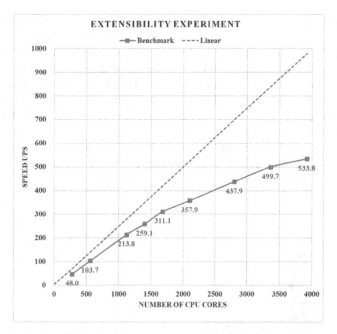

Fig. 2. Acceleration ratio of the strong expansion experiment parallel program

shown in Fig. 3. Taking 140 cores as an example, the horizontal axis takes time as a unit. Through the improvement of IB Fig. 3 b), it can be seen that the MPI waiting time of the model has been significantly reduced, the overall running time has been reduced to 33% of the original running time Fig. 3 a), and the running efficiency of the model has been significantly improved.

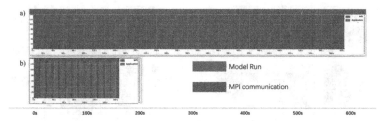

Fig. 3. Model operation and communication time distribution

3.3 Mesh Generation and Vectorization Optimization

ROMS parallel needs to define NtileI and NtileJ. For MPI programs, Cpu process number = NtileI * NtileJ is needed. Figure 4 [17] shows the communication and exchange of data by taking 9 MPI processes of 3 * 3 as an example. Data exchange first occurs in the east–west direction, followed by the north–south direction. The gray imaginary points are used for data exchange. Data communication mainly occurs in the virtual points of the boundary. The updated input is shown in Fig.4 c) and d). As shown in Fig. 4 a) and b), each tile is a discrete computing unit in the MPI process, OpenMP thread, or serial operation. The title contains enough information to calculate the time step of all internal points, so the number of ghost points is determined by the footsteps using the maximum number of adjacent points algorithm. In ROMS, generally speaking, the overlapping area is two grid points wide, unless the MPDATA advection scheme is used; in this case, three grid points are needed.

Because of the column priority feature of the Fortran language, with the same total number of cores, the allocation of larger values in the J direction is conducive to the vectorization of patterns and can achieve higher performance.

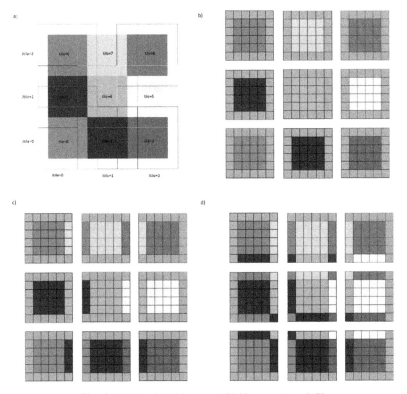

Fig. 4. Tiled grid with some ROMS parameters [17]

3.4 Parallel I/O Optimization

ROMS 4.0 supports the use of the PIO library and SCORPIO library promotion to improve the I/O performance of the model in clusters with parallel file systems (such as GPFS and Lustre). ROMS, a derivative of ROMS_Argif and CROCO support XIOS to improve parallel I/O performance. Because PIO is universal enough to be used in a wider range of applications, this paper takes the Parallel I/O (PIO) library as an example to improve the I/O capability of the model.

PIO development focuses on the use of back-end tools in the netCDF file format. PIO currently supports netCDF and PnetCDF as back-end libraries, both of which can be linked and used with runtime option control for a given file. PIO supports the classic version of netCDF and its parallel library PnetCDF, netCDF4 (HDF5), which subsequently supports parallel reading and writing. The installation of the PIO parallel version requires the installation of zlib, szip, hdf5, netCDF C, netCDF Fortran and other libraries. Fortran support needs to be enabled during the compilation process.

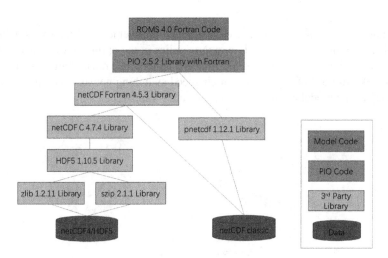

Fig. 5. ROMS, PIO architecture diagram and version

Figure 5 shows the PIO architecture for the ROMS model used in this paper. This article uses the PIO library and adopts the synchronous I/O mode, that is, all CPU cores perform I/O and calculation, and specify the total number of I/O and the distribution in the cluster nodes. The implementation of asynchronous I/O mode is difficult and requires comprehensive calculation and I/O resource allocation.

4   Result

In the actual sea area example, taking 140 CPU cores commonly used in commercialization as the benchmark experiment, the efficiency of the vectorization method was improved by 8.64%, and the efficiency of the IB switching network was improved by 204.45% through the optimization of vectorization, IB switching network, and parallel IO. The operation efficiency could be increased by 271.08% by the comprehensive use of vectorization, IB switching network, and parallel IO, as shown in Table 3.

It can be seen from Table 3 that the use of the IB switching network has the most significant effect on the efficiency of model operation.

Figure 6 shows the radar chart of data extracted by Paramon and analysed by Paratune. It can be seen from Fig. 6a) b) comparison that the mode can make full use of the IB

Table 3. Model parallel run time

Test Case	I	J	time (s)	Efficiency improvement
Baseline	28	5	605.917	
Vector	5	28	557.748	8.64%
IB	28	5	199.02	204.45%
IB&Vettor	5	28	166.218	264.53%
IB&Vettor&PIO	5	28	163.284	271.08%

switching network, and Fig. 6b) c) can obtain the improvement of vectorization, but the effect is not obvious. Figure 6c) d) comparison lies in the optimization of parallel I/O, which cannot significantly improve the analysis index in the radar chart. The optimized IB can make full use of the cluster's high-speed switching network, and the operation efficiency of the model has also been significantly improved.

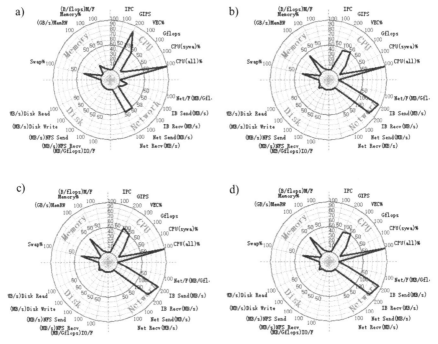

Fig. 6 Parallel feature analysis a) benchmark experiment b) IB optimization c) vectorization + IB optimization d) vectorization + IB + PIO optimization

5 Conclusion

Because of the huge amount of data and computation, the ROMS coupled marine eco-logical model needs to optimize the parallel computation to improve the computational efficiency. To meet the timeliness requirements of ecological operational forecasting systems, this paper proposes several parallel optimization methods to improve the efficiency of the model. According to the characteristics of the ROMS model, from the benchmark test and the actual sea area example, the analysis mode has a good acceleration ratio by optimizing the aspects of quantization, high-speed switching network, parallel I/O, and expanding to 3920 cores in the strong expansion experiment; In the actual business example, with the same computing resources, the operation efficiency is increased by 270%, and the actual business operation time is reduced to 27% of the original operation time. The relevant optimization methods provide reference for the optimization of other atmospheric and ocean models.

Acknowledgment. This research is supported by the National Natural Science Foundation of China (41976200), the project of Guangdong Ocean University (060302032106) and Southern Marine Science and Engineering Guangdong Laboratory (Zhuhai) (SML2022SP301). We acknowledge the comments of anonymous reviewers.

References

1. Fennel. K., Wilkin, J., Previdi, M., Najjar, R.: Denitrification effects on air-sea CO2 flux in the coastal ocean: Simulations for the northwest North Atlantic. Geophys. Res. Lett. **35**, L24608 (2008)
2. Fennel, K., Wilkin, J.L., Levin, J., Moisan J., O Reilly, J.E., Haidvogel, D.B.: Nitrogen cycling in the Middle Atlantic Bight: Results from a three-dimensional model and implications for the North Atlantic nitrogen budget. Global Biogeochem. Cycles **20**, B3007 (2006)
3. Liu G., Chai F.: Seasonal and interannual variation of physical and biological processes during 1994–2001 in the Sea of Japan/East Sea: a three-dimensional physical–biogeochemical modeling study. J. Marine Syst. **78**, 265 (2009)
4. Chai F., Dugdale R.C., Peng T.H., Wilkerson F.P., Barber R.T.: One-dimensional ecosystem model of the equatorial Pacific upwelling system. Part I: model development and silicon and nitrogen cycle. Deep Sea Res. Part II: Topical Stud. Oceanography **49**, 2713 (2002)
5. Kishi M.J.: NEMURO—a lower trophic level model for the North Pacific marine ecosystem. Ecological Model. **202**, 12 (2007)
6. Kishi, M.J., Ito, S., Megrey, B.A., Rose, K.A., Werner, F.E.: A review of the NEMURO and NEMURO.FISH models and their application to marine ecosystem investigations. J. Oceanography **67**, 3 (2011)
7. Moll, A., Radach, G.: Review of three-dimensional ecological modelling related to the North Sea shelf system: Part 1: models and their results. Progress Oceanography **57**, 175 (2003)
8. Edwards, K.P., Barciela, R., Butenschön, M.: Validation of the NEMO-ERSEM operational ecosystem model for the North West European Continental Shelf. Ocean Sci. **8**, 983 (2012)
9. Liu, T., et al.: Parallel implementation and optimization of regional ocean modeling system (ROMS) based on sunway SW26010 many-core processor. IEEE Access, 1 (2019)
10. Shchepetkin, A.F., McWilliams, J.C.: The regional oceanic modeling system (ROMS): a split-explicit, free-surface, topography-following-coordinate oceanic model. Ocean Model. **9**, 347 (2005)

11. Budgell, W.P.: Numerical simulation of ice-ocean variability in the Barents Sea region. Ocean Dynam. **55**, 370 (2005)
12. Moore, A.M., Arango, H.G., Di Lorenzo, E., Cornuelle, B.D., Miller, A.J., Neilson, D.J.: A comprehensive ocean prediction and analysis system based on the tangent linear and adjoint of a regional ocean model. Ocean Model. 7, **227** (2004)
13. Haidvogel D.B., et al.: Ocean forecasting in terrain-following coordinates: formulation and skill assessment of the Regional Ocean Modeling System. J. Comput. Phys. **227**, 3595 (2008)
14. Large, W.G., McWilliams, J.C., Doney, S.C.: Oceanic vertical mixing: a review and a model with a nonlocal boundary layer parameterization. Rev. Geophy. **32**, 363 (1994)
15. Fasham, M.J.R., Ducklow, H.W., Mckelvie, S.M.: A nitrogen-based model of plankton dynamics in the oceanic mixed layer. J. Marine Res. **48**, 591 (1990)
16. Egbert, G.D., Erofeeva, S.Y.: efficient inverse modeling of barotropic ocean tides. J. Atmospheric Oceanic Technol. **19**, 183 (2002)
17. Wang Y., Zhang T., Yin Z., Hao S., Wang C., Lin B.: Data analyses and parallel optimization of the tropical-cyclone coupled numerical model. In: Wang Y., Zhu G., Han Q., Wang H., Song X., Lu Z. (eds) ICPCSEE 2022. Springer Nature Singapore, Singapore (2022). https://doi.org/10.1007/978-981-19-5194-7_2

Comparison of Two Grey Models' Applicability to the Prediction of Passenger Flow in Sanya Airport

Yinan Chen[✉] and Yuanhui Li

Sanya Aviation and Tourism College, Sanya 572000, Hainan, China
305780161@qq.com

Abstract. The city airport serves as a fundamental infrastructure for air transportation, with its development planning largely contingent upon the prognostication of future airport activities' busyness. Given the multifarious factors that influence passenger flow, such as population size, economic structure, industrial policy, geographical location, and comprehensive transportation, grey system has the characteristic of incomplete information, and the passenger traffic at the airport conforms to this feature. In this article, the GM(1,1) grey prediction model and the GM(2,1) grey prediction model are respectively applied to predict the passenger flow of Sanya Airport, and the suitability of each model is compared. The outcomes show that the GM(2,1) grey model outperforms the GM(1,1) grey model in relation to the average relative error rate, the single point maximum error, the mean square error of relative error, and the average relative accuracy.

Keywords: airport · Passenger flow · Grey models · predict

1 Introduction

Sanya Airport officially opened to traffic in 1994 and was built in three phases. The construction of the airport further improved the infrastructure construction of Sanya City, provided more comfortable travel experience for the majority of Hainan people, and contributed to the construction of Hainan Free Trade Port. Studying and observing the change law of airport passenger flow and making scientific and reasonable predictions about its development trend holds immense significance for the future development planning of the airport. It is more conducive to the planning and development of Sanya Airport to choose a forecasting model with better effect and improve the accuracy of the airport throughput prediction [1].

Scholars have used various models to predict passenger traffic. Zhu Jiayue (2019) selected the passenger throughput of Chengdu Shuangliu Airport as the data. First, ARIMA model and grey prediction model were established and predicted. Then, the weighting method is used to combine the two models and form the gray-ARIMA combined prediction model. Finally, a comparison is made between the predicted results of the three models and the actual values. According to the findings, the combined model

Z. Yu et al. (Eds.): ICPCSEE 2023, CCIS 1880, pp. 225–236, 2023.
https://doi.org/10.1007/978-981-99-5971-6_17

exhibits superior forecasting efficacy as compared to the other two models, as indicated by its lower average absolute error [6]; As noted by Cheng Maolin (2022), the prevailing models used for grey system prediction encompass GM(1,1), GM(1,N), among others. It has been observed that the whitening equations for these models are characterized by single ordinary differential equations, which may not always adequately capture the intricate interconnections between variables. Attempts were made to establish a simultaneous grey model, and it was found that the extended simultaneous grey model had higher accuracy [12]; Qiu Hui (2019) used the GM(1,1) grey prediction model to estimate the airport passenger throughput. The prediction model was reliable and highly accurate [8]; Li Zhichao (2020) made improvements to the traditional GM(1,1) grey model. Firstly, an optimal dimension GM(1,1) model was established. However, the results showed that the optimal dimension GM(1,1) grey model did not enhance the precision of the forecast. Secondly, a interval GM(1,1) grey model was established. The concept of normalization and linear programming was utilized when assigning weights to the upper and lower bound sequences within the interval. The findings indicated that the interval GM(1,1) grey model can significantly improve the forecast precision and is more suitable for predicting oscillating sequences with small samples [19]; Li Jie (2022) used the GM(1,1) grey model to forecast the port data transfer rate demand of Wuhu Port in the next five years. After the residual test, the overall accuracy was high, which confirmed that the GM(1,1) grey model had good forecast accuracy and the predicted value was relatively reliable. This paper provides some reasonable suggestions for the future development planning of Wuhu Port [5]; Wang Qi (2023) selected six indicators closely related to the airport cargo throughput based on relevant studies, then determined the weight of each index affecting the air logistics throughput by entropy method, and finally predicted the airport cargo throughput by combining GA-BP neural network model [9].

Although there are various models for predicting passenger traffic, long-term and continuous observation data are generally required to obtain better prediction results [2]. For short data sequences, due to the lack of information and weak regularity, most models cannot be used effectively. In this case, the grey prediction model has significant advantages [11].

Gray prediction theory, introduced by the esteemed Chinese scholar Deng Julong, is an invaluable system prediction theory that operates under conditions of uncertainty [4]. This theory utilizes the generation and modeling of established sequences in order to extract crucial information, facilitating a precise and efficient understanding of the system's operational and evolutionary principles [20]. Among the most important models is GM(1,1), which Professor Deng Julong constructed based on the incomplete information conditions of data background and developmental laws. This model adeptly leverages partially known information to explore the laws of the system within uncertain systems [13, 15].The gray prediction model includes the univariate GM(N,1) grey model and the multivariate GM(1,N) grey model. For the GM(N,1) grey model, as N increases, the computational complexity also increases; however, there is no guarantee that the accuracy will improve [3]. Therefore, N is generally taken to be below 3. The multivariate GM(1,N) model can introduce other factors closely related to the research object, increasing the white information of the prediction system [7, 11].

This article is based on the study of the GM(1,1) grey prediction model and the GM(2,1) grey prediction model, and takes the prediction process of passenger traffic at Sanya Airport as a case study, predicting the passenger traffic for 2018 and 2019. The prediction results and accuracy of the two grey models are compared, and the applicability of each model in prediction is preliminarily analysed based on their characteristics. Considering the severe impact of the epidemic on the civil aviation industry in 2020, 2021, and 2022, data for these three years are not predicted.

2 The Procedural Steps for Constructing the Fundamental Structure of the GM(1,1) Grey Model

2.1 Assessment of Modelling Feasibility

For a given data series $x^{(0)} = \left(x^{(0)}(1), x^{(0)}(2), \cdots, x^{(0)}(n)\right)$, calculate its ratio-to-trend [20]

$$\lambda(k) = \frac{x^{(0)}(k-1)}{x^{(0)}(k)}, k = 2, 3, \cdots, n \tag{1}$$

Provided that the entirety of the ratio-to-trend values $\lambda(k)$ falls within the acceptable coverage $\Theta = \left(e^{-\frac{2}{n+1}}, e^{\frac{2}{n+1}}\right)$, the sequence $x^{(0)}$ can be effectively utilized as data for the purpose of grey prediction through application of the GM(1,1) model.

2.2 Generate Cumulative Sequence $x^{(1)}$

For the original data $x^{(0)} = \left(x^{(0)}(1), x^{(0)}(2), \cdots, x^{(0)}(n)\right)$, perform a first-order cumulative operation, and the result

$$x^{(1)} = \left(x^{(1)}(1), x^{(1)}(2), \cdots, x^{(1)}(n)\right) \tag{2}$$

is obtained after the cumulative operation, which $x^{(1)}(k) = \sum_{t=1}^{k} x^{(0)}(t), k = 1, 2, \cdots, n$

Upon cumulative operation, the volatility and randomness of the raw data are substantially attenuated, thereby effectuating a transformation into a more discernible, increasingly ordered sequence [10, 20]. This preparatory step is integral to establishing a differential equation-based prediction model and serves to streamline the application process [8].

2.3 Establishing the GM(1,1) Grey Model

Drawing upon the sequence $x^{(1)}$ that has been generated, a differential equation model

$$\frac{dx^{(1)}(t)}{dt} + ax^{(1)}(t) = b \tag{3}$$

is established, where in the unknown parameters a and b are determined. The model is a first-order differential equation with one variable, which is commonly denoted GM(1,1) [20].

Denoted as

$$u = \left[a, b\right]^T, \; Y = \left[x^{(0)}(2), x^{(0)}(3), \cdots, x^{(0)}(n)\right]^T,$$

$$B = \begin{bmatrix} -\frac{1}{2}\left(x^{(1)}(1) + x^{(1)}(2)\right) & 1 \\ -\frac{1}{2}\left(x^{(1)}(2) + x^{(1)}(3)\right) & 1 \\ \vdots & \vdots \\ -\frac{1}{2}\left(x^{(1)}(n-1) + x^{(1)}(n)\right) & 1 \end{bmatrix} \tag{4}$$

In accordance with the principles of the least squares method, estimates of the u that minimizes the value of $J(u) = (Y - Bu)^T(Y - Bu)$ is ascertained.

$$\hat{u} = [\hat{a}, \hat{b}]^T = \left(B^T B\right)^{-1} B^T Y \tag{5}$$

Solving the equation yields.

$$\hat{x}^{(1)}(t) = \left(x^{(0)}(1) - \frac{\hat{b}}{\hat{a}}\right) e^{-\hat{a}t} + \frac{\hat{b}}{\hat{a}} \tag{6}$$

The predicted value can be obtained

$$\hat{x}^{(1)}(k+1) = \left(x^{(0)}(1) - \frac{\hat{b}}{\hat{a}}\right) e^{-\hat{a}t} + \frac{\hat{b}}{\hat{a}}, \; k = 0, 1, 2, \cdots, n \tag{7}$$

Do cumulative subtraction and reduction operations, $\hat{x}^{(0)}$ can be obtained based on

$$\hat{x}^{(0)}(k+1) = \hat{x}^{(1)}(k+1) - \hat{x}^{(1)}(k), \; k = 1, 2, \cdots, n \tag{8}$$

2.4 Test Model

The residual test is one among several commonly employed techniques, including the posterior difference test, ratio deviation test, and correlation test, that are utilized for model validation [14]. However, in the present study, the residual test was selected as the method of choice. The value of

$$\delta(k) = \frac{\left|x^{(0)}(k) - \hat{x}^{(0)}(k)\right|}{x^{(0)}(k)}, \; k = 1, 2, \cdots, n \tag{9}$$

is calculated accordingly. Here $\hat{x}^{(0)}(1) = x^{(0)}(1)$

On the basis of the model validation results, it is inferred that $\delta(k) < 0.2$ satisfies the general criterion for model accuracy, whereas $\delta(k) < 0.1$ fulfils the more stringent criterion for model accuracy [16].

3 Establishment Steps of the Second-Order Univariate Grey Model GM(2,1)

3.1 Generate Sequence

For the original data $x^{(0)} = \left(x^{(0)}(1), x^{(0)}(2), \cdots, x^{(0)}(n)\right)$, perform an addition operation, generate a 1-AGO cumulative sequence

$$x^{(1)} = \left(x^{(1)}(1), x^{(1)}(2), \cdots, x^{(1)}(n)\right) \tag{10}$$

which $x^{(1)}(k) = \sum_{t=1}^{k} x^{(0)}(t), k = 1, 2, \cdots, n$

Then, perform a subtraction operation on $x^{(0)}$ and generate a 1-IAGO cumulative subtraction sequence

$$\alpha^{(1)}x^{(0)} = \left(\alpha^{(1)}x^{(0)}(2), \alpha^{(1)}x^{(0)}(3), \cdots, \alpha^{(1)}x^{(0)}(n)\right) \tag{11}$$

which $\alpha^{(1)}x^{(0)}(k) = x^{(0)}(k) - x^{(0)}(k-1), k = 2, 3, \cdots, n$
Mean sequence of $x^{(1)}$

$$z^{(1)} = \left(z^{(1)}(2), z^{(1)}(3), \cdots, z^{(1)}(n)\right) \tag{12}$$

which $z^{(1)}(k) = \left[x^{(1)}(k) + x^{(1)}(k-1)\right]/2, k = 2, 3, \cdots, n$

3.2 Establishing the Model

Establish a GM(2,1) model for the generated sequence

$$\alpha^{(1)}x^{(0)}(k) + a_1 x^{(0)}(k) + a_1 z^{(1)}(k) = b, k = 2, \cdots, n \tag{13}$$

Its whitening equation is

$$\frac{d^2 x^{(1)}(t)}{dt^2} + a_1 \frac{dx^{(1)}(t)}{dt} + a_2 x^{(1)}(t) = b \tag{14}$$

Denoted as

$$B = \begin{bmatrix} -x^{(0)}(2) & -z^{(1)}(2) & 1 \\ \vdots & \vdots & \vdots \\ -x^{(0)}(n) & -z^{(1)}(n) & 1 \end{bmatrix},$$

$$Y = \begin{bmatrix} \alpha^{(1)}x^{(0)}(2) \\ \alpha^{(1)}x^{(0)}(3) \\ \vdots \\ \alpha^{(1)}x^{(0)}(n) \end{bmatrix} = \begin{bmatrix} x^{(0)}(2) - x^{(0)}(1) \\ x^{(0)}(3) - x^{(0)}(2) \\ \vdots \\ x^{(0)}(n) - x^{(0)}(n-1) \end{bmatrix} \tag{15}$$

The estimate for the parameter sequence $u = [a_1, a_2, b]^T$ of the GM(2,1) model, obtained through the least squares method, is equivalent to the value of

$$\hat{u} = \left(B^T B\right)^{-1} B^T Y \tag{16}$$

By substituting the value of $\hat{u} = \left(\hat{a}_1, \hat{a}_2, b\right)^T$ and the boundary conditions $x^{(1)}(1)$ and $x^{(1)}(n)$ into the whitening equation of the GM(2,1) model, $x^{(1)}(t)$ 及 $\hat{x}^{(1)}$ can be obtained. Performing IAGO restoration, $\hat{x}^{(0)}$ can be obtained based on

$$\hat{x}^{(0)}(k + 1) = \hat{x}^{(1)}(k + 1) - \hat{x}^{(1)}(k), k = 1, 2, \cdots, n \tag{17}$$

3.3 Test Model

The residual test is one among several commonly employed techniques, including the posterior difference test, ratio deviation test, and correlation test, that are utilized for model validation. However, in the present study, the residual test was selected as the method of choice. The value of

$$\delta(k) = \frac{\left| x^{(0)}(k) - \hat{x}^{(0)}(k) \right|}{x^{(0)}(k)}, k = 1, 2, \cdots, n \tag{18}$$

is calculated accordingly. Here $\hat{x}^{(0)}(1) = x^{(0)}(1)$

On the basis of the model validation results, it is inferred that $\delta(k) < 0.2$ satisfies the general criterion for model accuracy, whereas $\delta(k) < 0.1$ fulfils the more stringent criterion for model accuracy [16].

4 Application Case

4.1 Historical Data

As per the data presented in Table 1, which have been sourced from the Statistical Yearbook available on the official website of Sanya Bureau of Statistics (http://tjj.sanya.gov.cn/tjjsite/), the passenger traffic of Sanya Airport has been recorded for the years spanning from 2008 to 2019 [21].

For training purposes, the passenger flow data from the years 2008 to 2017 will be selected, and for testing purposes, the passenger flow data from the years 2018 to 2019 will be chosen [21].

Recording the original data

$$x^{(0)} = (600.62, 794.14, 929.39, 1036.18, 1134.32, 1286.68, 1494.23, 1619.20,$$
$$1736.11, 1938.99)$$

Table 1. The passenger traffic volume measured in ten thousand passengers of Sanya Airport from 2008 to 2019.

Year	Passenger Flow	Year	Passenger Flow
2008	600.62	2014	1494.23
2009	794.14	2015	1619.20
2010	929.39	2016	1736.11
2011	1036.18	2017	1938.99
2012	1134.32	2018	2003.90
2013	1286.68	2019	2016.37

4.2 The Utilization of the GM(1,1) Model for Predicting Passenger Traffic Volume is Being Implemented

In accordance with Eq. (1), the range of the ratio of levels is (0.8563,0.9827), which is entirely encompassed within the range covered by

$$\Theta = \left(e^{-\frac{2}{11+1}}, e^{\frac{2}{11+1}} \right) = (0.8338, 1.1994)$$

This indicates the feasibility of constructing a GM(1,1) model.

Calculate $\hat{a} = -0.1068$, $\hat{b} = 727.6927$ from Eq. (5) and substitute them into Eq. (7) to obtain the GM(1,1) model.

$$\hat{x}(k+1) = 7413.2205e^{0.1068k} - 6812.6005, k = 0, 1, 2, \cdots, 10$$

Therefore,

$$\hat{x}^{(1)} = (600.62, 1436.31, 2366.20, 3400.92, 4552.28, 5833.43, 7259.01, 8845.29,$$
$$10610.39, 12574.47)$$

Performing cumulative reduction restoration can obtain

$$\hat{x}^{(0)} = (600.62, 835.69, 929.89, 1034.72, 1151.36, 1281.15, 1425.58,$$
$$1586.28, 1765.10, 1964.08)$$

Table 2 reveals that the relative error tests conducted from 2008 to 2017 all produced results that are less than 0.10, thus signifying an exceptional level of accuracy for the model. Based on the predicted average relative error for the passenger traffic of Sanya Airport spanning the years 2018 to 2019, which is measured at 0.148, it can be deduced that the level of prediction accuracy attained is of a sufficiently qualified standard.

4.3 Application of the GM(2,1) Model to Predict Passenger Traffic Volume

The first-order accumulated sequence of series $x^{(0)}$ is

$$x^{(1)} = (600.62, 1394.76, 2324.15, 3360.33, 4494.65, 5781.33,$$
$$7275.56, 8894.76, 10630.87, 12569.86)$$

Table 2. A comparison between the actual values and the GM(1,1) Grey predicted values spanning the years 2008 to 2019 is being conducted.

Year	Actual Value	Predicted Value	Residual Error	Relative Error
2008	600.62	600.62	0.00	0.00%
2009	794.14	835.69	−41.55	5.23%
2010	929.39	929.89	−0.50	0.05%
2011	1036.18	1034.72	1.46	0.14%
2012	1134.32	1151.36	−17.04	1.50%
2013	1286.68	1281.15	5.53	0.43%
2014	1494.23	1425.58	68.65	4.59%
2015	1619.2	1586.28	32.92	2.03%
2016	1736.11	1765.10	−28.99	1.67%
2017	1938.99	1964.08	−25.09	1.29%
2018	2003.90	2185.49	−181.59	9.06%
2019	2016.37	2431.86	−415.49	20.61%

The sequence 1-IAGO is
$$\alpha^{(1)}x^{(0)} = (193.52, 135.25, 106.79, 98.14, 152.36, 207.55, 124.97, 116.91, 202.88)$$
The mean-generated sequence of mean-generated sequence of $x^{(1)}$

$$z^{(1)} = (997.69, 1859.455, 2842.24, 3927.49, 5137.99, 6528.445,$$
$$8085.16, 9762.815, 11600.365)$$

Denoted as

$$B = \begin{bmatrix} -x^{(0)}(2) & -z^{(1)}(2) & 1 \\ \vdots & \vdots & \vdots \\ -x^{(0)}(n) & -z^{(1)}(n) & 1 \end{bmatrix} = \begin{bmatrix} -794.14 & -997.69 & 1 \\ \vdots & \vdots & \vdots \\ -1938.99 & -11600.365 & 1 \end{bmatrix},$$

$$Y = \begin{bmatrix} \alpha^{(1)}x^{(0)}(2) \\ \alpha^{(1)}x^{(0)}(3) \\ \vdots \\ \alpha^{(1)}x^{(0)}(n) \end{bmatrix} = \begin{bmatrix} 193.52 \\ 135.25 \\ \vdots \\ 202.88 \end{bmatrix},$$

Therefore,

$$\hat{u} = \begin{pmatrix} \hat{a}_1 \\ \hat{a}_2 \\ b \end{pmatrix} = (B^T B)^{-1} B^T Y = \begin{pmatrix} -0.1740 \\ 0.0164 \\ 9.7943 \end{pmatrix}.$$

The whitening model of GM(2,1) can be obtained as

$$\frac{d^2 x^{(1)}(t)}{dt^2} - 0.1740 \frac{dx^{(1)}(t)}{dt} + 0.0164 x^{(1)}(t) = 9.7943$$

The time response of the whitening equation is solved through the utilization of the boundary conditions $x^{(1)}(1) = 600.62$, $x^{(1)}(n) = 12569.86$

$$x^{(1)}(t) = [7302.6583 \sin(0.0940t) + 3.9538 \cos(0.0940t)]*e^{0.0870t} + 596.6662$$

Therefore, the time response of GM(2,1) is obtained

$$x^{(1)}(k+1) = [7302.6583 \sin(0.0940t) + 3.9538 \cos(0.0940t)]*e^{0.0870t} + 596.6662$$

Therefore,

$$\hat{x}^{(1)} = (600.62, 1348.99, 2226.15, 3240.90, 4401.20, 5713.96, 7184.66, 8816.98, 10612.43, 12569.86)$$

By performing IAGO reduction

$$\hat{x}^{(0)} = (600.62, 748.37, 877.17, 1014.74, 1160.30, 1312.76, 1470.70, 1632.32, 1795.45, 1957.43) \text{ can be obtained.}$$

Table 3 shows that the relative error tests conducted from 2008 to 2017 all produced results that are less than 0.10, thus signifying an exceptional level of accuracy for the model. Moreover, the predicted average relative error for the passenger traffic of Sanya Airport spanning the years 2018 to 2019 is measured at 0.089, which serves as a testament to the excellent level of prediction accuracy achieved.

Table 3. A comparison between the actual values and the GM(2,1) Grey predicted values spanning the years 2008 to 2019 is being conducted.

Year	Actual Value	Predicted Value	Residual Error	Relative Error
2008	600.62	600.62	0.00	0.00%
2009	794.14	748.37	45.77	5.76%
2010	929.39	877.17	52.22	5.62%
2011	1036.18	1014.74	21.44	2.07%
2012	1134.32	1160.30	−25.98	2.29%
2013	1286.68	1312.76	−26.08	2.03%
2014	1494.23	1470.70	23.53	1.57%
2015	1619.2	1632.32	−13.12	0.81%
2016	1736.11	1795.45	−59.34	3.42%
2017	1938.99	1957.43	−18.44	0.95%
2018	2003.90	2115.16	−111.26	5.55%
2019	2016.37	2265.01	−248.64	12.33%

5 Comparison of Forecast Results

It can be seen from Fig. 1 that both types of grey models can fit the actual data very well.

Table 4 provides evidence that the employment of GM(1,1) grey model and GM(2,1) grey model for prediction purposes has resulted in a reduction in the maximum error rate of a single point, from 20.61% to 12.33%, as well as a decrease in the average relative error rate, from 3.88% to 3.53%. Additionally, there has been an improvement in the average relative accuracy, which has increased from 96.12% to 96.47%. These findings suggest that the prediction accuracy has been enhanced.

Simultaneously, it can be observed that the mean square error of the average relative error has undergone a decline from 2.92% to 0.96%, thus indicating that the latter has exhibited greater stability.

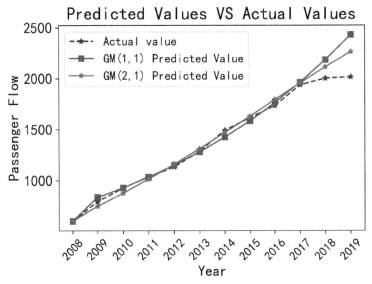

Fig. 1. An examination is undertaken to compare the actual and predicted passenger traffic at Sanya Airport over the period of 2008 to 2019. Unit: ten thousand people

Table 4. An analysis is carried out to draw a comparison between the two models.

Prediction Model	Single Point Maximum Error	Average Relative Error	Average Relative Accuracy	The Mean Square Error of The Mean Relative Error
GM(1,1)	20.61%	3.88%	96.12%	2.92%
GM(2,1)	12.33%	3.53%	96.47%	0.96%

Upon comprehensive evaluation and comparison of the predicted data and actual data spanning from 2008 to 2019 [21], it is evident that the performance indicators of the GM(2,1) grey model are superior to those of the GM(1,1) grey model.

6 Conclusion

Upon the utilization of two distinct grey models to predict and scrutinize the errors in passenger traffic at Sanya Airport, it was discerned that the implementation of GM(1,1) grey model and GM(2,1) grey model for short-term airport passenger traffic prediction yields predictions that are relatively proximate to the actual data. In the course of development of any grey system [17], the system is subjected to a continuous influx of random disturbances or driving factors over time, thereby exerting an influence on the system's development. Hence, the GM(1,1) grey prediction model is only appropriate for making short and medium-term forecasts [20].

Forecasts that extend beyond a duration of 10 years are prone to manifest substantial deviations, thereby necessitating modifications and optimizations of the prediction model. In the two models explored in this paper, relatively speaking, the GM(2,1) grey model has a higher forecast accuracy. Therefore, the GM(2,1) grey model should be considered for mid-to-long-term forecasting. [11].

The present study, owing to spatial constraints, has refrained from delving into the ways in which passenger traffic prediction at Sanya Airport can bolster the supply capacity and service level of the management department [20], facilitate the attainment of equilibrium between supply and demand, and optimize the service level of Hainan Free Trade Port, all of which embody the ultimate objective of passenger traffic prediction. These aspects, however, can be expounded upon in subsequent research endeavors [18].

Acknowledgments. Project supported by the Education Department of Hainan Province, project number: Hnky2022ZD-25.

References

1. Chen, S., Wei, W.: Grey neural network forecasting for traffic flow. J. Southeast Univ. (Nat. Sci. Ed.) **34**(04), 541–544 (2004)
2. Lyu, X., Yin, K., Li, X.: Grey multivariate variable weight combination prediction model and its application. Stat. Decis. **38**(14), 25–29 (2022)
3. Ye, L., Dang, Y., Wang, J.: Impact effect-based grey multivariable time delay model and its application. Syst. Eng.-Theory Pract., 1–25 (2023)
4. Duan, H., He, C., Wang, S.: Grey differential dynamic multivariate forecasting model and its application. Syst. Eng.-Theory Pract. **42**(05), 1402–1412 (2022)
5. Li, J., Ning, S.: Prediction and analysis of container throughput of Wuhu Port based on grey prediction model. China Storage Transp. **263**(08), 95–96 (2022)
6. Zhu, J., Fang, Y., Li, X.: Prediction research on passenger throughput of Chengdu Shuangliu airport. Technol. Econ. Areas Commun. **21**(01), 48–51+69 (2019)
7. Li, H., Zeng, B., Zhou, W.: Forcasting domestic waste clearing and transporting volume by employing a new grey parameter combination optimization model. Chin. J. Manag. Sci. **30**(04), 96–107 (2022)

8. Qiu, H., Xie, R., Liu, E.: Research on passenger throughput prediction of Yuncheng Airport based on grey theory. J. Yuncheng Univ. **33**(04), 33–36 (2015)

9. Wang, Q., Zhang, H.: Cargo throughput prediction of Nanjing Lukou international airport based on combined prediction model. Logist. Eng. Manag. **45**(02), 83–86 (2023)

10. Hou, R., Xu, X.: Population prediction based on improved multidimensional grey model and support vector machine. Stat. Decis. **37**(18), 41–44 (2021)

11. Peng, A., Zang, Z., Yang, W.: Preliminary study on applicability of different grey models for civil aviation passenger forecast. J. Civil Aviat. Flight Univ. China **27**(03), 62–65 (2015)

12. Cheng, M., Liu, B.: Parameter optimization method and model application of extended simultaneous grey model. Stat. Decis. **38**(21), 41–47 (2022)

13. Liu, S., Zeng, B., Liu, J.: Several basic models of GM(1,1) and their applicable bound. Syst. Eng. Electron. **36**(03), 501–508 (2014)

14. Cheng, M., Han, Y.: Grey model GM(1,1) based on variable generation coefficient and its applications. Stat. Decis. **36**(04), 15–18 (2020)

15. Deng, J.: Basic Methods of Grey Systems, 2nd edn., Chinese-English Version. Huazhong University of Science and Technology Press, Wuhan (2005)

16. Si, S., Sun, X.: Mathematical Experimentation and Modelling with Python. Science Press, Beijing (2020)

17. Xie, M., Wu, L.: Short-term traffic flow prediction based on GM(1, N) power model optimized by rough set algorithm. Math. Pract. Theory **51**(09), 241–249 (2021)

18. Tian, M., Zhou, F.: Study on tourism development and economic growth based on grey relational analysis: take Shandong Province as an example. J. Chongqing Univ. Technol. (Nat. Sci.) **33**(02), 208–215 (2019)

19. Li, Z., Liu, S.: Prediction of oscillation sequence based on grey interval GM (1,1) model – Take Shanghai Consumer confidence Index as an example. Stat. Decis. **36**(14), 145–148 (2020)

20. Li, Y., Han, H.: Forecast and analysis of passenger flow at sanya airport based on gray system theory. In: Wang, Y., Zhu, G., Han, Q., Zhang, L., Song, X., Zeguang, L. (eds.) Data Science: 8th International Conference of Pioneering Computer Scientists, Engineers and Educators, ICPCSEE 2022, Chengdu, China, August 19–22, 2022, Proceedings, Part II, pp. 427–434. Springer Nature Singapore, Singapore (2022). https://doi.org/10.1007/978-981-19-5209-8_29

21. Li, Y., Zhao, W., Han, H.: Comparison of ARIMA model and GM(1,1) model in passenger flow prediction of Sanya airport. In: 2022 International Conference on Cyber-Enabled Distributed Computing and Knowledge Discovery (2022)

Research on High Precision Autonomous Navigation of Shared Balancing Vehicles Based on EKF-SLAM

Xinyu Cheng, Yanqing Wang$^{(\boxtimes)}$, Dingdong Guo, Xuewei Li, and Yiming Gao

Nanjing Xiaozhuang University, NanJing 211171, China
wyq0325@126.com

Abstract. In view of the technical difficulties of autonomous navigation in local areas, this paper proposes a high-precision autonomous navigation shared balancing bike system based on EKF-SLAM. This system uses the EKF-SLAM algorithm in robot localization to achieve simultaneous localization and map construction using the extended Kalman filter. At the same time, GPS and IMU are also employed for absolute positioning, and point cloud matching is used for relative positioning to achieve multisensor fusion positioning. For the convenience of users, this system uses the RNN-T model for speech recognition destinations. Through experimental verification, the EKF-SLAM-based autonomous navigation technology proposed in this paper can meet the accurate localization service and can realize the function of high precision autonomous navigation and voice recognition of destinations for shared balancing vehicles in a local area.

Keywords: Shared balancing vehicle · EKF-SLAM · Autonomous navigation · GPS · IMU

1 Introduction

The development of the sharing economy can meet the growing sociocultural needs of the population while effectively mobilizing social resources and providing convenience for residents. The sharing economy contributes to a low-carbon economic structure and sustainable social development [1]. From the perspective of transportation, shared transport can be a good solution to traffic congestion and lack of parking spaces. From the perspective of sustainable development, shared transport can be safer and more environmentally friendly than traditional transport, enabling greener travel. At the same time, due to the rapid development of driverless technology, this paper proposes a shared balance vehicle with autonomous navigation in local areas. The specific positioning and navigation system and the algorithms used will vary according to the needs and environment of the application. Location navigation in places such as parks and airports relies heavily on GPS or other geolocation services such as WiFi, BLE, and NFC. to provide navigation, geolocation systems typically employ a range of algorithms, such as triangulation, Kalman filtering, and SLAM; however, it is important to note that each of these technologies has its own limitations. For example, the signals of WiFi and

Z. Yu et al. (Eds.): ICPCSEE 2023, CCIS 1880, pp. 237–251, 2023.
https://doi.org/10.1007/978-981-99-5971-6_18

BLE can be blocked by walls or other objects, while NFC has a very short effective range. Therefore, a combination of these positioning systems and algorithms can be used in practical applications to provide optimal positioning and navigation services [2]. In this paper, we mainly combine existing technologies to solve the problem of high-precision autonomous navigation of shared balance bikes in local areas by using high-precision maps, high-precision positioning, sensor fusion and redundancy to achieve high-precision map-based sensing, positioning and decision making with EKF-SLAM.

2 Related Work

2.1 The Application and Development of a Shared Balancing Vehicle

With the prevalence of sharing culture, shared bicycles have become popular in everyday life, but there are still many problems and flaws with ordinary shared bicycles. For example, they do not come with a navigation function, which greatly reduces their practical value. It is dangerous for users who are not familiar with the route to plan their route while riding, and for users who are already in a state of fatigue, shared bikes do not help to a great extent to relieve fatigue while catching up.

Compared to existing bicycle sharing, the improved technology of balancing bikes is the perfect solution to these problems. Bikes are mainly used on campuses, parks and airports, where users have short distances to travel. The future balance bike comes with a navigation system that allows for voice interaction and automatic route planning based on real-time road conditions, solving the problem of users having to navigate the route while using the balance bike. It can achieve high-precision local area autonomous navigation, which provides great convenience for users who are not familiar with the local environment. At the same time, the balance bike requires less manpower than a bicycle, and when the user is tired, the balance bike only requires a small amount of energy to achieve the effect of a fast journey, which will also become a reference indicator for the user. To a certain extent, balanced bikes are also more fun than traditional bicycle sharing.

2.2 High Precision Positioning and Navigation Technology

Based on different application scenarios, autonomous navigation balance vehicles may require different positioning accuracies. For example, navigation in a park may only need to be accurate to a few metres, while navigation in an airport may require higher accuracy, as one may need to find a specific gate or baggage claim. Therefore, the specific positioning navigation system and the algorithm used will vary depending on the needs and environment of the application. In 2019, GPS-SLAM was proposed [3], and experimental results showed that the improved SLAM algorithm can be successfully used on low or high frame rate datasets. In 2020, Kalman filter-based IMU and UWB integration for high-accuracy indoor positioning and navigation techniques was proposed [4], and experimental results have shown that this technique can improve robustness and accuracy, resulting in smoother and more stable positioning results. Inspired by previous research, this paper combines GPS and IMU to form an integrated positioning system by extending Kalman filtering to improve the accuracy and tracking capability of autonomous navigation shared balance bikes in indoor and outdoor positioning and navigation.

3   Overall Framework of the System

3.1   Software Design

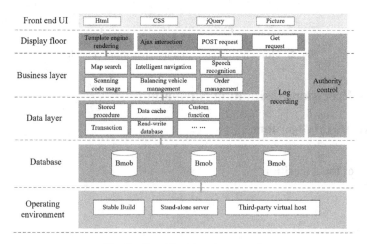

Fig. 1. Overall system framework

As shown in Fig. 1, the overall design is based on Stable Build and uses the Bmob back-end cloud management database, which realizes the functions of map search, intelligent navigation, speech recognition, scanning code usage, balanced vehicle management, order and user management.

3.2   Key Technology Design

The key technologies of the high precision autonomous navigation shared balancing vehicle system include high precision autonomous navigation design, QR code generation and speech recognition technology. The system is based on EKF-SLAM for map creation and autonomous positioning navigation. As local navigation requires more precise positioning, this system combines absolute positioning with relative positioning, implementing absolute positioning using GPS and IMU data fusion and relative positioning using point cloud matching and visual positioning. In addition, the system uses Quick Response Code to generate QR codes and the RNN-T model for speech recognition (Fig. 2).

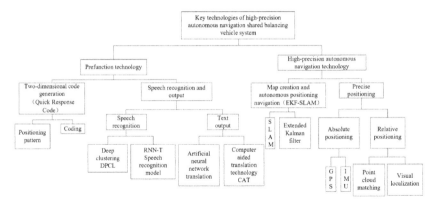

Fig. 2. Flow chart of autonomous navigation technology

4 High Precision Positioning Navigation

4.1 EKF-SLAM Algorithm

The Flow of the Algorithm

Fig. 3. EKF-SLAM algorithm flow

When a moving robot is in an unknown operating environment, the control information is first derived from the motion measurements, the positional information from the positional measurements and the observation information from the environmental measurements. The SLAM motion model, the SLAM observation gain and the SLAM observation model are then used to calculate the specific information about the motion and observation, respectively, and the digital information is processed through subfilters. The information associated with the data is fused with the positional information for real-time map updates. Finally, all known data are checked to see if the run is complete, and if it is, the process is ended; if not, the process starts again from the next step of initializing the data (Fig. 3).

Results of the Algorithm Experiment. The general Kalman filter has very significant limitations, only being able to estimate linear processing and measurement models accurately. It does not achieve optimal estimation in nonlinear scenarios, so EKF-SLAM (Extended Kalman SLAM) has emerged as a filtering algorithm capable of being used extensively in the case of nonlinear systems. EKF-SLAM is able to introduce linearization steps for nonlinear systems, which is practically applied to EKF-SLAM-guided autonomous mobile robots in outdoor environments. Maps can only be constructed for fixed space-time and space, whereas real-time localization is possible with the EKF-SLAM algorithm [5] (Fig. 4).

As shown in the 4th set of diagrams, the green line represents the original route, and purple asterisks represent landmarks. EKF-SLAM (Extended Kalman SLAM) enables autonomous navigation in four scenarios: straight line, mesh, dense landmarks and ultra-dense landmarks, which improves the accuracy of positioning by a large margin compared to relying solely on sensors and model predictions. Furthermore, the accuracy of the EKF-SLAM algorithm increases rather than decreases when the number of landmarks increases.

4.2 Fusion Positioning System

GPS + IMU Data Fusion Absolute Positioning. This system uses GPS + IMU data fusion for absolute positioning. The following section will explain the advantages and disadvantages of GPS and IMU and show the experimental results of GPS + IMU data fusion (Table 1).

The real-time GPS positioning speed is fast and provides continuous global ground coverage, but the frequency is only approximately 10 Hz, which is below the response requirements of the automatic navigation balancing vehicle. However, the frequency of IMU is as high as 1 kHz, so the combination of GPS and IMU can make it possible to solve the problem of response speed of balancing vehicle positioning. Due to the existence of the problem of GPS signal noise and signal loss, it is easy to affect the accuracy of balanced vehicle positioning. IMU belongs to inertial navigation and can autonomously navigate without relying on external information, which easily affects the balance of the car positioning accuracy. On the other hand, the IMU is an inertial navigation unit that can navigate independently without relying on external information. However, IMUs also have some disadvantages. For example, the calculation error will

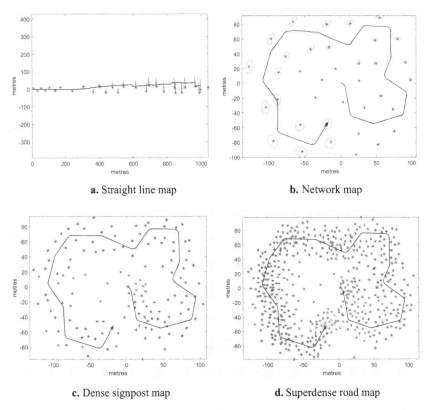

a. Straight line map   **b.** Network map

c. Dense signpost map   **d.** Superdense road map

Fig. 4. a. Straight line map **b.** Network map **c.** Dense signpost map **d.** Superdense road map

Table 1. Comparison of advantages and disadvantages of GPS and IMU.

Advantages/disadvantages	GPS	IMU
Advantages	Fast real-time positioning speed	High frequency (approximately 1 kHz)
	Continuous coverage of the global ground	Inertial navigation (autonomous navigation without reliance on external information) [4]
Disadvantages	Signal noise and loss [6]	Calculation error increases with time
	Low frequency (approximately 10 Hz)	High cost

increase with time, so it cannot realize stable positioning for a long time. However, GPS can compensate for this disadvantage. In summary, the system uses GPS + IMU for fusion positioning to improve the positioning accuracy of local areas.

In mobile road detection, the combination of measurable stereo image GPS and IMU data can greatly improve the accuracy and reliability of mobile data detection [7]. There are different characteristics between GPS and IMU in terms of positioning. The Kalman filter algorithm is usually used to fuse the data of different sensors. Therefore, this system combines GPS and IMU for absolute positioning and simulates the algorithm based on the Kalman filter fusion positioning of GPS + IMU, in which IMU inertial navigation is used for state prediction and GPS is used for filtering correction.

Relative Positioning Combined with Point Cloud Matching and Visual Positioning. Relative position localization is based on the current driving scenario as a reference system. It can be divided into two technical routes: point cloud matching (Lidar) and visual positioning. Point cloud matching is based on LIDAR as the core, where LIDAR emits laser pulses outwards and multiple echoes from the ground or the surface of an object are returned for matching to achieve high accuracy positioning of the current scene of the balancing vehicle. Visual positioning is usually low cost but is susceptible to environmental changes such as weather and lighting, which is suitable for specific positioning scenarios. Therefore, combining the advantages of point cloud matching and visual localization, this system uses a combination of both for relative localization.

5 Speech Recognition

5.1 Speech Recognition Technology

The speech search destination function of this system can recognize Mandarin, Cantonese and English, effectively combining speech recognition technology with translation. The system's voice search destination function can recognize Mandarin, Cantonese and English by combining speech recognition technology with translation effectively. The diagram below shows the flow chart of the speech recognition technology of this system (Fig. 5).

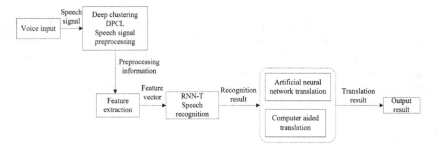

Fig. 5. Flow chart of speech recognition technology

In the classical speech recognition model, the acoustic model, pronunciation model and language model are trained separately, which makes speech recognition training more complicated. In contrast, the three models are jointly optimized for training in the point-to-point recognition model during which the neural network can be used to convert the input audio information directly into text [8]. The speech recognition technology of this system uses deep clustering DPCL as the speech preprocessing method. The input discourse of this system is relatively short rather than sentences with a strong correlation between the upper and lower levels. The training discourse of the end-to-end model is short, and the evaluation model is from the same field [9]. At the same time, this system uses RNN-T as the language recognition model because of its small size and accuracy [10]. Considering the convenience of users who adopt different languages, artificial neural network translation technology and computer-aided translation technology are added to this system [11].

5.2 Speech Recognition Test

Table 2. Speech recognition test results.

Language	Input speech	Output text
Mandarin	Nan2 shi2 tang2	南食堂(South canteen)
	Su4 she4 B qu1	宿舍B区(Dormitory B)
English	/ saʊθ//kænˈtiːn/	南食堂(South canteen)
	/ ˈdɔːmətri//ˈeəriə/	宿舍B区(Dormitory B)
Cantonese	naan4 faan6 tong4	南食堂(South canteen)
	suk1se5 B keoi1	宿舍B区(Dormitory B)

As shown in Table 2, the system's speech recognition function can accurately output the destination names in different languages for the two destinations "South Canteen" and "Dormitory B". Through training, the system's speech recognition technology can recognize Cantonese, Mandarin, English and other languages, and the accuracy rate of speech recognition can reach 95%, which can meet the requirements of the system's speech recognition accuracy.

6   Experiment

6.1   Experimental Results of GPS + IMU Data Fusion

As shown in the figure, the 6th set of plots is a comparison of the original standard route, angle, and speed with the real data generated by GPS and IMU data fusion. Figure 6a shows the trajectory generated by the fusion of GPS and IMU data, with

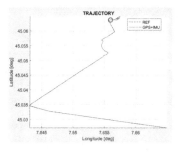

a. Trajectory comparison diagram

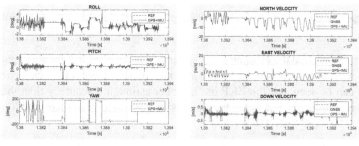

b. Angular contrast diagram   **c.** Velocity comparison diagram

Fig. 6. a. Trajectory comparison diagram **b.** Angular contrast diagram **c.** Velocity comparison diagram

the horizontal coordinates indicating longitude and the vertical coordinates indicating latitude. Figure 6b shows the comparison between the reference roll angle, pitch angle and yaw angle vectors and the real data generated by GPS and IMU data fusion, with the x-coordinate indicating time and the y-coordinate indicating degree. Figure 6c shows the comparison between the velocity in each direction and the real data generated by GPS and IMU data fusion, with the x-coordinate indicating time and the y-coordinate indicating velocity. As seen from the comparison figures, the trajectory, angular volume and velocity are consistent with the reference value. Therefore, the use of GPS + IMU data fusion can achieve more accurate absolute positioning.

6.2 Positioning and Navigation Simulation Experiment

Simulation experiments comparing the estimation performance of KF, EKF and UKF were carried out based on MATLAB. The covariance of the process state was set to 10, the measurement noise covariance to 1, the initial estimation variance to 5, and the number of particle filter particles to 500. State values were first calculated from the initial state values, process noise variance values and other data, and observed values were calculated from the initial state values and measurement noise variance values data, followed by estimation and estimation updates. Then, it goes to the next iteration of parameter changes, and particle resampling is performed by averaging the likelihood of each estimate through individual particles for weight normalization. Finally, a Kalman

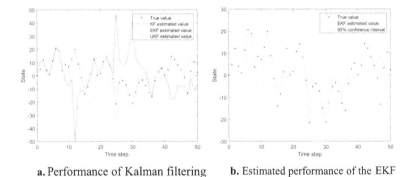

a. Performance of Kalman filtering   b. Estimated performance of the EKF

Fig. 7. **a.** Performance of Kalman filtering **b.** Estimated performance of the EKF

gain is performed, and the estimates and variances are updated. As shown in panel 7, the X-axis indicates the step size, and the Y-axis indicates the state. From the experimental results, it can be seen that the EKF has a higher estimation accuracy than the KF and UKF and is suitable for use in the high-precision construction of maps and localization (Fig. 7).

6.3   Positioning and Navigation Simulation Experiment

As shown in the figure, the 8th set of plots shows real-time positioning results based on EKF-SLAM, with optional IMU and GPS data using point cloud matching and a monocular camera. A continuous trajectory is generated by interpolating the existing attitude, and angular rates and linear accelerations are given by differentiating the continuous trajectory to simulate noisy IMU data given a realistic ground attitude. Noise is added to these real IMU samples given the error specifications for a particular IMU model. GPS and IMU complement each other to realize stable positioning of balanced vehicles, solving the problem of positioning refresh frequency. It enables map updates, real-time high precision positioning and navigation with real-time data detection based on motion measurements, posture measurements and environmental measurements (Fig. 8).

7   System Demonstration

7.1   Customer Usage Process

The customer use flow chart of this high-precision autonomous navigation shared balancing vehicle system based on EKF-SLAM is shown in Fig. 9. First, the user opens the system page and authorizes the system avatar, nickname and location of the system. Users can view the location of the balancing vehicle, their own location and personal center information. The user clicks to scan the QR code of the balancing vehicle, and the camera is aroused to recognize the QR code. If the recognition fails, the screen displays the mismatch of the QR code. Otherwise, the balancing vehicle is unlocked successfully by the user. After successfully unlocking the lock, the user can use speech recognition

Fig. 8. EKF-SLAM and GPS + IMU fusion localization simulation results

of the destination to select the correct destination name. Based on route planning and positioning algorithms, the balancing vehicle makes accurate local positioning and successfully delivers the user to the destination. Users settle the charges, and the system generates an order based on the calculation.

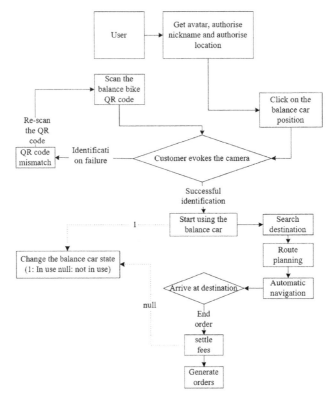

Fig. 9. Customer usage flow chart

7.2 System Test Results

(See Fig. 10).

8 Summary and Prospects

This system is a map-based localization solution that uses EKF-SLAM in robot local-ization to achieve simultaneous localization and map construction using an extended Kalman filter. GPS and IMU are used with multiple sensors for combined navigation positioning and fused positioning. The experimental results show that the autonomous navigation technique and speech recognition technology proposed in this paper can realize speech recognition of destinations, precise positioning and autonomous naviga-tion. However, the computational speed of the EKF-SLAM algorithm decreases when observing a larger number of landmarks [12]. At the same time, the performance of the EKF-SLAM algorithm degrades in a noisy state. These questions still require further study.

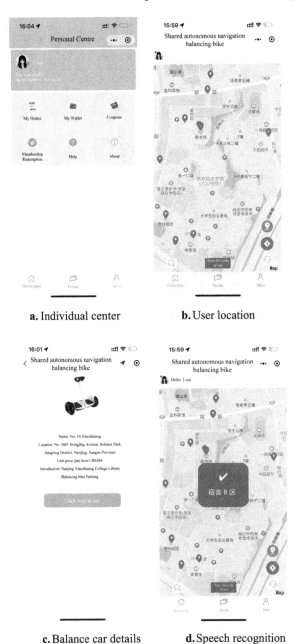

a. Individual center **b.** User location

c. Balance car details **d.** Speech recognition

Fig. 10. **a.** Individual center **b.** User location **c.** Balance car details **d.** Speech recognition **e.** Destination selection **f.** Pedestrian navigation route **g.** Driving navigation route **h.** Dynamic navigation route

e. Destination selection **f.** Pedestrian navigation route

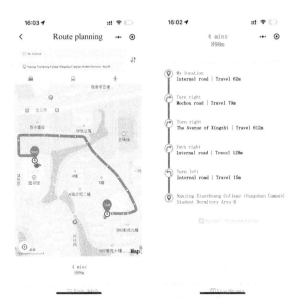

g. Driving navigation route **h.** Dynamic navigation route

Fig. 10. (*continued*)

References

1. Zhifu, M., D'Maris, C.: The sharing economy promotes sustainable societies. Nat. Commun. **10**(1), 1214–1217 (2019)
2. Pascacio, P., et al.: Collaborative indoor positioning systems: a systematic review. Sensors **21**(3), 1002 (2021)
3. Kiss-Illés, D., Barrado, C., Salamí, E.: GPS-SLAM: an augmentation of the ORB-SLAM algorithm. Sensors **19**(22), 4973 (2019)
4. Feng, D., et al.: Kalman filter based integration of IMU and UWB for high-accuracy indoor positioning and navigation. IEEE Internet Things J. **7**(4), 3133–3146 (2020)
5. Kokovkina, V.A., et al.: The algorithm of EKF-SLAM using laser scanning system and fisheye camera. In: 2019 Systems of Signal Synchronization, Generating and Processing in Telecommunications (SYNCHROINFO), pp. 1–6 (2019)
6. Shen, L., Stopher, P.R.: Review of GPS travel survey and GPS data-processing methods. Transp. Rev. **34**(3), 316–334 (2014)
7. Girbes-Juan, V., et al.: Asynchronous sensor fusion of GPS, IMU and CAN-based odometry for heavy-duty vehicles. IEEE Trans. Veh. Technol. **70**(9), 8617–8626 (2021)
8. Gojcic, Z., et al.: The perfect match: 3D point cloud matching with smoothed densities. In: 2019 IEEE/CVF Conference on Computer Vision and Pattern Recognition (CVPR), pp. 5540–5549 (2019)
9. Park, D.S., et al.: SpecAugment: a simple data augmentation method for automatic speech recognition, pp. 98–105 (2019). arXiv preprint arXiv:1904.08779
10. Saon, G., et al.: Advancing RNN transducer technology for speech recognition. In: ICASSP 2021 - 2021 IEEE International Conference on Acoustics, Speech and Signal Processing (ICASSP), pp. 5654–5658 (2021)
11. Chen, G.: Exploration and research on computer aided translation technology and English translation practice. J. Phys: Conf. Ser. **1213**(4), 042011 (2019)
12. Abdulredah, S.H., Kadhim, D.J.: Developing a real time navigation for the mobile robots at unknown environments. Indon. J. Electr. Eng. Comput. Sci. **20**(1), 500–509 (2020)

Social Media and Recommendation Systems

Link Prediction Based on the Relational Path Inference of Triangular Structures

Xin Li, Qilong Han[✉], Lijie Li, and Ye Wang

College of Computer Science and Technology, Harbin Engineering University, Harbin, China
{lixin,hanqilong,lilijie,wangye2020}@hrbeu.edu.cn

Abstract. Link prediction is used to complete the knowledge graph. Convolutional neural network models are commonly used for link prediction tasks, but they only consider the direct relations between entity pairs, ignoring the semantic information contained in the relation paths. In addition, the embedding dimension of the relation is generally larger than that of the entity in the ConvR model, which blocks the progress of downstream tasks. If we reduce the embedding dimension of the relation, the performance will be greatly degraded. This paper proposes a convolutional model PITri-R-ConvR based on triangular structure relational inference. The model uses relational path inference to capture semantic information, while using a triangular structure to ensure the reliability and computational efficiency of relational inference. In addition, the decoder R-ConvR improves the initial embedding of the ConvR model, which solves the problems of the ConvR model and significantly improves the prediction performance. Finally, this paper conducts sufficient experiments in multiple datasets to verify the superiority of the model and the rationality of each module.

Keywords: Link Prediction · Triangular Structure · Relational Path Inference · Attention Mechanism · Convolution Neural Network Model

1 Introduction

Knowledge graphs are usually used to represent various things in the world and the relations between them. With the deepening of research, knowledge graphs have shown a wealth of application value. At present, there are multiple knowledge graphs, such as Freebase [1], DBpedia [2], YAGO [3–5], and WordNet [6], but in Freebase and DBpedia, more than 66% of individual entries lack a birthplace. Link prediction aims to complete a knowledge graph.

The translation model optimizes the model by using the distance between entity pairs after relational translation as a scoring function, such as TransE [7], TransH [8], TransR [9], TransD [10], and TransM [11]. In addition to translation transformations, other models have more complex transformations, such as RotatE [12], HAKE [13], MuRP [14], and TorusE [15]. The semantic matching model utilizes a scoring function based on similarity, such as RESCAL [16], DistMult [17], ComplEx [18] and Hole [19].

© The Author(s), under exclusive license to Springer Nature Singapore Pte Ltd. 2023
Z. Yu et al. (Eds.): ICPCSEE 2023, CCIS 1880, pp. 255–268, 2023.
https://doi.org/10.1007/978-981-99-5971-6_19

People have introduced the idea of convolution into link prediction tasks, resulting in many convolutional neural network models, such as ConvE [20], ConvR [21], AcrE [22], InteractE [23], ConvKB [24], and other models.Due to the excellent performance of CNN models, they have become a common model method for link prediction tasks.

However, there is a problem with the ConvR model. In this model, the convolution kernel is obtained by splitting the relation embedding vector, which leads to a difference in the initial embedding dimensions of the entity and relation.This situation is not friendly to downstream tasks. In addition, the ordinary convolution model only considers the single-step relations between entities, ignoring the semantic information in the relation path.

We propose the PITri-R-ConvR model,and the way to solve the above problems is as follows.

- Based on the triangular structure and relational inference, this paper proposes an encoder model PITri. The triangular structure consists of three entity pairs and the relations between them. PITri utilizes the uniqueness of this structure to ensure the reliability and computational efficiency of relational inference while capturing the rich semantic information present in relational paths for the convolutional neural network model.
- This paper utilizes an attention mechanism to comprehensively utilize the information of the path set for multiple relation paths existing in the same entity sequence.
- The R-ConvR model uses FNN to adjust the relational initial embedding structure of the ConvR model and improve prediction performance while solving the problem of its inability to adapt to downstream tasks
- The comparison experiments and ablation experiments on the datasets FB15k-237, WN18RR, YAGO3–10, and kinship have been performed to verify the superiority of the PITri-R-ConvR model.

2   Related Work

2.1   Convolution Model

Introducing convolution to the link prediction task was first proposed by ConvE, which proposed to reshape the vectors of the head entity and the relation and then randomly generate convolution kernels for convolution. ConvR takes the relation as the convolution kernel, which increases the interactivity of the entity-relation relative to ConvE. The gap between the interactivity of the two models can be seen in Fig. 1.

AcrE expands the receptive field by adding dilated convolutions, further increasing the interaction between entities and relational embeddings to link head entities and relations more closely. After stacking entities and relations, InteractE further improves interactivity by circular convolution. The difference between ConvKB and the previous one is that it adds tail entities to the convolution. However, these convolutional models only deal with each triple separately, ignoring the topology and connection between each triple, which has a great impact on the prediction task.

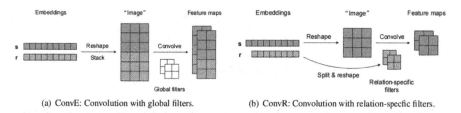

(a) ConvE: Convolution with global filters.   (b) ConvR: Convolution with relation-specfic filters.

Fig. 1. Convolution model interaction diagram

2.2 Relational Path Inference

As shown in Fig. 2, the Path-RNN [25] model utilizes the PRA algorithm to discover relation paths. Then, the model inputs relation vectors into the RNN model to obtain the path vector. Through training, the model learns a potential relation prediction model for entity pairs.

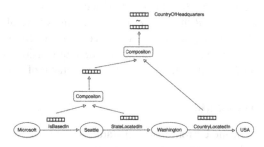

Fig. 2. Path-RNN model diagram

It should be noted that not all relation paths are reliable and meaningful for knowledge graph representation learning, and there are many relation paths in the knowledge graph traversing them will require a significant amount of calculation. These issues will be addressed in this paper.

3 PITri-R-ConvR

3.1 Motivation for Model Proposition

Many previous works have proven the reliability and effectiveness of inference based on relational paths, but there are still the following problems.

As shown in Fig. 3, (h, father, a)(a,father,t) implies (h, grandfather, t), but not all multiple-step relational paths can infer a reliability connection, such as (h, friend, a) (a, friend,t). It is difficult to say what the relation in (h, t). It can be seen that not all relational paths are reliable and meaningful for knowledge graph representation learning. Moreover, when the length of the relation path is too large, the inferred relation is weaker, and the inference becomes less significant.

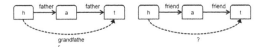

Fig. 3. Inference reliability analysis diagram

There may be multiple paths between entity pairs, which contain varying levels of semantic information and have varying impacts on the prediction performance of the model. Therefore, it is particularly important to process the path set. Existing models typically adopt methods such as calculating the similarity between the relation vector inferred by different paths and the target relation representation vector, taking the maximum, average, or LSE of the similarities. However they cannot effectively use the information of the path set. Because only one path is used for inference by the maximum value method, the information of other paths is ignored. The method of averaging will make the model affected by noise; lSE is a smooth approximation of the maximum value and cannot effectively integrate information from multiple paths. The model uses the triangle structure to ensure the reliability of inference.

In addition, since the ConvR model uses the embedding vector of the relation as the convolution kernel, the convolution operation has requirements for the size and number of convolution kernels, which will cause the embedding dimension of the relation to be much larger than that of the entity, hindering the development of downstream tasks.

3.2   Overall Design of the Model

In this paper, in view of the current research status and existing problems in the field of convolutional models and relational path reasoning algorithms, a convolution model PlTri-R-ConvR based on triangular structure relational path inference is proposed. In Fig. 4, the overall structure of the model is introduced.

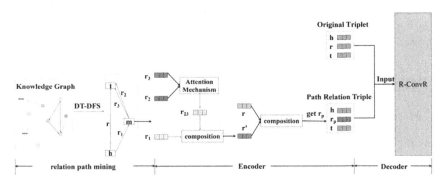

Fig. 4. Encoder diagram

This paper uses the path mining algorithm based on direct relation triples (DT-DFS) to obtain a set of relation paths with a length of 2 between entity pairs in each direct relation triple. The relation path and the original triplet form a triangular structure.

Then, we use an attention mechanism to address the phenomenon that there are multiple relation paths in the same entity sequence and comprehensively utilize the information of the path set. At the same time, the RNN-based relation vector combination algorithm is used for relation path inference, and the obtained path relation r_p is combined with the head h and tail entity t into a path relation triplet (h, r_p, t), which is put into the R-ConvR model together with the original triplet.

3.3 Triangular Structure Analysis

This paper proposes the encoder PITri to capture the semantic information for the convolutional model. To fully utilize this semantic information, it is necessary to mine the most valuable relation paths. Therefore, this paper proposes a path mining algorithm, DT-DFS, based on direct relation triples to obtain the relation path. The specific algorithm is shown in 3.1.

Algorithm 3-1 DT-DFS(T)

---

input: dataset T

output: relation path collection $pathC$

1: Statement $e2es$;//A collection of neighbors of an entity;
1: Statement $es2rs$;//A collection of relations between entities pair;
3: Statement $pathC$;//The collection of paths obtained;
4: **for** (h, r, t) in T **do**
5:   $e2es[h].append(t)$;
6:   $es2rs[(h, t)].append(r)$;
7: **end for**
8: **for** (h, r, t) in T **do**
9:   neighs $= e2es[h]$;
10:    **for** m in neighs **do**
11:     if t in $e2es[m]$ then//
12:       $r1s = es2rs[(h, m)]$;
13:       $r2s = es2rs[(m, t)]$;
14:       $rs = es2rs[(h, t)]$;
15:     break;
16:     end if
17:    $pathC.add(h, r1s + r2s, t)$
18:    **end for**
19: **end for**
20: return $pathC$;

Through Algorithm 3.1, we have obtained relation path $r_1 + r_2$ and the intermediate entity m. As shown in Fig. 5, the triangular structure is formed by the direct relation triples (h, r, t), relation paths $r_1 + r_2$, and intermediate entities m.

The triangle structure contains a relational path $(r_1 + r_2)$. In general, whether there is a connection between (h, t) is questionable, which leads to some irrationality in the inference of the relational path. In the triangular structure, since there is a direct relation

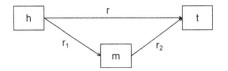

Fig. 5. Convolution model interaction diagram

r between entity pairs (h, t), it proves that there is a connection between the entity pairs, thus providing further evidence for the rationality of the relation path inference. Moreover, the relation paths in the triangular structure have a length of 2, which ensures high computational efficiency while providing the most abundant semantic information.

3.4 Application of the Attention Mechanism

In the knowledge graph, there may be multiple paths between two entities, and multiple paths may also pass through the same intermediate entity. As shown in Fig. 6, there are still two paths (r_1, r_2), (r_1, r_3) between the entity sequences (h, m, t). In this case, if the information about the path set is not effectively used, it may lead to errors in the application of the triangular structure.

When faced with the problem of multiple paths between two entities, existing models typically calculate the similarity between the relation vectors inferred by different paths and the given relation vector, and choose the path with the highest similarity. Alternatively, they may adopt methods such as taking the average or LSE of multiple paths. However, these methods have difficulty to effectively utilizing the information of the path set. Because the method of taking the maximum value usually only selects one relation path, information about other paths is lost. Similarly, taking the average of multiple paths treats each relation path equally, making it vulnerable to noise (unimportant information). To solve the shortcomings of existing methods, this paper uses the attention mechanism to fuse multiple relations between entity pairs and performs path inference on the fused relation, solving the problems mentioned above.

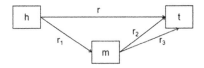

Fig. 6. Multirelation path diagram

The specific solution is analysed in the following relational path inference algorithm.

3.5 Relational Path Inference Algorithm

The triangular structure is processed by the relational path inference algorithm, as shown in Fig. 7, which includes the following steps.

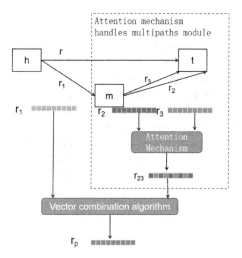

Fig. 7. Relation path inference diagram

A relational path can be obtained through the triangular structure. If (h, t) obtains multiple relational paths, the attention mechanism is used to fuse the paths into one relation path. As shown in Fig. 7, there are two relations r_2, r_3 between the entity pair (m, t), and then the relation paths between (h, t) include $(h, r_1 + r_2, t)$, and $(h, r_1 + r_3, t)$. Then, it is necessary to use the attention mechanism to fuse r_2, r_3 into a relation embedding vector r_{23}. The calculation formula is as follows, where the vector dimension is $lam \epsilon R^d$; $r_2 \epsilon R^d$; $r_3 \epsilon R^d$:

$$att_2 = \frac{exp(r_2^T lam)}{exp(r_2^T lam) + exp(r_3^T lam)} \tag{1}$$

$$att_3 = \frac{exp(r_3^T lam)}{exp(r_2^T lam) + exp(r_3^T lam)}; exp(x) = \frac{1}{e^{-x} + 1} \tag{2}$$

$$r_{23} = att_2 \cdot r_2 + att_3 \cdot r_3 \tag{3}$$

We fuse multiple paths existing in the relational path into one path through the attention mechanism, the information of multiple paths will be integrated, and the more important relational path will occupy a larger weight.

The composition operation is realized using a matrix **W**:

$$r_p = f(W[r_1; r_2]) \tag{4}$$

Additionally, the inference method described above is only the basic content and does not effectively utilize our triangular structure scene. In our scene, in addition to r_1 and r_2 on the path, we also have the original relation r. We also add the direct relation r to the combinatorial algorithm, which is a better use of the triangular structure. By adding the direct relation, the inference result can be closer to the direct relation r and still retain the path semantic information obtained during the process of relational inference.

$$r\prime = f(W[r_1; r_2]); r_p = f(W[r\prime; r]) \tag{5}$$

3.6 R-ConvR

Because of the special mechanism of ConvR, the high probability of entity embedding and relation embedding is different, which may have a certain impact on the use of downstream tasks. However, the number and size of convolution kernels required by the ConvR model are relatively large, and if forcibly adjusted to be consistent, it will lead to performance degradation. Therefore, we adjust the embedded vector dimension of relations to be consistent with entities. As shown in Fig. 8, we use FNN technology to adjust the dimension of the relation to a size that is sufficient to segment the appropriate convolution kernel. After sufficient experiments, the reconstruction performance has been improved and can better adapt to downstream tasks.

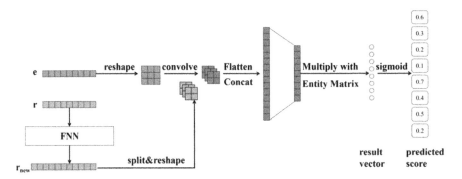

Fig. 8. R-ConvR diagram

In this paper, it should be noted that the loss involved in the PITri-R-ConvR model needs to consider two parts, namely the loss caused by the direct relation triplet and the loss caused by the path relation triplet. We need to add the two parts of the loss and then optimize the model.

4 Experimental Analysis

4.1 Dataset

This paper utilizes a series of general link prediction datasets to evaluate the PITri-R-ConvR model. A brief introduction of the datasets is provided.

- FB15k-237: The FB15k-237 dataset prevents the testing leakage problem that can occur after reversing the relations in the FB15k dataset.
- WN18RR: The WN18RR dataset is processed in the same way as the WN18 dataset.
- YAGO3–10: YAGO3–10 dataset is characterized by at least 10 relations for each entity.
- kinship: The kinship dataset describes the relations between family members.

4.2 Evaluation Protocol

To verify the superiority of the PITri-R-ConvR model, we compare it with multiple baseline models on the evaluation metrics MRR and Hits@N.

The model in this paper will be compared with the following link prediction models:

• Convolution model: Our PITri-R-ConvR model introduces the triangle structure into the convolution model and then uses the idea of path inference to address the triangle structure, so we need to compare it with the convolution model. We use the classical models ConvE, ConvR, and InteractE.

- Convolutional Models Based on Graph Neural Networks: SCAN, CompGCN.
- Other models: DistMult, ComplEx, ComplEx(w/LS) [26], RotatE, GenKGC [27].

4.3 Comparison Experiment

In the comparison experiment, the PITri-R-ConvR model was compared with the seven baseline models of DistMult, ComplEx, ConvE, RotatE, ConvR, SCAN and CompGCN. As shown in Table 1, the bold data are the experimental results of the PITri-R-ConvR model in the WN18RR dataset and FB15k-237 dataset.

Table 1. Performance of PITri-R-ConvR in WN18RR and FB15k-237.

Model	WN18RR				FB15k-237			
	MRR	Hits@1	Hits@3	Hits@10	MRR	Hits@1	Hits@3	Hits@10
DistMult	0.43	0.39	0.44	0.49	0.241	0.155	0.263	0.419
ComplEx	0.44	0.41	0.46	0.51	0.247	0.158	0.275	0.428
RotatE	0.476	0.428	0.492	0.571	0.338	0.241	0.375	0.533
ConvE	0.46	0.39	0.43	0.48	0.316	0.239	0.35	0.491
SCAN	0.47	0.43	0.48	0.54	0.35	0.26	0.39	0.54
CompGCN	0.479	0.443	0.494	0.546	0.355	0.264	0.39	0.535
ComplEx(w/LS)	0.477	0.441	0.491	0.546	0.318	0.231	0.348	0.493
GenKGC	0.40	0.192	0.355	0.439	0.352	0.287	0.403	0.535
ConvR	0.475	0.443	0.489	0.537	0.35	0.261	0.385	0.528
PITri-R-ConvR	**0.482**	**0.454**	**0.50**	**0.557**	**0.357**	**0.28**	**0.41**	**0.54**

Experimental results prove the superiority of the PITri-R-ConvR model in all performance metrics. In the WN18RR dataset, compared with the ConvR model, the model in this paper has an MRR increase of approximately 1.5%, Hits@1 increase of approximately 2.5%, and other indicators have improved to varying degrees. In the FB15k-237 dataset, the model has an MRR increase of approximately 2.0%, Hits@1 increase of approximately 7.3%, and other indicators have improved to varying degrees.

The experimental results of our model and other baseline models are shown in Fig. 9 and Fig. 10. The PITri-R-ConvR model has obvious advantages in link prediction performance compared with other models.

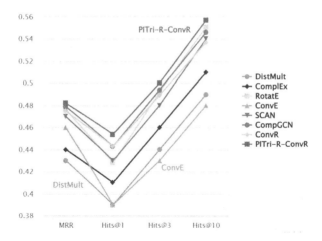

Fig. 9. Comparison of the experimental results in the WN18RR diagram

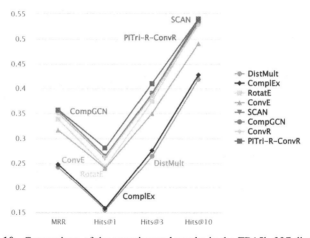

Fig. 10. Comparison of the experimental results in the FB15k-237 diagram

The PITri-R-ConvR model consists of PITri and R-ConvR. The PITri model uses the path inference algorithm to infer the relation paths, obtains rich semantic information and uses an attention mechanism to comprehensively utilize the information of the path set between entity pairs. In addition, the R-ConvR model improves the initial relational embedding structure of the ConvR model, which not only solves the problems that it cannot adapt to downstream tasks but also improves the predictive performance of the model.

Through comparison experiments and effectiveness analysis, it can be shown that the PITri-R-ConvR model is superior.

4.4 Encoder Ablation Experiment

The ablation experiment is mainly to verify whether the encoder PITri can improve the convolutional neural network model. In this paper, ConvE and InteractE models are selected as the decoder to verify the effectiveness of PITri. The experimental datasets are WN18RR, YAGO3–10 and kinship datasets. In addition to the verification of the encoder, this experiment also needs to verify whether there is an improvement effect after adding the direct relation r to the relation path inference algorithm.The results are shown in the following tables.

In the tables provided, the three models are defined as follows:Model represents the original model, including ConvE and InteractE; PITri-Model means adding PITri on the basis of Model; PITri-Model + r means adding direct relation r to the relational path inference algorithm.

As shown in Tables 2, 3 and 4, when the decoders are ConvE and InteractE, the performance metrics of the three models show a step-by-step improvement. The results of PITri-Model in each evaluation metric are higher than those of Model, and the results of PITri-Model + r in each evaluation metric are higher than those of PITri-Model.

The results of the WN18RR dataset show that the MRR of PITri-Model + r increased by more than 1.4% in ConvE, and the MRR of PITri-Model + r increased by more than 2.1% in InteractE.

The results of the YAGO3–10 dataset shows that the MRR of PITri-Model + r increased by more than 27% in ConvE.

The results of the kinship dataset show that the MRR of PITri-Model + r increased by more than 3.7% in ConvE, and the MRR of PITri-Model + r increased by more than 2.3% in InteractE.

Table 2. Performance of PITri on ConvE and InteractE in WN18RR.

WN18RR	ConvE				InteractE			
	MRR	Hits@1	Hits@3	Hits@10	MRR	Hits@1	Hits@3	Hits@10
Model	0.43	0.40	0.43	0.52	0.463	0.43	0.475	0.528
PITri-Model	0.433	0.40	0.44	0.512	0.471	0.44	0.482	0.538
PITri-Model + r	0.436	0.40	0.45	0.512	0.473	0.44	0.483	0.538

Through sufficient experiments on multiple datasets and models, we can conclude that the encoder PITri significantly improves on the predictive performance of the convolutional model.At the same time, the relational path inference algorithm after adding the direct relation r performs better than the original algorithm.

4.5 Decoder Ablation Experiment

The ablation experiment is mainly to verify whether there is a performance improvement after using FNN to improve the relational embedding structure of ConvR, in addition to solving the problem that it cannot adapt to downstream tasks (Table 5).

Table 3. Performance of PITri on ConvE and InteractE in YAGO3–10.

YAGO3–10	ConvE				InteractE			
	MRR	Hits@1	Hits@3	Hits@10	MRR	Hits@1	Hits@3	Hits@10
model	0.44	0.35	0.50	0.62	0.541	0.462	0.582	0.687
PITri-Model	0.552	0.477	0.597	0.69	0.541	0.461	0.583	0.689
PITri-Model + r	0.56	0.487	0.60	0.694	0.543	0.463	0.588	0.69

Table 4. Performance of PITri on ConvE and InteractE in kinship.

kinship	ConvE				InteractE			
	MRR	Hits@1	Hits@3	Hits@10	MRR	Hits@1	Hits@3	Hits@10
model	0.83	0.75	0.90	0.95	0.862	0.785	0.93	0.98
PITri-Model	0.85	0.776	0.92	0.983	0.876	0.806	0.936	0.985
PITri-Model + r	0.861	0.781	0.931	0.986	0.882	0.816	0.94	0.986

Table 5. Performance of R-ConvR on WN18RR and FB15k-237.

Model	WN18RR				FB15k-237			
	MRR	Hits@1	Hits@3	Hits@10	MRR	Hits@1	Hits@3	Hits@10
ConvR	0.475	0.443	0.489	0.537	0.35	0.261	0.385	0.528
R-ConvR	0.48	0.45	0.495	0.542	0.355	0.265	0.391	0.535

The experimental results show that the R-ConvR model outperforms the ConvR model in all performance metrics. In the WN18RR dataset, compared with the ConvR model, the R-ConvR has an MRR increase of approximately 1.1%, Hits@1 increase of approximately 1.6%, and other indicators have improved to varying degrees. In the FB15k-237 dataset, the model PITri-R-ConvR has an MRR increase of approximately 1.5%, Hits@1 increases of approximately 1.6%, and other indicators have improved to varying degrees.

In summary, we can conclude that the improving the relational embedding structure of ConvR leads to significant performance improvements.

5   Conclusion

Based on the insufficiency of the convolution in capturing the semantic information of the relation path and the inconsistency of entity and relation embedding dimensions in the ConvR model, which affects downstream tasks, this paper proposes the PITri-R-ConvR model. The model utilizes the direct relation r existing in the triangular structure to ensure the reliability of relational path inference. In addition, the attention mechanism is used to comprehensively process the path set, and the relational path inference algorithm is optimized. It obtains rich semantic information for the convolutional model, and the decoder improves the relational embedding structure of the ConvR model, which solves the problem and improves the prediction performance. Finally, sufficient comparison and ablation experiments are performed to verify the effectiveness and rationality of the PITri-R-ConvR model.

Acknowledgment. This work was supported by the National Key R&D Program of China under Grant No. 20201710200.

References

1. Bollacker, K., Evans, C., Paritosh, P., et al.: Freebase: a collaboratively created graph database for structuring human knowledge. In: Proceedings of the 2008 ACM SIGMOD International Conference on Management of Data, pp. 1247–1250 (2008)
2. Auer, S., Bizer, C., Kobilarov, G., Lehmann, J., Cyganiak, R., Ives, Z.: Dbpedia: a nucleus for a web of open data. In: Aberer, K., et al. (eds.) ASWC/ISWC -2007. LNCS, vol. 4825, pp. 722–735. Springer, Heidelberg (2007). https://doi.org/10.1007/978-3-540-76298-0_52
3. Suchanek, F.M., Kasneci, G., Weikum, G.: Yago: a core of semantic knowledge. In: Proceedings of the 16th International Conference on Worldwide Web, pp. 697–706 (2007)
4. Hoffart, J., Suchanek, F.M., Berberich, K., et al.: YAGO2: a spatially and temporally enhanced knowledge base from Wikipedia. Artif. Intell. **194**, 28–61 (2013)
5. Mahdisoltani, F., Biega, J., Suchanek, F.: Yago3: a knowledge base from multilingual wikipedias. In: 7th Biennial Conference on Innovative Data Systems Research. CIDR Conference (2014)
6. Miller, G.A.: WordNet: a lexical database for English. Commun. ACM **38**(11), 39–41 (1995)
7. Bordes, A., Usunier, N., Garcia-Duran, A., et al.: Translating embeddings for modelling multirelational data. Adv. Neural Inf. Process. Syst. **26**, 1–9 (2013)
8. Wang, Z., Zhang, J., Feng, J., et al.: Knowledge graph embedding by translating on hyperplanes. In: Proceedings of the AAAI Conference on Artificial Intelligence, vol. 28, no. 1 (2014)
9. Lin, Y., Liu, Z., Sun, M., et al.: Learning entity and relation embeddings for knowledge graph completion. In: Proceedings of the AAAI Conference on Artificial Intelligence, vol. 29, no. 1 (2015)
10. Ji, G., He, S., Xu, L., et al.: Knowledge graph embedding via dynamic mapping matrix. In: Proceedings of the 53rd Annual Meeting of the Association for Computational Linguistics and the 7th International Joint Conference on Natural Language Processing, vol. 1: Long papers, pp. 687–696 (2015)
11. Fan, M., Zhou, Q., Chang, E., et al.: Transition-based knowledge graph embedding with relational mapping properties. In: Proceedings of the 28th Pacific Asia Conference on Language, Information and Computing, pp. 328–337 (2014)
12. Sun, Z., Deng, Z.H., Nie, J.Y., et al.: Rotate: knowledge graph embedding by relational rotation in complex space (2019). arXiv preprint arXiv:1902.10197
13. Zhang, Z., Cai, J., Zhang, Y., et al.: Learning hierarchy-aware knowledge graph embeddings for link prediction. In: Proceedings of the AAAI Conference on Artificial Intelligence, vol. 34, no. 03, pp. 3065–3072 (2020)

14. Balazevic, I., Allen, C., Hospedales, T.: Multirelational poincaré graph embeddings. Adv. Neural Inf. Process. Syst. **32**, 1–11 (2019)
15. Ebisu, T., Ichise, R.: Toruse: knowledge graph embedding on a lie group. In: Proceedings of the AAAI Conference on Artificial Intelligence, vol. 32, no. 1 (2018)
16. Nickel, M., Tresp, V., Kriegel, H.P.: A three-way model for collective learning on multirelational data. In: ICML, vol. 11, no. 10.5555, pp. 3104482–3104584 (2011)
17. Yang, B., Yih, W., He, X., et al.: Embedding entities and relations for learning and inference in knowledge bases (2014). arXiv preprint arXiv:1412.6575
18. Trouillon, T., Welbl, J., Riedel, S., et al.: Complex embeddings for simple link prediction. In: International Conference on Machine Learning, pp. 2071–2080. PMLR (2016)
19. Nickel, M., Rosasco, L., Poggio, T.: Holographic embeddings of knowledge graphs. In: Proceedings of the AAAI Conference on Artificial Intelligence, 30, no. 1 (2016)
20. Dettmers, T., Minervini, P., Stenetorp, P., et al.: Convolutional 2d knowledge graph embeddings. In: Proceedings of the AAAI Conference on Artificial Intelligence, vol. 32, no. 1 (2018)
21. Jiang, X., Wang, Q., Wang, B.: Adaptive convolution for multirelational learning. In: Proceedings of the 2019 Conference of the North American Chapter of the Association for Computational Linguistics: Human Language Technologies, vol. 1 (Long and Short Papers), pp. 978–987 (2019)
22. Ren, F., Li, J., Zhang, H., et al.: Knowledge graph embedding with atrous convolution and residual learning (2020). arXiv preprint arXiv:2010.12121
23. Vashishth, S., Sanyal, S., Nitin, V., et al.: Interacte: improving convolution-based knowledge graph embeddings by increasing feature interactions. In: Proceedings of the AAAI Conference on Artificial Intelligence, vol. 34, no. 03, 3009–3016 (2020)
24. Nguyen, D.Q., Nguyen, T.D., Nguyen, D.Q., et al.: A novel embedding model for knowledge base completion based on convolutional neural network (2017). arXiv preprint arXiv:1712.02121
25. Neelakantan, A., Roth, B., McCallum, A.: Compositional vector space models for knowledge base completion (2015). arXiv preprint arXiv:1504.06662
26. Kamigaito, H., Hayashi, K.: Unified interpretation of softmax cross-entropy and negative sampling: With case study for knowledge graph embedding (2021). arXiv preprint arXiv:2106.07250
27. Xie, X., Zhang, N., Li, Z., et al.: From discrimination to generation: knowledge graph completion with generative transformer. In: Companion Proceedings of the Web Conference 2022, pp. 162–165 (2022)

Research on Link Prediction Algorithms Based on Multichannel Structure Modelling

Gege Li[1,2,3,4], Lin Zhou[1,2,3,4], Zhonglin Ye[1,2,3,4]([✉]), and Haixing Zhao[1,2,3,4]

[1] College of Computer, Qinghai Normal University, Xining, Qinghai, China
zhonglin_ye@foxmail.com
[2] The State Key Laboratory of Tibetan Intelligent Information Processing and Application, Xining, Qinghai, China
[3] Tibetan Information Processing and Machine Translation Key Laboratory of Qinghai Province, Xining, Qinghai, China
[4] Key Laboratory of Tibetan Information Processing, Ministry of Education, Xining, Qinghai, China

Abstract. Today's link prediction methods are based on the network structure using a single-channel approach for prediction, and there is a lack of link prediction algorithms constructed from a multichannel approach, which makes the features monotonous and noncomplementary. To address this problem, this paper proposes a link prediction algorithm based on multichannel structure modelling (MCLP). First, the network is sampled three times to construct its three subgraph structures. Second, the node representation vectors of the network are learned separately for each subgraph on a single channel. Then, the three node representation vectors are combined, and the similarity matrix is calculated for the combined vectors. Finally, the performance of the MCLP algorithm is evaluated by calculating the AUC using the similarity matrix and conducting multiple experiments on three citation network datasets. The experimental results show that the proposed link prediction algorithm has an AUC of 98.92%, which is better than the performance of the 24 link prediction comparison algorithms used in this paper. The experimental results sufficiently prove that the MCLP algorithm can effectively extract the relationships between network nodes, and confirm its effectiveness and feasibility.

Keywords: Link Prediction · Subgraph Sampling · Matrix Factorization · Similarity Matrix · Multichannel

1 Introduction

The complex network is an abstract model describing the real world. Nodes are used to describe entities in reality, and the relationships between entities are described by the connected edges between nodes. For the study of social analysis networks [1–3], individuals can be considered nodes and relationships between people can be considered edges. In the real world, everything is constantly moving with the passage of time, so the complex network can be regarded as a dynamic evolutionary model. There are hot issues in current research, for example, the accurate study of network evolution mechanisms [4]

Z. Yu et al. (Eds.): ICPCSEE 2023, CCIS 1880, pp. 269–284, 2023.
https://doi.org/10.1007/978-981-99-5971-6_20

and mining the relationship between nodes and between connected edges. The research object of link prediction is the relationship between nodes in a complex network, which is also a hot problem. In 2007, Liben-Nowel et al. [5] first defined the link prediction, which predicts the probability of links between unlinked nodes in a network using known network properties as a reference basis. There are two types of link predictions: predicting unknown links, and predicting future links. Predicting unknown links means predicting link relationships that already exist but are temporarily undiscovered. Predicting future links means predicting link relationships that may arise between nodes at a future moment [6].

For the study of link prediction, researchers have proposed various types of algorithms, which can be summarized into three major categories: link prediction algorithms based on node attributes, link prediction algorithms based on network topology, and link prediction algorithms based on entropy and maximum likelihood estimation [7]. Bu et al. [8] used the number of fans, followers, tweets, and retweets as node attributes and combined them with the common neighbor-based algorithm for link prediction, which made node prediction achieve good results. While the methods based on node attributes and network topology are simple, using node attributes cannot accurately reflect the characteristics of the network; thus, the network nodes of unconnected edges cannot be accurately predicted. The disadvantage of link prediction algorithms based on entropy and maximum likelihood estimation is that they are not suitable for link prediction on large-scale networks, because they cannot accurately explore the structural features of large-scale networks, which in turn leads to low prediction accuracy [9] and high algorithm complexity [9]. Currently, link prediction research is not only limited to small-scale networks but also aims to explore the intrinsic mechanisms of large-scale networks. However, there is a scarcity of methods to achieve efficient link prediction on large-scale networks, so it is necessary to combine neural networks and link prediction. Due to the complexity of the graph structure, the idea of convolution is used in graphs with good results. In the graph convolutional neural networks (GCNs) [10], as the number of layers deepens, the problem of an exponential increase in the number of neighbors arises, which leads to many methods to accelerate the training complexity of the graph convolutional network. Subgraph sampling (Graph SAINT) [11] is one method to accelerate the depth of convolutions. Subgraph sampling is a process of randomly removing the connected edges while keeping the nodes connected on the networks, which can effectively extract the features of networks and accelerate the training speed. Therefore, it is necessary to apply subgraph sampling in this paper. Then, a single subgraph sampling of the original network does not optimize its effect, and it cannot accurately represent the characteristics of the whole network. Therefore, subgraph sampling of the network is performed three times to fully extract the features of the original network, which further conducts the basis for the construction of a link prediction algorithm based on multichannel structure modelling.

In the field of word vector representation learning, various methods already exist, among which methods based on neural networks usually show excellent results, such as Word2Vec [12, 13]. The Word2Vec was proposed in 2013, which relies on co-occurrence relationships between words and achieves word vector representations by two language models, namely, CBOW and Skip-gram. The major difference between the CBOW model and the skip-gram model is that CBOW predicts the central word from the surrounding words, while skip-gram predicts the words around the central word from the central word. Later, some scholars put forward the DeepWalk [14] network representation learning algorithm based on Word2Vec. Network representation learning is actually the transformation from network features into vectors, and the network nodes are now displayed visually, as shown in Fig. 1.

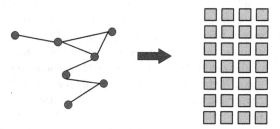

Fig. 1. Network representation learning visualization.

DeepWalk is a classical network representation learning algorithm. In a network, if the nodes have similarity to each other, the node representation vector obtained by Deep-Walk will have similarity. Yang et al. [15] proved that the DeepWalk method is equivalent to the factorization network structure feature matrix M. Then the network representation features obtained by these two methods must be equal. The representation vector obtained by using the factorized target matrix M not only inherits the advantages of DeepWalk, but also effectively avoids the problem of obtaining contextual node relationships using the DeepWalk random walk strategy and the problem of undertraining due to network sparsity.

The main works of this paper are to enrich the network structure features by subgraph construction and modelling the subgraph structure based on network representation learning algorithms. The MCLP algorithm is not only a simple and efficient model for predicting edge relationships between nodes, but can also be applied to many types of complex network tasks. In summary, the main contributions of this paper are follows.

(1) In this paper, the method of accelerating graph convolution, subgraph sampling, is applied to link prediction for the first time, which effectively extracts the features of the network and reduces the complexity of the algorithm. At the same time, the idea of multichannel is incorporated into the link prediction algorithm, which effectively avoids the unstable problem algorithm effect due to occasionality.

(2) In this paper, DeepWalk is introduced as the link prediction algorithm. The node representation vector based on matrix factorization is equivalent to the DeepWalk-based algorithm, which avoids the problem of using the DeepWalk random walk strategy to obtain contextual node relationships and insufficient training set due to sparse network structure. This method can fully exploit the network structure features and help to accomplish link prediction-related tasks.

(3) In this paper, a link prediction algorithm with multichannel structure modelling is proposed to effectively fuse subgraph sampling and network structure feature matrix.

To effectively extract the network features to different views, subgraph sampling is performed on the original network; hence, a new link prediction feature fusion framework is built, called the Multi-Channel Structure Modelling-based Link Prediction Algorithm (MCLP), and it further improves the accuracy of link prediction.

2 Related Work

The link prediction problem has important research value, and among the existing methods, there are three major categories: link prediction algorithms based on node attributes, link prediction algorithms based on network topology, and link prediction algorithms based on machine learning. Initially, researchers studied link prediction from the perspective of node attributes, and if the node attributes of two nodes are more similar, the probability of their link is higher [16]. However, this approach is difficult to implement because most of the information about node attributes is not publicly available.

Subsequently, more researchers have studied from the perspective of network topology, which includes three approaches: local information similarity-based approach, path-based approach, and random wander-based approach. (1) In the local information similarity-based approach, the metrics used are RA [17], AA [17], CN [17], etc. In RA (resource allocation), it is assumed that each node of the network will have its resources, the common neighbor acts as a transfer bridge between two nodes that are not directly connected, and each node takes an equal distribution to allocate its resources to the common neighbor. Finally, the similarity of the two nodes is the number of resources accepted by the target node. In AA (adamic-adar), the similarity value of node pairs is obtained by summing the node weights to all common neighbour, where the weights of each node are based on the node degree of the common neighbour. The CN (common neighbors) is judged based on the common neighbors of nodes, and a higher similarity indicates the existence of more common neighbours between nodes. The difference between the above three metrics is whether they take the common neighbor role into account. In addition, there are other basic metrics, such as Jaccard [18], Sorenson [19], and Salton [17]. Although the computational complexity of the method based on local information similarity is lower, it limits the mining of the information contained in the network, thus reducing the accuracy of the algorithm. (2) The path-based approach uses three metrics: LP (local paths) [20], Katz [21], and LHZ-II [22]. There is a certain connection between the three metrics, where the LP expands the range of neighbours when calculating the influence of neighbours on the occurrence of links between nodes, namely, it adds neighbours with path 3 to the CN. If the path length of LP is continuously increased, it can be equated with the Kata algorithm considering all the path information. (3) The method based on random walk can be measured by cos+ [23], SRW [20], etc. The basis of cos+, which calculates the similarity of node vectors, is the Mahalanobis distance. The above methods of link prediction based on network topology are also commonly used methods, which have lower computational complexity, but they cannot fully exploit the network structure features, which will further affect the link prediction effect.

In addition, there is machine learning-based link prediction, such as support vector machine (SVM), Bayesian network (BN) and neural network (NN). The use of such methods makes great progress in link prediction. Liang et al. [24] proposed a support vector machine-based link prediction for opportunity networks based on the weighted fusion of node pairs of spatial similarity and event features. Cao et al. [25] proposed a new method for plain Bayesian link prediction combining modalities, which fully takes the influence of modal density on link formation into account, thus constructing a role contribution function, while combining plain Bayesian and then inferring a new method for calculating similarity. Ye et al. [6] proposed a method based on matrix factorization from DeepWalk that can be applied in link prediction, which not only cleverly inherits the advantages of DeepWalk but also has a simple model framework. Although the machine learning-based link prediction methods summarized above are very effective, they expose their disadvantages when dealing with large data, which is the problem of longer training time and large memory consumption.

In summary, the three main types of link prediction algorithms have limitations, and all of them perform link prediction from a single channel perspective. Due to the higher quality of the network feature vectors obtained through neural networks, it has become one of the means to solve problems in various fields. In the field of graph representation learning, data enhancement by manipulating the graph structure can play a role in preventing feature loss. Zhu et al. [26] proposed a novel graph comparison representation learning method with adaptive enhancement. Therefore, we first use subgraph sampling three times to achieve graph enhancement and effectively extract the structural features of different views to accomplish multichannel construction. Second, we combine the matrix factorization based on DeepWalk to factorize the target matrix to obtain the node representation vector, which is further applied to link prediction.

3 Link Prediction Algorithm Based on Multichannel Structure Modelling

When studying link prediction, the data need to be transformed into a network structure for processing. Obtaining the representation vector and similarity matrix of each node is crucial for solving the link prediction problem. The description of the proposed link prediction algorithms based on multichannel structure modelling is shown in Fig. 2.

As shown in Fig. 2, the link prediction algorithm based on multichannel structure modelling consists of four steps, which are shown below.

Step 1, Subgraph sampling: The original network is subjected to edge deletion operations while ensuring network connectivity. We perform subgraph sampling three times on the original network to construct multiple channels.

Step 2, Node representation vector acquisition: For the three obtained subgraphs, the corresponding node representation vectors are calculated using DeepWalk's method based on matrix factorization.

Step 3, Similarity matrix calculation: The node representation vectors corresponding to the three subgraphs are combined, and then the similarity of the combined vectors is calculated to construct the similarity matrix of the network nodes.

Step 4, AUC to evaluate link prediction: According to the similarity matrix of network nodes, AUC values are calculated to verify the feasibility and effectiveness of the MCLP algorithm.

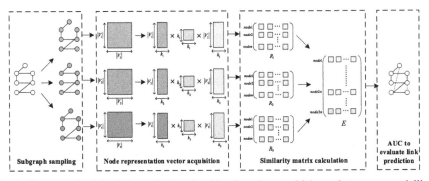

Fig. 2. Framework of the link prediction algorithm based on multichannel structure modelling. (*nodern* represents the node vector, here $r \in [1, 3]$, r represents the number of channels, which also corresponds to the number of subgraph samples.)

3.1 DeepWalk-Based Network Node Representation Vector Acquisition

In 2014, DeepWalk was first applied to the field of representation learning, and a new network representation learning model was proposed to obtain node representation vectors through co-occurrence relationships between nodes. The process of DeepWalk is as follows. First, the initial representation vector is defined for each node, and the walk path length is set. Second, the initial representation vector of the node is input into the neural network, the neighboring nodes of the current node are selected by uniform sampling, the random walk is continuously performed until the walk path length is reached, and then the walk is stopped once. Then, by repeating the above operation, multiple walk sequences are generated, which dynamically adjust the node representation vector. Finally, the final node representation vector is obtained.

However, the DeepWalk-based method for obtaining node representation vectors suffers from the problem of undertraining or slow training due to network sparsity. Yang et al.[15] prove that the DeepWalk algorithm is equivalent to factorizing the target matrix M. Therefore, this idea can be used to generate the node representation vector by factorizing the target matrix M, which effectively avoids the drawback of obtaining the node representation vector based on DeepWalk. Ye et al. [6] defined this target matrix M as shown in Eq. 1.

$$M_{i,j} = \log \frac{(e_i(A + A^2 + A^3 + ... + A^t))_j}{t} \tag{1}$$

Assuming a random walk from node i, e_i represents the initial vector, and e_i is a $|V|$ dimensional vector. A denotes the structural feature matrix, and the specific expression is shown in Eq. 2. The value of column j in $e_i A^t$ represents the probability value from node i to node j when the random walk step is t,

$$A_{ij} = \begin{cases} 1/d_i, & (i, j \in E). \\ 0, & otherwise. \end{cases} \tag{2}$$

Here, d_i represents the degree of node i.

From Eq. 1, we can see that the complexity of computing the target matrix M is $O(|V|^3)$, the matrix contains a large number of nonzero elements, and removing the *log* will achieve a reduction in complexity. Then, Yang et al.[15] combined the algorithm time complexity, space complexity and accuracy and proposed the factorized target matrix $M = (A + A^2)/2$, and the complexity of computing the target matrix is $O(n^2)$, which effectively improves the performance of the algorithm. In summary, the target matrix chosen for matrix factorization is

$$M = \frac{A + A^2}{2} \tag{3}$$

Algorithm1: Subgraph sampling (G, trainRatio)

Input: Network G,   Training ratio *trainRatio*.
Output: Training set network *train*, Test set network *test*.
Begin
    num_testlinks = [(1-*trainRatio*) * The number of edges of G]. # [ ] is the integer.
    test = an edge-free subgraph of G.
    while (The number of edges of test < num_testlinks) **then**
        uid1, *uid2* = The endpoint of the random, unselected side of G.
        $g = G.$ delete([*uid1*, *uid2*]).
        if *uid1* and *uid2* are still connected in g **then**
            test. add([*uid1*, *uid2*]).
            $G = g$.
        end
    end
    return *train* = G, *test*;
End

3.2 Network Structure Subgraph Sampling

Subgraph sampling is the process of deleting the edges from the original network while keeping the number of nodes and the connectivity between nodes constant. This method is used to accelerate the graph convolution, which is done by deleting edges. In addition, multiple subgraph sampling can extract different features from different perspectives, which enables feature complementation between subnetworks and effective mining of the overall network features, as well as making the model lightweight. Therefore subgraph sampling is introduced into the algorithmic framework to conduct multichannel construction, and the number of subgraph samples is the number of channels. By comparing experiments based on the number of channels in Subsect. 4.4, it has been demonstrated that the subgraph sampling module is effective for the model framework of this paper and that the multichannel link prediction-based approach outperforms the single-channel link prediction-based algorithm.

The introduction of the MCLP algorithm is based on a three-channel example, so the network structure is sampled three times to generate three subgraphs G_r, where $r \in [1, 3]$,

r is the number of subgraphs and the number of channels. The detailed procedure of one subgraph sampling is shown in algorithm1, where the training set network *train* is the subgraph generated by sampling once.

3.3 Node Representation Vector Acquisition

The node representation vector is obtained from the DeepWalk-based representation learning in Subsect. 3.1. The essence of this method is to obtain the network node representation vector by matrix factorization, where the target matrix M is defined as $M = (A + A^2)/2$.

Based on the above contents, the three subgraphs generated by the original network G_r must be transformed into the corresponding structural feature matrices as $A_{i,j}^{(r)}$, whose expressions are defined as

$$A_{i,j}^{(r)} = \frac{1}{d_i^{(r)}}, (i, j \in |V_r|, r = 1, 2, 3) \tag{4}$$

Here $d_i^{(r)}$ are the degrees of node i corresponding to subgraphs G_r, $r \in [1, 3]$, r is the number of subgraphs and the number of channels.

The target matrices M_r are obtained by $A_{i,j}^{(r)}$ in Eq. 4, and the specific expressions are shown in Eq. 5.

$$M_r = \frac{A_{i,j}^{(r)} + A_{i,j}^{(r)2}}{2} \tag{5}$$

For the target matrices M_r, the SVDS method is chosen in this paper. Based on the SVD method, SVDS makes great improvements. For example, the SVDS method is less complex, highly customizable, and highly moldable, and it can derive specific maximum eigenvalues as well as a feature row vector and feature column vector. The proposed method factorizes the target matrices M_r into the specific form of three matrix multiplications, and the whole factorization process is shown in Eq. 6.

$$M_r = U_{V_r \times k_r}^{(r)} \times S_{k_r \times k_r}^{(r)} \times V_{k_r \times V_r}^{(r)} \tag{6}$$

The singular vector corresponding to the target matrix M is matrix U, the diagonal matrix is matrix S, and the diagonal elements in matrix S are the singular values of matrix M. $|V_r|$ represents the number of nodes corresponding to each subgraph. k_r represents the number of feature values corresponding to target matrices M_r, $r \in [1, 3]$, r is the number of subgraphs and the number of channels.

Then we multiply the factorized $U_{|V_r| \times k_r}^{(r)}$ with $S_{k_r \times k_r}^{(r)}$, and the corresponding nodes of the three subgraphs will be represented by the vectors R_r, as shown in Eq. 7.

$$R_r = U_{V_r \times k_r}^{(r)} \times (S_{k_r \times k_r}^{(r)}{}^T)^{1/2} \tag{7}$$

3.4 Similarity Matrix Calculation and AUC to Evaluate MCLP

With the above descriptions, the node representation vectors R_r of the three subgraph structures have been obtained, and we combine them to obtain the combined vector E, whose expression is shown in Eq. 8.

$$E = R_1 \oplus R_2 \oplus R_3 \tag{8}$$

"$\oplus$" is defined as the combination symbol, and the combination vector E is a matrix with $|V_1| + |V_2| + |V_3|$ rows and $|k_1|$ columns, which can also be considered to be a matrix with $3|V_1| = 3|V_2| = 3|V_3|$ rows and $|k_1| = |k_2| = |k_3|$ columns.

Based on the combined vector E obtained above, the similarity matrix $W_{i,j}$ is calculated and its expression is shown in Eq. 9.

$$W_{i,j} = E(i, *) \times E(j, *) \tag{9}$$

Here i is the i-th row vector in the combined vector E and j is the j-th row vector in the combined vector E.

The similarity matrix $W_{i,j}$ obtained above is applied to the link prediction algorithm. The AUC is used to evaluate link prediction algorithms based on multichannel structure modelling (MCLP).

4 Experimental Results and Analysis

4.1 Experimental Data

In this paper, we conduct extensive experiments on the Citeseer, DBLP and Cora datasets to demonstrate the validity of the proposed model (MCLP). All three datasets are publicly available citation network datasets. The Citeseer dataset is composed of the world's top conferences, whose data link is http://www.cs.umd.edu/~sen/lbc-proj/data/citeseer.tgz. The DBLP dataset consists of author partnerships, whose data link is http://dblp.uni-trier.de/xml/. The Cora dataset is composed of scientific publications, whose data link is http://www.cs.umd.edu/~sen/lbc-proj/data/cora.tgz. The relevant specific information is shown in Table 1. In Table 1, $N.$ represents the number of nodes, $S.$ represents the number of edges, $C.$ represents the number of categories, $A.$ represents the average degree, and $D.$ represents the density.

Table 1. Dataset description.

Dataset	N	S	C	A	D
Citeseer	3312	4732	6	2.857	0.001
DBLP	3119	39516	4	21.07	0.005
Cora	2708	5429	7	4.01	0.001

From Table 1, the following can be seen:

(1) The Citeseer dataset has 3312 nodes, the DBLP dataset has 3119 nodes, and the Cora dataset has 2708 nodes. The difference in the number of nodes is not very large, but there is a large difference in the number of edges. The dataset with the largest difference in the number of edges is the DBLP dataset, which also has the smallest number of categories.

(2) Under almost equal numbers of nodes, the values corresponding to the DBLP dataset are several times larger than those corresponding to the Citeseer and Cora datasets based on the average degree and density, which is caused by the different numbers of edges. Therefore, it can be further inferred that the DBLP network is dense, the Citeseer, and Cora networks are sparse.

4.2   Comparison Algorithm

In this paper, 24 existing link prediction algorithms are used as comparison algorithms to facilitate effective evaluation and analysis of the performance of the proposed MCLP algorithms. In total, these can be summarized into two types: traditional link prediction algorithms and the latest link prediction algorithms.

Traditional link prediction algorithms are as follows:

① Link prediction algorithms based on local information: CN, AA, PA, RA, HPI, HDI, LHN-I, Salton, Jaccard, LNBAA, LNBCN, LNBRA.

② Link prediction algorithms based on a path: LP, Katz, LHN-II.

③ Link prediction algorithms based on random walk: ACT, Cos+, LRW, SRW.

④ Link prediction algorithm based on matrix forest theory: MFI.

⑤ Link prediction algorithm based on self-consistent similarity: TSCN.

The latest link prediction algorithms are as follows:

① LPMF: This algorithm uses a matrix factorization equivalent to DeepWalk for link prediction, which was proposed by Ye et al. in 2018 [6].

② TELP: This algorithm is a network node text enhancement algorithm that fully takes into account the textual information related to the target network, which was proposed by Cao et al. in 2019 [25].

③ NRLP: This algorithm is a link prediction algorithm that fuses node labels, strong relationships and weak relationships, which considers the correlation properties and different relationship strengths of known paths and was proposed by Wang et al. in 2022 [27].

4.3   Analysis of Experimental Results

To ensure the effectiveness of the link prediction algorithm based on multichannel structure modelling (MCLP), several experiments are performed on three citation network datasets: Citeseer, DBLP, and Cora. The results of the three-channel experiments are compared with 24 existing link prediction algorithms. For the evaluation metrics of the algorithms, the AUC evaluated on the testset is selected in this paper. A higher the represents higher accuracy and better performance of the algorithm. The principle of this metric is that (1) if the similarity value of the existing edge is greater than that of nonexistent edges, then 1 point can be added to the AUC, and (2) if the similarity value of the existing edge is equal to the similarity value of the nonexistent edge, then 0.5 points can be added to the AUC. The formula for calculating AUC is shown in Eq. 10.

$$AUC = \frac{n' + 0.5n''}{n} \tag{10}$$

Here, n denotes the number of independent comparisons, n' denotes the number of times that the similarity value of the existing edge is greater than that of the nonexistent edge, and n'' denotes the number of times that the similarity value of the existing edge is equal to the similarity value of the nonexistent edge.

To ensure the fairness of the comparison experiments, the length of the representation vector obtained by training is set to 300, and the training ratios are 0.7, 0.8 or 0.9. The results of the comparison experiments with 24 link prediction algorithms are shown in Table 2.

As shown in Table 2, the multichannel structure modelling-based link prediction algorithm (MCLP) is compared with 24 link prediction algorithms in this paper, and the following analytical results are obtained.

(1) The MCLP algorithm obtains the optimal performance on both the DBLP and Cora datasets obtaining a maximum AUC of 98.92% especially on the Cora dataset, which shows that the MCLP algorithm can effectively extract the structural features of the networks.
(2) The range of AUC for the MCLP algorithm on the Citeseer, DBLP and Cora datasets is from 95.98% to 98.92%, which shows that the AUC of the algorithm fluctuates very little, reflecting the stability of the model.
(3) The performance of the MCLP algorithm is slightly lower than that of MFI, Katz and NRLP algorithms on the Citeseer dataset. This is because the MFI algorithm is a global algorithm, the Katz algorithm considers the global path, which is more suitable for application to the network with a shorter average path, and the NRLP algorithm considers the node-related attributes and different relationship strengths. However, our MCLP algorithm obtains the subgraphs by subgraph sampling and it builds a multichannel structure modelling method, that fully extracts the network node features. Therefore, the MCLP algorithm works better than the MFI, Katz and NRLP algorithms on both the DBLP and Cora datasets, which further proves that the performance of the MCLP algorithm is excellent in general.

In summary, the performance of the MCLP algorithm proposed in this paper is better than that of 24 comparison algorithms. This is the result integrating subgraph sampling and factorizing matrix M. Multichannel subgraph sampling enables the network features to be fully extracted and utilized, and then combining factorization matrix M enables deeper learning for the structural features and deeper training for node representation vectors.

4.4   Channel Number Comparison Experiments

To prove the influence of the number of channels on the MCLP model, this paper conducts channel comparison experiments. The basis of constructing multiple channels

Table 2. Comparison of experimental results table.

Dataset	Citeseer			DBLP			Cora		
Training ratio	0.7	0.8	0.9	0.7	0.8	0.9	0.7	0.8	0.9
CN	68.13	72.08	74.67	85.49	88.40	90.68	69.50	72.38	78.19
Salton	66.32	72.73	74.44	86.00	87.92	90.74	69.38	72.13	77.89
Jaccard	66.51	72.25	74.33	85.92	88.26	90.98	69.25	72.00	77.09
HPI	66.29	72.18	74.42	85.61	88.95	90.77	69.38	72.44	77.93
HDI	66.03	72.52	74.17	85.72	88.31	90.84	69.52	72.53	76.67
LHN-I	66.47	72.93	74.46	85.80	87.87	89.95	69.19	72.16	77.30
AA	66.37	72.22	74.33	86.00	88.22	90.95	69.35	72.66	77.60
RA	66.37	72.12	74.63	86.56	88.50	90.81	69.47	72.47	77.97
PA	78.98	79.06	79.53	76.39	77.13	77.54	71.50	71.91	71.50
LP	81.06	86.83	88.45	92.96	93.65	94.94	80.12	82.97	87.90
Katz	**96.89**	**97.98**	97.19	93.45	94.18	94.83	90.89	92.14	94.44
LHN-II	95.76	96.85	96.20	90.86	91.80	92.80	89.41	90.37	93.64
LNBAA	66.37	72.64	74.52	86.07	88.42	91.12	69.42	72.50	78.01
LNBCN	66.70	72.27	74.25	85.60	88.47	90.80	69.50	72.19	77.79
LNBRA	66.05	72.23	74.27	85.86	88.91	91.23	69.32	72.84	77.74
ACT	75.88	75.59	73.79	79.00	80.07	80.84	74.11	73.67	74.00
Cos+	88.57	89.38	88.49	91.53	93.47	95.08	90.25	90.98	93.22
LRW	87.21	90.13	91.25	92.75	93.35	94.09	88.48	90.58	93.63
SRW	86.34	90.05	90.47	90.50	92.25	94.06	88.40	90.50	93.62
MFI	96.68	98.00	**97.80**	95.13	96.00	97.07	93.13	94.25	95.60
TSCN	84.26	85.68	86.27	91.25	91.03	92.34	88.35	90.64	92.98
TELP	95.00	96.65	96.81	93.75	94.79	95.77	91.21	92.43	94.40
LPMF	87.18	90.64	94.98	93.42	94.70	95.13	89.57	92.13	93.93
NRLP	96.71	97.83	97.15	94.43	95.42	96.40	92.32	93.57	94.53
MCLP	95.98	96.16	96.21	**97.41**	**97.64**	**97.63**	**98.04**	**98.34**	**98.92**

is subgraph sampling, and then the number of subgraph samplings is the number of channels. In this paper, we set the number of channels as 1, 2, 3, 4 or 5 and the training ratios as 0.7, 0.8 or 0.9, respectively, and the experimental results are shown in Table 3.

From Table 3, the highest AUC corresponds to the channel on the Citeseer dataset of 5 when the training ratio is not 0.8. On the DBLP and Cora datasets, the highest AUC corresponds to 5 channels when the training ratio is 0.7, 0.8 or 0.9, and the corresponding AUC shows an increasing trend as the number of channels increases. In summary, the premise of constructing a multichannel is to perform subgraph sampling, and the number of channels represents the number of subgraph samplings. It can be proven that the

Table 3. Experimental results of channel number comparison.

Dataset	Citeseer			DBLP			Cora		
Training ratio	0.7	0.8	0.9	0.7	0.8	0.9	0.7	0.8	0.9
1	84.73	89.61	90.53	93.52	94.66	95.20	85.39	88.85	90.94
2	92.78	96.23	96.76	96.62	96.85	97.04	95.69	96.95	98.26
3	95.98	96.16	96.21	97.41	97.64	97.63	98.04	98.34	98.92
4	93.51	**97.25**	97.04	97.58	97.62	98.16	98.90	99.06	99.00
5	**96.74**	96.91	**98.28**	**97.74**	**98.00**	**98.23**	**99.01**	**99.26**	**99.14**

subgraph sampling module has an influence on this model, and it further proves that the multichannel-based link prediction algorithm has superior performance to that of the single-channel-based link prediction algorithm.

4.5 Tuning Participation Analysis

In the tuning experiments, the results of the three-channel experiments are also used as an example, and two parameters are set as the representation vector length k and the training ratio *trainingRatio* in the training set. The original network is sampled three times and the structural feature matrices of the three subgraphs are obtained. The corresponding target matrices are obtained by $M = (A + A^2)/2$, and the node representation vectors are obtained by SVDS factorization. Finally, the three node representation vectors are combined into one node representation vector, and then the similarity matrix of the node representation vectors is calculated and evaluated by the AUC. Then, to study the effects of vector length k and training ratio *trainingRatio* on the performance of the MCLP algorithm, this paper sets five different representation vector lengths as 60, 90, 150, 210, and 300 and six different training ratios as 0.7, 0.75, 0.8, 0.85, 0.9 and 0.95, and the experimental results are shown in Fig. 3.

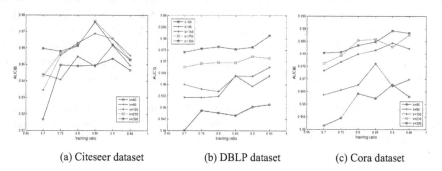

(a) Citeseer dataset   (b) DBLP dataset   (c) Cora dataset

Fig. 3. Relationships between training ratio, vector length and AUC.

An analysis of Fig. 3 shows that different training ratios and representation vector lengths have different effects on the performance of the MCLP algorithm.

(1) On the Citeseer, DBLP and Cora datasets, when the AUC is the highest, it means that the vector lengths are all 300, and when the AUC is the lowest, it means that the vector lengths are all 60. While the corresponding training ratios are very different, the MCLP has the best performance when the corresponding training ratio is set as 0.85, 0.95 and 0.9 on the Citeseer, DBLP and Cora datasets.

(2) The DBLP dataset is a dense network, and the AUC does not vary very much and it is stable within a certain range except that the training ratio is 0.7 or 0.75, and the vector length is 60, the training ratio is 0.8 or 0.9, and the vector length is 90 or 150.

(3) The Citeseer and Cora datasets are sparse networks, and their corresponding AUCs are more variable.

In summary, both the training ratio and the length of the representation vector affect on the performance of the MCLP algorithm. The impact is smaller for dense networks and larger for sparse networks.

5 Summary

In this paper, we propose a link prediction algorithm based on multichannel structure modelling (MCLP). To effectively extract features, the algorithm first samples the network three times to construct multiple channels. Second, the node representation vectors are derived by matrix factorization. Then the node representation vectors are combined and the similarity matrix is calculated. Finally, the performance of the MCLP algorithm is evaluated on three real citation networks the Citeseer, DBLP, and Cora datasets based on the AUC. The experimental results show that (1) The AUC of the link prediction algorithm based on the multichannel structure modelling is 98.92%, and its AUC has a 1%-30% improvement over the 24 link prediction algorithms compared in this paper. (2) The number of channels comparison experiments show that, when the training ratio is 0.7, 0.8 or 0.9, the AUC shows an increasing trend with the increasing number of channels on the DBLP and Cora datasets. Therefore, the performance of the multichannel-based link prediction algorithm is better than that of the single-channel-based link prediction algorithm for the algorithmic model in this paper. (3) By adjusting the representation vector length k, it is known that the algorithm performs best on the three datasets when the combined vector length is 300. (4) By adjusting the training ratio, it is known that the training ratio for the best performance is 0.85 on the Citeseer dataset, 0.9 on the DBLP dataset, and 0.95 on the Cora dataset. Then, in future studies, different matrix factorization algorithms can be chosen for decomposing factorizing the target matrix, thus allowing the structural features of the network to be fully extracted.

Acknowledgement. This article was supported by the National Key Research and Development Program of China (No. 2020YFC1523300) and the Innovation Platform Construction Project of Qinghai Province (2022-ZJ-T02).

References

1. Xie, X., Li, Y., Zhang, Z., Han, S., Pan, H.: A joint link prediction method for social network. In: Wang, H., et al. (eds.) ICYCSEE 2015. CCIS, vol. 503, pp. 56–64. Springer, Heidelberg (2015). https://doi.org/10.1007/978-3-662-46248-5_8

2. Schafer, L., Graham, J.W.: Missing data: our view of the state of the art. Psychol Methods **7**(2), 147–152 (2016)
3. Kossinets, G.: Effects of missing data in social networks. Social Netw. **28**(3), 247–268 (2003)
4. Albert, R., Barabasi, A.L.: Statisrical mechanics of complex networks. Rev. Mod. Phys. **74**(51), 47–97 (2002)
5. Liben-nowell, D., Kleinberg, J.: The link-prediction problems for social networks. J. Am. Soc. Inform. Sci. Technol. **58**(7), 1019–1031 (2007)
6. Ye, Z.L., Cao, R., Zhao, H.X., Zhang, K., Zhu, Y.: Link prediction based on matrix factorization for DeepWalk. Appl. Res. Comput. **37**(02), 424–429 (2020)
7. Bianconi, G.: Entropy of network ensembles. Phys. Rev. E: Stat., Nonlin, Soft Matter Phys. **79**(2), 036114 (2009)
8. Bu, X.Y., Chen, M.L.: Research on link prediction in social networks. Researches Libr. Sci. (17), 17–21 (2016)
9. Liu, W.P., Lu, L.Y.: Link prediction based on local random walk. Europhys. Lett. **89**(5), 58007–58012 (2010)
10. Kipf, T.N., Welling, M.: Semi-supervised classification with graph convolutional networks (2016). arXiv: 1609.02907
11. Zeng, H., Zhou, H., Srivastavs, A, et al.: Graphsaint: graph sampling based inductive learning method (2019). arXi: 1907.04931
12. Mikolov, T., Chen, K., Corrado, G., et al.: Efficient estimation of word representations in vector space (2013). arXiv: 1301.3781
13. Mikolov, T., Sutskever, I., Chen, K., et al.: Distributed representations of words and phrases and their compositionality. Adv. Neural. Inf. Process. Syst. **26**(14), 3111–3119 (2013)
14. Perozzi, B., Al-Rfou, R., Skiena, S.: DeepWalk: online learning of social representations. In: Proceedings of the 20th ACM SIGKDD Conference on Knowledge Discovery and Data Mining, pp. 701–710 (2014)
15. Yang, C., Liu, Z.: Comprehend deepwalk as matrix factorization (2015). arXi:1501.00358
16. Zhang, Y.X., Feng, Y.X.: A summary of the methods and development of link prediction. Meas. Control Technol. **38**(02), 8–12 (2019)
17. Zhou, T., Lu, L., Zhang, Y.C.: Predicting missing links via local information. Eur. Phys. J. B **71**(4), 623–630 (2009)
18. Jaccard, P.: Etude de la distribution florale dans une portion des Alpes et du Jura. Bull. Soc. Vaud. Sci. Nat. **37**(142), 547–579 (1901)
19. Sorenson, T.: A method of establishing groups of equal amplitude in plant sociology based on similarity of species and its application to analyses of the vegetation on Danish commons. Biologiske Skrifter **5**(4), 1–34 (1957)
20. Lu, L., Jin, C.H., Zhou, T.: Similarity index based on local paths for link prediction of complex networks. Phys. Rev. E **80**(4), 046122 (2009)
21. Katz, L.: A new status index derived from sociometric analysis. Psychometrika **18**(1), 39–43 (1953)
22. Leicht, E.A., Holme, P., Newman, M.E.: Vertex similarity in networks. Phys. Rev. E: Stat., Nonlin, Soft Matter Phys. **73**(2), 026120 (2006)
23. Fouss, F., Pirotte, A., Renders, J.M., et al.: Random-walk computation of similarities between nodes of a graph with application to collaborative recommendation. IEEE Trans. Knowl. Data Eng. **19**(3), 355–369 (2007)
24. Liao, L., Zhang, H.F.: Link prediction of opportunistic networks based on support vector machine. Inf. Commun. **09**, 23–25 (2018)

25. Cao, R., Zhao, H.X., Ye, Z.L.: Link prediction algorithm based on network node text enhancement. Comput. Appl. Softw. **36**(3), 227–235+242 (2019)
26. Zhu, Y., Xu, Y., et al.: Graph contrastive learning with adaptive augmentation. In: Proceedings of the Web Conference 2021, pp. 2069–2080 (2021)
27. Wang, S.Y., Gong, J.Y.: Link prediction algorithm fusing node label and strength relationship. Comput. Eng. Appl. **58**(18), 71–77 (2022)

Modal Interactive Feature Encoder for Multimodal Sentiment Analysis

Xiaowei Zhao[1,2], Jie Zhou[1], and Xiujuan Xu[1(✉)]

[1] School of Software, Dalian University of Technology, Dalian 116620, China
{xiaowei.zhao,xjxu}@dlut.edu.cn
[2] Key Laboratory for Ubiquitous Network and Service Software of Liaoning Province,
Dalian 116620, China

Abstract. Multimodal Sentiment analysis refers to analyzing emotions in information carriers containing multiple modalities. To better analyze the features within and between modalities and solve the problem of incomplete multimodal feature fusion, this paper proposes a multimodal sentiment analysis model MIF (Modal Interactive Feature Encoder For Multimodal Sentiment Analysis). First, the global features of three modalities are obtained through unimodal feature extraction networks. Second, the inter-modal interactive feature encoder and the intra-modal interactive feature encoder extract similarity features between modalities and intra-modal special features separately. Finally, unimodal special features and the interaction information between modalities are decoded to get the fusion features and predict sentimental polarity results. We conduct extensive experiments on three public multimodal datasets, including one in Chinese and two in English. The results show that the performance of our approach is significantly improved compared with benchmark models.

Keywords: Multimodal Sentiment Analysis · Modal Interaction Feature · Encoder

1 Introduction

The modality of information means the source or form of information [1], such as text, video, voice, etc. Compared with a single modality, which only describes one aspect of things, multimodal data contains multiple modalities that can present more abundant and diversified content. However, the modalities in multimodal data are not isolated, and there are various associations between different modalities. Therefore, in studying multimodal data, effectively using the correlation information of inter-modalities and solving the contradictory information between each modality is the key to improving the performance of multimodal sentiment analysis.

In the methods of unimodal feature extraction, researchers usually extract the unimodal internal space by convolution neural network, and extract time series information by recurrent neural network. For text characteristics, there are some pre-trained

Z. Yu et al. (Eds.): ICPCSEE 2023, CCIS 1880, pp. 285–303, 2023.
https://doi.org/10.1007/978-981-99-5971-6_21

language models, such as ELMo [2], GPT [3], and BERT [4], which can map high-dimensional sparse features to low-dimensional dense semantic representation and have better text semantic understanding. The key to multimodal fusion feature extraction lies in effectively extracting the internal structural features of modalities and analyzing the interactive features between modalities. Nowadays, many feature fusion methods and models have been proposed by researchers, such as TFN [5], LMF [6], MFN [7], Graph-MFN [8], MULT [9], MISA [10], Self-MM [11], etc. These multimodal fusion models all discuss how to better integrate internal features of each unimodal and interactive information between modalities to synchronize multimodal information sequences and improve the accuracy of multimodal sentiment analysis tasks.

Although there are many optimization and improvement models for multimodal sentiment analysis tasks, there are still significant areas for improvement in the trade-off between the inter-modal and inter-modal interactive features. For example, ignoring whether the introduction of multimodal information will cause redundancy. How to solve the emotional conflicts between modalities, whether each modality has the same contribution weight to the final results, and whether to consider the temporal context of sequence information while paying attention to the characteristics of the unimodal itself. Therefore, it is still a vital issue of multimodal sentiment analysis to consider the inter-modal features and the inter-modal features, remove redundant information to obtain valuable features, and weigh the contribution of each modality. According to the above analysis, this paper proposes a multimodal sentiment analysis model based on an interactive modal feature (Modal Interactive Feature Encoder for Multimodal Sentiment Analysis, MIF). The main contributions can be summarized as follows:

- We propose a multimodal sentiment analysis model for a video that follows hierarchical fusion and considers interactions in intra-modal and inter-modal. We introduce the multi-head attention mechanism and bi-directional long and short-term memory neural network (BiLSTM) into the Encoder function. Multi-head attention focuses on the multi-dimensional features of the modalities, reducing the redundant information and noise interference, and BiLSTM fully extracts the context series information of the modalities.
- We introduce cosine similarity and difference loss into task loss, which acts on the inter-modal Encoder and intra-modal Encoder, respectively. Inter-modal Encoder captures common features with a high similarity between modalities and fully uses inter-modal interaction information. Intra-modal Encoder extracts intra-modal special features and global information.
- We conduct extensive experiments on three public datasets, one is in Chinese, and the others are in English, and the performance exceeds the benchmark models on most evaluation Metrics.

The rest of this paper is organized as follows. Section 2 gives a brief overview of the related literature. Section 3 proposes a multimodal sentiment analysis model MIF and introduces its framework and algorithms. Section 4 presents detailed information about the data source used in our study and reports the experimental results. Finally, we draw our conclusions in Sect. 5.

2   Related Work

Social media (such as Tiktok, Weibo, Youtube, etc.) use short video interaction as the mainstream form. Voice and visual modality could make up for the missing information in text sentiment analysis. These two data forms are crucial to identifying opinion holders' sentiments. Effective integration of text, voice, and facial expression information can improve the performance of sentiment analysis tasks [12].

2.1   Multimodal Fusion Embedding

In recent years, researchers have proposed many optimized and improved models in multimodal feature fusion methods. They are committed to designing a new feature fusion network that can thoroughly learn intra-modal representation and inter-modal dependency and capture better multimodal fusion features for better performance.

Morency et al. (2011) [13] first proposed a multimodal sentiment analysis problem integrating text, voice, and visual features proving that the joint model can be effectively used to identify emotions in online video, and developed an English multimodal sentiment data set YouTube. Before multimodal fusion, it is an important research topic in multimodal sentiment analysis to extract the context relationship between adjacent utterances and consider the importance of cross-modal utterances. Poria et al. (2017) [14] proposed the Bi-direction Contextual Long Short-Term Memory (BC-LSTM) model, which uses two-way LSTM to extract the context semantic information of each sample feature, and then considers the context relationship between the utterances in each video. Chen et al. (2017) [15] proposed the Gated Multimodal Embedding Long Short-Term Memory with Temporal Attention (GME-LSTM (A)) model with temporal attention, embedding LSTM into the gating mechanism and combining temporal attention for word-level fusion. Majumder et al. (2018) [16] proposed a hierarchical fusion and context modeling model to associate bimodal and tri-modal information data.

However, the above methods ignore the combination of the internal information of each modality and the interaction between modalities.

2.2   Modal Interaction Feature Learning

The extraction of modal interaction features is not only necessary to ensure the unique features and semantic integrity of each modality, but also to achieve effective extraction of interrelated information between different modalities. With the introduction of an attention mechanism, it can weigh the impact of various modalities and extract important features [17].

Zadeh et al. (2017) proposed the Tensor Fusion Network (TFN) [5], which creates a multi-dimensional tensor by calculating the outer product to combine the interaction between modalities. However, its computational space complexity is high. In 2018, Zadeh et al. proposed the Multi-Attention Recurrent Network (MARN) model [18]. The interaction within and between modalities is simulated by allocating multiple attention scores. Liu et al. [6] (2018) used low-rank matrix decomposition to extract cross-modal information, effectively reducing the complexity of tensor calculation. Tsai et al. (2019) proposed a cross-modal transformer model [9], which uses the cross-modal Transformer to model all modalities, and repeatedly strengthen the target modality with the low-level features of the other modality. Huddar et al. (2020) proposed a multi-level context feature extraction technology based on BiLSTM to consider the context information at the utterance level [19]. The inter-modal fusion based on attention is used to adapt to the connection between modalities in multimodal fusion. Yu et al. (2021) proposed a single modality label generation strategy based on self-monitoring to assist multimodal sentiment classification and prediction tasks [11].

3   Method

Since interaction features between multiple modalities lead to different information contribution weights, this paper proposes a multimodal sentiment analysis model MIF (Modal Interactive Feature Encoder for Multimodal Sentiment Analysis) to solve the video multimodal sentiment analysis task. In the model, text (T), acoustic (A), and visual (V) are used as inputs. The overall framework of the MIF model is shown in Fig. 1. From Fig. 1, the model proposed in this paper consists of three layers. The bottom layer is the unimodal representation learning, the middle layer is the modality representation encoder, and the top layer is feature fusion prediction.

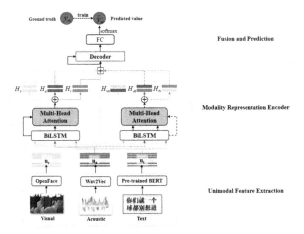

Fig. 1. Workflow of the Proposed Modal Interactive Feature Encoder for Multimodal Sentiment Analysis (MIF). Arrows indicate the direction of data transmission, a dotted line is a common data pipeline, and a full line is a unique data pipeline.

3.1   Unimodal Representation Learning

Text, acoustic and visual modalities are drawn from the video, and each video is divided into small utterances, and each utterance contains three unimodal features. Given a multimodal signal $U_i = \{U_i^t, U_i^a, U_i^v\}$, U_i^m, $m \in \{t, a, v\}$ represents the original unimodal sequence obtained from the video fragment i, $m \in \{t, a, v\}$ means text, acoustic, and vision respectively, and i ∈ N indicates the number of N samples. The final purpose of the multimodal sentiment analysis task is to obtain the predicted value $y_i^r \in R$, which is used to reflect the sentiment intensity value.

The pre-trained Chinese and English BERT-base [4] model is used for a textual modality to obtain 768-dimensional word embedding. Because the sentence length differs, we adopt padding and truncation to let the final length be L. In short sentences, specific characters are used to fill in the end of the sentence. In long sentences, the first L words are selected to constitute word embedding. The final hidden layer output obtained by BERT is used as a word embedding vector $u_a \in R^{S_a \times d_a}$, where S_a, d_a is the sentence length L and word embedding dimension, respectively.

For acoustic modality, wav2vec 2.0 [20] extracts the general features of audio. This tool is based on CNN and Transformer to predict the future frame with the current frame on the feature level. The principle of wav2vec is similar to the pre-training model BERT. We obtain the acoustic feature vector through the wav2vec2-base and train the model by the fine-tuning mechanism. Finally, we get 768-dimensional general features, the final acoustic feature vector $u_a \in R^{S_a \times d_a}$, S_a, d_a represents the acoustic length and embedding dimension, respectively.

For visual modality, FFmpeg [21] frames the videos at a specific rate, and then MTCNN [22] is applied to detect the aligned face. Finally, Multi Comp OpenFace2.0 [23] is used to extract facial landmark features, and 709-dimensional frame-level visual features are obtained, the final visual representation $u_v \in R^{S_v \times d_v}$, where S_v, d_v is the sequence length and embedding dimension, respectively.

After the unimodal feature extractor, we get each utterance's unimodal representation vector u_t, u_a, u_v. u_t', u_a', u_v' is obtained after layer normalization, showed in Eq. (1). Then we input u_t', u_a', u_v' to the two modal interaction encoders.

$$
\begin{aligned}
u_t' &= LayerNorm(u_t) \\
u_a' &= LayerNorm(u_a) \\
u_v' &= LayerNorm(u_v)
\end{aligned}
\tag{1}
$$

3.2 Intra-Modal Interaction Encoder.

In the inter-modal interaction encoder, we use a BiLSTM network to compute the output of u_t', u_a', u_v'. We consist of the concatenated output of the forward and reverse hidden layer states at each time step in the sequence o_m^{intra}, $m \in \{t, a, v\}$. Meanwhile, the concatenation of the final forward and reverse hidden states output h_m^{intra}, $m \in \{t, a, v\}$, are shown in Eq. (2), where θ_m^{intra} denotes the trainable parameters.

$$
o_m^{intra}, h_m^{intra} = BiLSTM\left(u'_m, \theta_m^{intra}\right)
\tag{2}
$$

To convert to the unified dimensions of the hidden layers and facilitate subsequent calculations, we use a linear layer to convert h_m^{intra}, and use the ReLU activation function for nonlinear mapping to increase the diversity of features. Equation (3) shows that W_m^{intra} is the learnable transformation matrix, and b_m^{intra} is the bias.

$$
\overline{h}_m^{intra} = ReLU\left(W_m^{intra} \times h_m^{intra} + b_m^{intra}\right)
\tag{3}
$$

After nonlinear transformation, three hidden layer features with the same dimensions are obtained. To further extract the deep features, they will be input into the fully connected layer, where the parameters are shared for training, and $\hat{h}_m^{intra} \in R^d$ is obtained, as shown in Eq. (4). Where FC refers to the fully connected linear layer, θ^{intra} is a learnable parameter matrix.

$$\hat{h}_m^{intra} = FC(\bar{h}_m^{intra}, \theta^{intra}) \in R^d \tag{4}$$

After passing through the BiLSTM network, o_m^{intra} and $\hat{h}_m^{intra}$ are input into the multi-head attention mechanism. In the multi-head attention mechanism, each head's Q, K, and V are obtained by a linear transformation of the input feature matrix, as shown in Eq. (5), where W_q^h, W_k^h, W_v^h is the parameter matrix, and h represents a specific attention head. The attention score is calculated as shown in Eq. (6). d is the dimension of the input X. The multi-head attention score is obtained by concatenating the attention scores of each head, as shown in Eq. (7), where W^O is the learnable parameter matrix, and h is the number of heads.

$$\begin{aligned} Q^h &= X \times W_q^h \\ K^h &= X \times W_K^h \\ V^h &= X \times W_v^h \end{aligned} \tag{5}$$

$$Att^h = softmax\left(\frac{Q^h \times (K^h)^T}{\sqrt{d}}\right) \times V^h \tag{6}$$

$$MHA(Q, K, V) = concat(Att^1, \ldots, Att^h) \times W^O \tag{7}$$

In the intra-modal interaction encoder, to better focus on the unique internal features of the uni-modality and find the correlation between internal elements, we transform Q, K, and V from the input o_m^{intra} linearly to obtain the multi-head attention scores. These scores are averaged for dimensionality reduction, then are concatenated with $\hat{h}_m^{intra}$ to obtain three intra-modal interaction feature vectors $H_t^{intra}, H_a^{intra}, H_v^{intra}$, as shown in Eq. (8). Among them, $m \in \{t, a, v\}$ represents the three modalities of text, audio and vision.

$$H_m^{intra} = concat\left(MHA_m^{intra}\left(o_m^{intra}, o_m^{intra}, o_m^{intra}\right), \hat{h}_m^{intra}\right) \tag{8}$$

According to the above calculation process, we pay more attention to the interactive features within the modalities and highlight the unique features in intra-modalities. So, we add the loss of the difference L_{diff} on this basis. This loss function will strengthen the sense of the difference between modalities during training and learning specific features between individual modalities.

3.3 Inter-Modal Interaction Encoder

In the inter-modal interaction encoder, we calculate the bimodal attention score in the inter-modal interaction encoder by changing the values Q, K, V of the multi-head attention mechanism. We map Q, K, and V from different unimodal representations and pay attention to the interaction between the bimodal representations when calculating the attention score. Hence, we can obtain text-visual (TV), audio-text (AT), and visual-audio (VA) three modal interaction features, H_{tv}^{inter}, H_{at}^{inter}, H_{va}^{inter}, respectively.

The unimodal representation u_t', u_a', u_v' is input in the intermodal interaction encoder. The BiLSTM network obtains the concatenated output of the forward and reverse hidden layer states at each time step in the sequence o_m^{inter}, $m \in \{t, a, v\}$. The concatenation of the final forward and reverse hidden states output h_m^{inter}, $m \in \{t, a, v\}$. In Eq. (9), θ_m^{inter} denotes the trainable parameters.

$$o_m^{inter}, h_m^{inter} = BiLSTM\,(u'_m, \theta_m^{inter}) \tag{9}$$

Next, we use a linear layer h_m^{inter} to transform and then apply a ReLU activation function to obtain $\overline{h}_m^{inter}$, as shown in Eq. (10), where W_m^{inter} is the learnable transformation matrix, and b_m^{inter} is the bias. On this basis, a fully connected layer with shared parameters is applied for further feature extraction. Finally, $\hat{h}_m^{inter}$ is obtained, as shown in Eq. (11), where θ^{inter} is a learnable parameter matrix.

$$\overline{h}_m^{inter} = ReLU\,(W_m^{inter} \times h_m^{inter} + b_m^{inter}) \tag{10}$$

$$\hat{h}_m^{inter} = FC(\overline{h}_m^{inter}, \theta^{inter}) \in R^d \tag{11}$$

Assuming the head of multi-head attention is h, we use the same parameter matrix W_O for different bimodal matrices to reduce the number of parameters and memory consumption. After obtaining the features of different mapping spaces, we encode the relationship between the two modalities by calculating the attention scores, thus obtaining the attention scores of TV, AT, and VA. The values of Q, K, and V come from the two modalities respectively, which are shown in Eq. (12), where W_q^i, W_k^i, W_v^i is the initialized mapping matrices, $i \in h$ is the i-th attention head.

$$\begin{aligned} Q_m^i &= o_m^{inter} \times W_q^i \\ K_m^i &= o_m^{inter} \times W_k^i \\ V_m^i &= o_m^{inter} \times W_v^i \end{aligned} \tag{12}$$

After obtaining the initial Q, K, and V matrices, the attention calculation Equations between the two modalities are as shown in Eq. (13).

$$\begin{aligned} Att_{tv}^i &= softmax(\frac{Q_v^i \times (K_t^i)^T}{\sqrt{d}}) \times V_t^i \\ Att_{at}^i &= softmax(\frac{Q_a^i \times (K_t^i)^T}{\sqrt{d}}) \times V_t^i \\ Att_{va}^i &= softmax(\frac{Q_a^i \times (K_v^i)^T}{\sqrt{d}}) \times V_v^i \end{aligned} \tag{13}$$

We connect the attention scores of each head separately and perform linear transformation to obtain the final multi-head attention scores, as shown in Eq. (14), which are the attention scores among TV, AT, and VA, respectively. Exchange the position between Q and K, V, and then calculate the score between VT, TA, and AV. Then the final attention scores between text-vision, text-audio, and vision-audio are the average score, where W_O is the weight parameter.

$$
\begin{aligned}
MHA_{tv}^{inter} &= \left[Att_{tv}^1, \ldots, Att_{tv}^h\right] \times W_O \\
MHA_{at}^{inter} &= \left[Att_{at}^1, \ldots, Att_{at}^h\right] \times W_O \\
MHA_{va}^{inter} &= \left[Att_{va}^1, \ldots, Att_{va}^h\right] \times W_O
\end{aligned}
\tag{14}
$$

After obtaining the interaction features between two modalities, it is concatenated with the previous hidden layer feature matrix to obtain three intra-modal interaction features $H_{tv}^{intra}, H_{at}^{intra}, H_{va}^{intra}$, as shown in Eq. (15).

$$
\begin{aligned}
H_{tv}^{inter} &= concat\left(MHA_{tv}^{inter}, \hat{h}_t^{inter}\right) \\
H_{at}^{inter} &= concat\left(MHA_{at}^{inter}, \hat{h}_a^{inter}\right) \\
H_{va}^{inter} &= concat\left(MHA_{va}^{inter}, \hat{h}_v^{inter}\right)
\end{aligned}
\tag{15}
$$

After calculating the attention score, the interaction information between the two modalities is well encoded into the interaction feature matrix. The multi-head attention mechanism is used to focus on the dimension of the feature from different feature subspaces. During training, we introduce the cosine loss function L_{sim}, focusing on the spatial similarity and learning the association information between pair modalities.

3.4 Fusion Decoder and Prediction

Six interactive feature matrices $H_t^{intra}, H_a^{intra}, H_v^{intra}, H_{tv}^{inter}, H_{at}^{inter}, H_{va}^{inter}$ from the modal interaction encoders,. These features are stacked and input to one Transformer Decoder Layer for decoding. The obtained fusion feature matrix is concatenated and then transferred to the 3-layer DNN for the output of the predicted value.

We perform better by minimizing the loss function when conducting multimodal sentiment analysis tasks. Loss function is expressed by Eq. (16), which includes three kinds of loss: task loss, cosine similarity loss, and difference loss. α, β represents the interaction weight, which determines the contribution of each interaction feature matrix to the overall loss.

$$
L = L_{task} + \alpha L_{sim} + \beta L_{diff}
\tag{16}
$$

For L_{task}, it is a loss function designed for a specific task to evaluate the model's prediction quality during training. Here, for the classification task, the standard cross-entropy loss function Equation is used as shown in Eq. (17), while for the regression task, using the mean square error as the loss function, the Equation is shown in Eq. (18). The loss function Equation is as follows for N samples as batch size.

$$
L_{task} = -\frac{1}{N}\sum_{i=0}^{N} y_i \cdot \log \hat{y}_i \, classification
\tag{17}
$$

$$
L_{task} = -\frac{1}{N}\sum_{i=0}^{N} \|y_i - \hat{y}_i\|_2^2 \, regression
\tag{18}
$$

For L_{sim}, the cosine loss is used here, and the absolute value of the cosine similarity between the two feature matrices is calculated as the loss function. The closer the cosine value is to 1, the closer the angle is to 0 degrees, that is, the more similar the two vectors are. The Equation for calculating the cosine similarity is shown in Eq. (19).

$$\cos(\theta) = \frac{\sum_{i=1}^{n}(x_i \times y_i)}{\sqrt{\sum_{i=1}^{n}(x_i)^2} \times \sqrt{\sum_{i=1}^{n}(y_i)^2}} \tag{19}$$

The paper uses the cosine similarity function as the similarity loss function to calculate the similarity between two modalities. The Equation is shown in Eq. (20):

$$L_{sim} = \frac{1}{3} \sum_{(m_1,m_2)\in\{(t,v),(a,t),(v,a)\}} \cos\left(H_{m1}^{inter}, H_{m2}^{inter}\right) \tag{20}$$

For L_{diff}, it's the difference loss. Here we restrict the soft orthogonal constraint between two feature matrices to obtain the non-redundant information of the modalities. Assuming two modal matrices $H_{m1}, H_{m2} \in R^{N \times d}$, is a two-dimensional matrix, each row represents each utterance's modal hidden layer vector in a sample with a batch size of N. Then the calculation Equation of the orthogonal constraint between the two modal vector pairs is as follows in Eq. (21) as shown:

$$\left\|H_{m_1}{}^T H_{m_2}\right\|_F^2 \tag{21}$$

$\|\cdot\|_F^2$ represents the Fibonacci norm. To comprehensively consider the redundant information of each interaction feature matrix, we calculate the orthogonal constraints of the interactive features within the modality. Meanwhile, we compute the orthogonality constraints between the intra-modal and inter-modal interaction matrices. Therefore, the calculation Equation of the different loss functions is shown in Eq. (22):

$$L_{diff} = \sum_{m\in\{t,v,a\}} \left\|\left(H_m^{intra}\right)^T H_m^{inter}\right\|_F^2 + \sum_{(m_1,m_2)\in\{(t,a),(t,v),(a,v)\}} \left\|\left(H_{m1}^{intra}\right)^T H_{m2}^{intra}\right\|_F^2 \tag{22}$$

4 Experiments

We conducted comparative experiments on CH-SIMS, CMU-MOSI, and CMU-MOSEI datasets. The experimental results showed that the MIF model performed well on these multimodal datasets. Moreover, it did better than the benchmark models on most of the evaluation metrics. According to the experimental results, error analyses are made on the performance of different models. Finally, the modal features are visualized with T-SNE visualization tool, which directly reflects the changes in modal distribution during network training, and verifies that the multimodal sentiment analysis model proposed in this paper can effectively capture the modal interaction features.

4.1 Datasets

We conducted experiments on three public datasets, CH-SIMS [24], CMU-MOSI [25], and CMU-MOSEI [8]. The first is in Chinese, and the latter two are in English, the statistical information of the dataset is shown in Table 1, and detailed information about the dataset is as follows.

CH-SIMS (A Chinese Multimodal Sentiment Analysis Dataset with Fine-grained Annotation of Modality) dataset was proposed by Yu et al. [24]. The dataset is collected from 60 original videos of variety shows of Chinese film and television dramas, and 2281 video clips are obtained. The average length of the clips is 3.67 s, and each video clip has three unimodal tags, including text, audio, and vision, in addition to multimodal tags. The sentiment polarity of the video sequence is marked as 11 polarity values from -1 to 1.

CMU-MOSI (CMU Multimodal Opinion-level Sentiment Intensity Corpus) dataset was proposed by Zadeh et al. [25] in 2016. They selected the monologue videos of users expressing opinions on different topics from the Youtube website and obtained 93 viewpoint videos from videos of 89 different speakers with an average age of 20–30. The subjectivity annotation of the dataset finally resulted in 2199 subjective segments and 1503 objective ones. For sentiment annotations, we only focus on 2199 subjective segments in this paper, and the annotators labeled each sample from -3 (strongly negative) to 3 (strongly positive) sentiment intensity score.

CMU-MOSEI (CMU Multimodal Opinion Sentiment and Emotion Intensity) dataset is a multimodal sentiment and emotion recognition dataset proposed by Zadeh et al. [8] to expand CMU-MOSI in 2018 with more sentences, more samples, and diversity of speakers and topics. It obtained 5,000 videos from YouTube, 1,000 speakers with different identities, selected 250 topics and collected 23,453 annotated video clips. The sentiment polarity label is similar to CMU-MOSI. Each sentence is scored according to the sentimental strength from -3 to 3, referring to a sentiment category from strongly negative to strongly positive.

Table 1. Statistical Information of CH-SIMS, CMU-MOSI, and CMU-MOSEI Dataset.

Dataset	Train	Valid	Test	Total
CH-SIMS	1368	456	457	2281
CMU-MOSI	1284	229	686	2199
CMU-MOSEI	16326	1871	4659	22856

4.2 Baselines

EF-LSTM. The EF-LSTM network [26] uses early fusion method to concatenate modality features and then extract modality context sequence information by LSTM.

LF-DNN. The LF-DNN model [24] uses a late fusion method with the three unimodal features first passing through the DNN network. The output feature vectors are concatenated to obtain the prediction result.

TFN. The TFN model [5] employs outer product calculations during feature fusion to create multi-dimensional tensors that combine interactions between modalities.

LMF. The LMF model [6] improves TFN, which uses low-rank matrix factorization to extract cross-modal information, reducing the complexity of tensor computation.

MFN. The MFN method [7] stores the modal's internal information and interactive information through the gated memory unit and introduces the dynamic fusion graph to model the modal's internal information and interactive information continuously.

Graph-MFN. The graph-MFN network [8] is improved based on MFN, and a graph memory fusion network is proposed on the structure of memory fusion network.

MulT. The MulT model [9] utilizes cross-modal attention to extract key information inside each modality and performs feature merging based on Transformer.

MISA. The MISA model [10] uses distributional similarity, orthogonal loss, reconstruction loss, and task loss to learn invariant and specific modality representation.

MLF-DNN, MTFN, MLMF. These are multi-task frameworks of LF-DNN, TFN, and LMF, using independent unimodal annotation labels for multi-task learning [24].

Self-MM. The Self-MM model [11] introduces a self-supervised unimodal label generation strategy and utilizes unimodal subtasks to learn modality representations.

ICCN. The ICCN model [27] exploits canonical correlations to analyze hidden relationships among text, audio, and vision to learn multimodal embedding features.

MAG-BERT. The MAG-BERT model [28] designed a multimodal fusion gating mechanism, enabling BERT and XLNet [29] to accept multimodal non-linguistic data for fine-tuning.

MMIM. The MMIM model [30] introduces hierarchical maximum mutual information in single-modal input pairs and multimodal fusion results and maintains task-related information through multimodal fusion.

4.3 Experiments Settings

During the training process, we adjust the following hyperparameters: batch size (bs), learning rate (lr), hidden layer size (hs), linear layer drop rate (ldrp), attention drop rate (adrp), difference loss weight (dw), cosine loss weight (sw), weight decay (wd), and the head number of Multi-head attention (h). The hyperparameter values of different data sets are shown in Table 2. The Adam optimizer with an initial learning rate and early stopping at 12 epochs is used during the experiments.

4.4 Evaluation Metrics

Evaluation metrics are different for different tasks. Mean Absolute Error (MAE) and Pearson Correlation (Corr) are used as evaluation criteria for regression tasks. MAE represents the average of the absolute error between the predicted value and the ground truth. Pearson Correlation is referred to as linear correlation coefficient, which means the degree of linear relationship between two vectors. Accurate (Accuracy) and F1 value (F1-score) are used in the binary classification task as evaluation standards. The standards include binary classification accuracy (Has0-acc-2) and F1 value (Has0- F1-score), and the two-category accuracy (Non0-acc-2) and F1 value (Non0-F1-score) without the

Table 2. Hyperparameters of MIF for various tasks on CH-SIMS, CMU-MOSI, and CMU-MOSEI Datasets.

Para	Dataset		
	CH-SIMS	CMU-MOSI	CMU-MOSEI
bs	64	64	8
lr	1e-4	5e-5	1e-5
hs	128	128	32
ldrp	0.5	0	0.2
adrp	0.2	0.5	0.2
dw	0.3	0.5	0.5
sw	0.8	0.8	0.8
wd	0	0.001	0.0002
h	8	8	2

neutral label (0). For multi-classification, only the accuracy (Accuracy) is used as an indicator, including three-category accuracy (Acc-3), five-category accuracy (Acc-5), and seven-category accuracy (Acc-7). Except for MAE, the higher the value of other indicators, the better the performance. For a more intuitive comparison, the evaluation indicators used in this paper are consistent with those in the baseline models.

4.5   Results

The experimental comparison results of the MIF model and other benchmark models on three datasets are shown as follows. The accuracy and F1 value of classification 2 include two experimental data of neutral label (left) and non-neutral label (right).

Table 3 shows the results of the CH-SIMS dataset. The performance of EF-LSTM is the worst because this model ignores the modal interaction features. The method of LF-DNN based on post-fusion is better than EF-LSTM because it considers the mutual information between modalities and obtains more useful fusion features. TFN introduces tensor fusion by calculating the outer product of different modal features, and its performance is better than in some models. MulT could have been better, although it uses the Transformer architecture. The MISA model could have performed better on the CH-SIMS dataset, possibly because Chinese task was relatively more complicated than that in English, so it did not perform well due to the influence of the dataset.

Multi-task learning improves the generalization performance of the whole task by using the information contained in different subtasks. Therefore, these multi-task learning frameworks performs better than the other eight methods. However, from the experimental results, it's known that the performance of Self-MM is inferior to MLF-DNN, MTFN, and MLMF. The main reason is that these three methods are based on manually annotated unimodal labels. And Self-MM continuously updates unimodal labels based on the self-supervised unimodal label generation strategy, which may lead to inaccurate

Table 3. The results on CH-SIMS Dataset for multimodal sentiment analysis (%).

Model	Acc-2	F1-score-2	Acc-3	F1-score-3	Acc-5	MAE	Corr
EF-LSTM	69.37/63.92	56.82/49.85	54.27	38.18	21.23	59.07	5.45
LF-DNN	79.87/56.70	79.97/55.27	70.02	65.29	39.74	44.63	55.51
TFN	75.32/53.56	75.66/52.79	65.95	62.04	39.30	43.22	59.10
LMF	77.99/57.06	77.59/53.83	66.87	**72.46**	40.53	44.12	57.59
MLF-DNN	80.79/58.19	80.59/55.55	70.37	65.94	40.22	39.58	66.52
MTFN	81.23/56.91	81.24/55.29	70.28	66.44	40.31	**39.54**	66.58
MLMF	81.45/56.60	**81.62/55.66**	**71.60**	70.45	41.05	40.91	63.90
MFN	78.25/56.96	78.08/54.14	67.57	64.20	39.47	43.49	58.24
Graph-MFN	79.21/57.99	78.92/54.66	68.44	63.44	39.82	44.50	57.84
MulT	78.07/56.34	78.07/54.26	68.27	64.23	37.94	45.32	56.41
MISA	78.07/57.27	77.70/53.99	67.05	60.98	24.95	55.79	19.29
Self-MM	80.04	80.44	65.47	-	41.53	42.50	59.52
MIF	**81.88/64.81**	81.26/54.28	70.76	70.11	**42.53**	40.09	**66.68**

prediction results. As we know, MIF makes the performance better than other models, especially in five categories. The model's design structure is more suitable for multi-classification tasks, and the learned features are more comprehensive. The concerned features have a high contribution to classification prediction, so it can well improve the accuracy of multi-classification tasks.

Table 4 shows the results of the CMU-MOSI dataset. The MIF model is superior to other models in almost all evaluation metrics and has significant improvement. The performance of the seven classifications is slightly lower than that of the self-MM model. The unimodal labels generation strategy of self-MM can help predict the multimodal sentiment polarity to a certain extent. Compared with other benchmark models, self-MM performance is better, followed by MMIM and MAG-BERT models, which are multimodal fusion strategies designed for MOSI and MOSEI datasets. The multimodal fusion feature is obtained by introducing a multimodal fusion gating mechanism and hierarchical maximum mutual information, which can improve the performance of the sentiment analysis task to a certain extent. The performance of other models are not good, which mainly focus on the simple fusion of multimodal features, more on the design of fusion methods, more on the fusion calculation process, and concatenating or tensor calculation, ignoring interaction features between the bi-modalities and the connection within modality, so the performances are relatively general. The model we proposed has a strong feature capture ability in this respect, so it has achieved good results.

Table 5 shows the results of the CMU-MOSEI dataset. The MIF model outperforms other models in most metrics. The accuracy of MIF model in five categories is slightly lower than that of Self-MM, which is negligible, perhaps due to the effect of multitasking learning. And MMIM model performs well in the accuracy of the seven categories and

Table 4. The results on CMU-MOSI Dataset for multimodal sentiment analysis (%).

Model	Acc-2	F1-score-2	Acc-3	F1-score-3	Acc-5	Acc-7	MAE	Corr
EF-LSTM	77.38/78.48	77.35/78.51	74.43	72.74	40.15	35.39	94.88	66.9
LF-DNN	77.52/78.63	77.46/78.63	74.64	72.99	38.05	34.52	95.48	65.84
TFN	77.99/79.08	77.95/79.11	73.44	71.86	39.39	34.46	94.73	67.33
LMF	77.90/79.18	77.8/79.15	74.11	72.50	38.13	33.82	95.04	65.1
ICCN	-/83.07	-/83.02	-	-	-	39.01	86.20	71.40
MAG-BERT	82.37/84.43	82.50/84.61	-	-	-	43.62	71.20	79.60
MMIM	84.14/86.06	84.00/85.98	-	-	-	46.65	70.00	80.00
MFN	77.67/78.87	77.63/78.90	66.59	64.31	40.47	35.83	92.68	67.02
Graph-MFN	77.14/78.35	77.08/78.35	66.39	64.00	38.63	34.64	95.57	64.86
MulT	79.71/80.98	79.63/80.95	74.99	73.06	42.68	36.91	87.99	70.22
MISA	81.84/83.54	81.82/83.58	76.30	74.57	47.08	41.37	77.65	77.81
Self-MM	83.44/85.46	83.36/85.43	-	-	53.47	**46.67**	70.80	79.63
MIF	**87.53/89.65**	**87.65/89.66**	**82.79**	**80.58**	**58.97**	44.79	**69.51**	**81.39**

Table 5. The results on CMU-MOSEI Dataset for multimodal sentiment analysis (%).

Model	Acc-2	F1-score-2	Acc-3	F1-score-3	Acc-5	Acc-7	MAE	**Corr**
EF-LSTM	77.84/80.79	78.34/80.67	66.09	63.68	51.16	50.01	60.05	68.25
LF-DNN	80.60/82.74	80.65/82.52	67.35	64.65	51.97	50.83	58.02	70.87
TFN	78.50/81.89	78.96/81.74	66.63	63.93	53.1	51.6	57.26	71.41
LMF	80.54/83.48	80.94/83.36	66.59	64.86	52.99	51.59	57.57	71.69
ICCN	-/84.20	-/84.20	-	-	-	51.60	56.50	70.40
MAG-BERT	82.51/84.82	82.77/84.71	-	-	-	52.67	54.30	75.50
MMIM	82.24/85.97	82.66/85.94	-	-	-	**54.24**	52.60	**77.20**
MFN	78.94/82.86	79.55/82.85	66.59	64.31	52.76	51.34	57.33	71.82
Graph-MFN	81.28/83.48	81.48/83.23	66.39	64.00	52.69	51.37	57.45	71.33
MulT	81.15/84.63	81.56/84.52	67.04	65.01	54.18	52.84	55.93	73.31
MISA	80.67/84.67	81.12/84.66	67.63	65.39	53.63	52.05	55.75	75.15
Self-MM	83.76/85.15	83.82/84.90	-	-	**55.53**	53.87	53.09	76.49
MIF	**84.31/85.12**	**84.72/85.01**	**68.04**	**67.04**	55.26	53.93	**52.31**	74.86

Pearson correlation coefficient. Since MMIM introduces maximum mutual information, which is to measure the degree of correlation between two features, i.e. linear or non-linear relationship, this model has great advantages in Pearson correlation coefficient and performs better than MIF on complex classification tasks. However, from the perspective of comprehensive evaluation metrics, MIF has the best all-around performance and plays a stable role in various metrics, which can prove that the method we propose in this paper performs better than other models in extracting modal interaction features and better adapt to various tasks.

4.6 Ablation Study

To verify the function of the overall structure of MIF and the impact of different modal feature combinations, we conduct ablation experiments on the proposed model from the following three aspects, as shown in Table 6. The first group is the experimental results of a single modality. In this group of experiments, we refer to using only text, audio, and vision for sentiment prediction; The second group of experiments is the combination of paired modalities, including the experimental results of four modal combinations: text-audio (T + A), text-vision (T + V), audio-vision (A + V) and text-audio-vision (T + A + V); The third group of experiments is the ablation experiment of the internal components of MIF, including the removal of the intra-modal interaction feature encoder (- w/o intra Encoder), the removal of the inter-modal interaction feature coder (- w/o inter Encoder), the removal of the cosine loss L_{sim} (- w/o sim_loss) and the removal of the difference loss L_{diff} (- w/o diff_loss). Unlike the above two groups of experiments, the third group uses four modal labels (M + T + V + A) for training, the final one is the MIF proposed in this paper, which contains all components, uses four modal labels (M + T + V + A) for training, and has the best performance in various metrics.

Table 6. The experimental results of different combinations on CH-SIMS Dataset (%). Ablation experiments on single modality (A, V, T), pair-wise modalities (AV, AT, VT), multi-modalities (AVT), and various modules without inter-modal Encoder (-w/o inter Encoder), an intra-modal encoder (-w/o intra Encoder), difference loss (-w/o diff_loss) or similarity loss (-w/**bo** sim_loss).

Num	Model	Acc-2	Acc-3	Acc-5	F1-score	MAE	Corr
1	A	70.18	48.46	20.39	58.53	58.09	15.28
2	V	68.64	47.59	19.08	57.35	59.51	-0.35
3	T	75.00	62.50	35.53	75.65	44.97	57.09
4	A + V	69.96	49.56	21.05	64.12	58.24	19.45
5	A + T	75.88	64.91	40.13	76.42	43.18	57.97
6	V + T	76.1	64.04	37.28	76.38	43.63	55.74
7	A + V + T	77.85	64.91	37.28	78	42.75	58.31
8	-w/o intra Encoder	76.97	64.91	36.18	76.90	43.05	59.82
9	-w/o inter Encoder	77.63	66.89	32.89	77.33	45.65	57.08
10	-w/o sim_loss	79.41	66.04	34.02	79.34	41.93	63.03
11	-w/o diff_loss	78.07	66.89	34.87	78.31	42.30	63.25
12	**MIF**	**81.88**	**70.76**	**42.53**	**81.26**	**40.09**	**66.68**

In Table 6, from the results of experiments 1, 2, and 3, the text modality is the most predictive, while the visual modality performs poorly, which is consistent with our cognition. Most of the data extracted from the original audio and visual data are transcribed from videos, including some noise and other interference. At the same time, the rich semantic information in the text is obtained using the pre-trained model, so the text has more advantages than vision and audio. In this paper, wav2vec is used for audio pre-training. These factors make the extracted phoneme, pitch, and other features relatively accurate. Therefore, in uni-modality experiments, audio features perform better than visual features. According to the experiments in groups 4, 5, 6, and 7, the performance of the three modal combinations is better than the bimodal combination. In the bimodal combination experiment, the classification accuracy of the bimodal combination containing text features is higher than that of the audio-vision combination, which explains that the sentiment information of text features dominates in the CH-SIMS dataset. In the final ablation experiments of the internal module, the inter-modal interaction feature is more critical than the intra-modal interaction feature. Removing the two kinds of interaction feature encoders significantly impacts on results, while removing the two loss functions has a relatively small impact.

4.7 Visualization

We select several samples from the test set to visualize the three modal features. We use the T-SNE [31] visualization tool to perform PCA dimensionality reduction on the text, audio, and visual modal features in the initial state, as shown in the left figure of Fig. 2. After the modal interaction feature encoding, the output features of the three modal features are visualized. The results are shown in the right figure of Fig. 2.

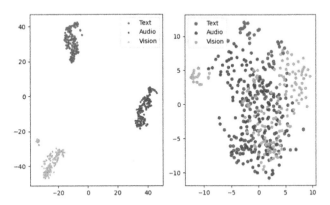

Fig. 2. The Distribution of initial states after encoding on text, audio, and visual modalities.

From Fig. 2, the initial state's distribution of text, audio, and visual features is relatively independent, and each modality is distributed intensively without intersection. After the modal interaction features encoding, the three modal features are fused. After the encoder function, the interaction between modalities is relatively apparent. Therefore, the modal interaction feature encoders can learn from the interaction features. They can get effective multimodal fusion feature vectors, proving that the modal interactive feature encoders have a noticeable effect and are conducive to improving the performance of sentiment analysis tasks.

5 Conclusions

Due to the difference between the interactive features within and between modalities, the weight contribution of multimodal fusion is different. This paper proposes a multimodal sentiment analysis model based on modal interactive features encoder(MIF). The intra-modal interaction features and the intermodal interaction features are encoded by two encoder functions, respectively. Compared with the previous models, MIF focuses on the interaction characteristics and fine-tuning the contribution weight of each modality, thus significantly improving the classification performance. Extensive comparative experiments have been conducted on three datasets. The experimental results show that MIF proposed in this paper performs well on these multimodal datasets and performs better than the benchmark model on most evaluation metrics.

In the future, we will be more committed to studying new multimodal fusion methods and conducting experiments on CH-SIMS datasets to provide more experimental results for multimodal sentiment analysis in Chinese. In addition, we will also improve the model according to the actual emotional scene and explore the impact of other dimensions on the results of multimodal sentiment analysis. For example, emotional duration and other physiological characteristics dimensions will be introduced, or better results will be obtained.

References

1. Bronstein, M.M., Bronstein, A.M., Michel, E.: Data fusion through cross-modality metric learning using similarity-sensitive hashing. In: CVPR, pp. 3594–3601 (2010)
2. Peters, M.E., Neumann, M., Iyyer, M.: Deep contextualized word representations. In: NAACL (2018)
3. Radford, A., Narasimhan, K., Salimans, T.: In: Improving language understanding by generative pre-training (2018)
4. Devlin, J., Chang, M.W., Lee, K.: Bert: Pre-training of deep bidirectional transformers for language understanding. arXiv preprint arXiv:1810.04805 (2018)
5. Zadeh, A., Chen, M., Poria, S.: Tensor fusion network for multimodal sentiment analysis. arXiv preprint arXiv:1707.07250 (2017)
6. Liu, Z., Shen, Y., Lakshminarasimhan, V.B.: Efficient low-rank multimodal fusion with modality-specific factors. arXiv preprint arXiv:1806.00064 (2018)
7. Zadeh, A., Liang, P.P., Mazumder, N.: Memory fusion network for multi-view sequential learning. In: Proceedings of the AAAI Conference on Artificial Intelligence, vol. 32(1), (2018)
8. Zadeh, A.A.B., Liang, P.P., Poria, S.: Multimodal language analysis in the wild: Cmu-mosei dataset and interpretable dynamic fusion graph. In: Proceedings of the 56th Annual Meeting of the Association for Computational Linguistics (Volume 1: Long Papers), pp. 2236–2246 (2018)
9. Tsai, Y.H.H., Bai. S., Liang. P.P.: Multimodal Transformer for unaligned multimodal language sequences. In: Proceedings of the Conference. Association for Computational Linguistics. Meeting. NIH Public Accessm vol. 2019, p. 6558 (2019)

10. Hazarika, D., Zimmermann, R., Poria, S.: Misa: Modality-invariant and-specific representations for multimodal sentiment analysis. In: Proceedings of the 28th ACM International Conference on Multimedia, pp. 1122–1131 (2020)

11. Yu, W., Xu, H., Yuan, Z.: Learning modality-specific representations with self-supervised multi-task learning for multimodal sentiment analysis. arXiv preprint arXiv:2102.04830 (2021)

12. He, J., Liu, Y., He, Z.: Research progress in multimodal emotion recognition. Appli. Res. Comput. **35**(11), 3201–3205 (2018)

13. Morency, L.P., Mihalcea, R., Doshi, P.: Towards multimodal sentiment analysis: Harvesting opinions from the web. In: Proceedings of the 13th International Conference on Multimodal Interfaces, pp. 169–176 (2011)

14. Poria S, Cambria E, Hazarika D.: Context-dependent sentiment analysis in user-generated videos. In: Proceedings of the 55th annual meeting of the association for computational linguistics, Volume 1: Long papers, pp. 873–883 (2017)

15. Chen, M., Wang, S,, Liang, P.P.: Multimodal sentiment analysis with word-level fusion and reinforcement learning. In: Proceedings of the 19th ACM International Conference on Multimodal Interaction, pp. 163–171 (2017)

16. Majumder, N., Hazarika, D., Gelbukh, A.: Multimodal sentiment analysis using hierarchical fusion with context modeling. Knowl.-Based Syst. **161**, 124–133 (2018)

17. Yu, J., Jiang, J., Xia, R.: Entity-sensitive attention and fusion network for entity-level multimodal sentiment classification. IEEE/ACM Trans, Audio, Speech, Language Process. **28**, 429–439 (2019)

18. Zadeh, A., Liang, P.P., Poria, S.: Multi-attention recurrent network for human communication comprehension. In: Proceedings of the AAAI Conference on Artificial Intelligence, vol. 32(1) (2018)

19. Huddar, M.G., Sannakki, S.S., Rajpurohit, V.S.: Multi-level context extraction and attention-based contextual inter-modal fusion for multimodal sentiment analysis and emotion classification. Inter. J. Multimedia Inform. Retrieval **9**(2), 103–112 (2020)

20. Baevski, A., Zhou, Y., Mohamed, A.: wav2vec 2.0: A framework for self-supervised learning of speech representations. In: Advances in Neural Information Processing Systems **33**, 12449–12460 (2020)

21. Tomar, S.: Converting video formats with FFmpeg. Linux J. **2006**(146), 10 (2006)

22. Zhang, K., Zhang, Z., Li, Z.: Joint face detection and alignment using multi-task cascaded convolutional networks. IEEE Signal Process. Lett. **23**(10), 1499–1503 (2016)

23. Baltrusaitis, T., Zadeh, A., Lim, Y.C.: Openface 2.0: Facial behavior analysis toolkit. In: 2018 13th IEEE international Conference on Automatic Face & Gesture Recognition (FG 2018), pp. 59–66. IEEE (2018)

24. Yu W, Xu H, Meng F.: Ch-sims: a Chinese multimodal sentiment analysis dataset with fine-grained annotation of modality. In: Proceedings of the 58th Annual Meeting of the Association for Computational Linguistics, pp. 3718–3727 (2020)

25. Zadeh A, Zellers R, Pincus E.: Mosi: multimodal corpus of sentiment intensity and subjectivity analysis in online opinion videos. arXiv preprint arXiv:1606.06259 (2016)

26. Williams, J., Kleinegesse, S., Comanescu, R.: Recognizing emotions in video using multimodal dnn feature fusion. In: Proceedings of Grand Challenge and Workshop on Human Multimodal Language (Challenge-HML), pp. 11–19 (2018)

27. Sun, Z., Sarma, P., Sethares, W.: Learning relationships between text, audio, and video via deep canonical correlation for multimodal language analysis. In: Proceedings of the AAAI Conference on Artificial Intelligence, vol. 34(05), pp. 8992–8999 (2020)

28. Rahman, W., Hasan, M.K., Lee, S.: Integrating multimodal information in large pretrained transformers. In: Proceedings of the conference. Association for Computational Linguistics. Meeting. NIH Public Access (2020)

29. Yang, Z., Dai, Z., Yang, Y.: Xlnet: Generalized autoregressive pretraining for language understanding. In: Advances in Neural Information Processing Systems 32 (2019)
30. Han. W., Chen, H., Poria, S.: Improving multimodal fusion with hierarchical mutual information maximization for multimodal sentiment analysis. arXiv preprint arXiv: 2109.00412 (2021)
31. Van der Maaten, L., Hinton, G.: Visualizing data using t-SNE. J. Mach. Learn. Res. **9**(11) (2008)

Type-Augmented Link Prediction Based on Bayesian Formula

Ye Wang[1], Enze Luo[1], Lijie Li[1(✉)], and Wenjian Tao[2]

[1] College of Computer Science and Technology, Harbin Engineering University, Harbin, China
{wangye2020,lilijie}@hrbeu.edu.cn
[2] Purification Equipment Research Institute of CSIC, Handan, China

Abstract. Knowledge graphs (KGs) play a pivotal role in various real-world applications, but they are frequently plagued by incomplete information, which manifests in the form of missing entities. Link prediction, which aims to infer missing entities given existing facts, has been mostly addressed by maximizing the likelihood of observed triplets at the instance level. However, they ignore the semantic information most KGs contain and the prior knowledge implied by the semantic information. To address this limitation, we propose a Type-Augmented Link Prediction (TALP) approach, which builds a hierarchical feature model, computes type feature weights, trains them to be specific to different relations, encodes weights into prior probabilities and convolutional encodes instance-level information into likelihood probabilities; finally, combining them via Bayes rule to compute the posterior probabilities of entity prediction. Our proposed TALP approach achieves significantly better performance than existing methods on link prediction benchmark datasets.

Keywords: Knowledge Graph · Link Prediction · Bayes Formula · Type Information

1 Introduction

As a popular structured information representation method, knowledge graphs have been put into practice for tasks such as question answering systems [1], recommendation [2], dialogue generation [3] and natural language reasoning [4]. However, due to the incompleteness of knowledge graphs, the task of link prediction [5] is needed to infer missing facts. As in Fig. 1, given two entities'Albert Einstein'and'unified field theory'in the knowledge graph, the objective of link prediction is to deduce a third entity by leveraging the association between entities within the graph.

At present, mainstream link prediction methods are based on the embedding of knowledge graphs. In this approach, the information contained in the knowledge graph is separated into type information and ontology information. The type information is represented by triples of relationally linked entities, such as Albert Einstein - > place of birth - > Germany in Fig. 1. Ontological information comprises meta-information pertaining to entities and relationships, such as entity type information. For example,

Albert Einstein's ontology types include {Doctor, physicist, human}. Most existing link prediction methods only use instance-level information to learn embeddings [6], and some methods also use instance-level information and ontology information. Ontology information, such as type information, can directly contribute to link prediction because most relationships connect different types of entities. For example, the acting relationship always connects the film and television entities with the actor entities. By incorporating ontology information, such as type information, into the training triples at the instance level, link prediction tasks can benefit, especially in the absence of sufficient training data for embedding learning.

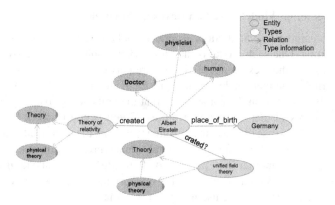

Fig. 1. A knowledge graph with type information.

One of the prevalent embedding-based approaches is centered around the convolutional neural network (CNN) [7]. The CNN-based methodology has also been employed in knowledge graphs to enhance the precision of link prediction [8]. Because the convolutional layer may learn the noise in training, the generalization performance of the model on the test data is suboptimal. As a result, CNN-based methods may have overfitting problems when dealing with small-scale knowledge graphs. This paper proposes an effective decision-level integration method. This technique encodes instance-level type information as prior probability and likelihood probability and subsequently fuses the two probabilities using the Bayesian formula. The advantage of this method is that it can adapt to new embedding techniques more flexibly. Moreover, this approach enables the adjustment of the trade-off between prior probability and likelihood probability to achieve a harmonious balance of the contributions of various types of information.

In this paper, we present a straightforward yet effective framework for expanding the existing type information-based embedding model, which yields the following significant contributions:

- Our proposed model leverages type information to its full potential by employing a decision framework that is independent of the embedding-based model. This flexibility enables the model to be easily adapted to extend different embedding-based models, thereby enhancing its suitability for diverse real-world applications.

- We introduce a training matrix in the training process of the prior model so that different types have different weights under different relationships, thus satisfying the specificity and improving the accuracy and reliability of the prediction results.
- Finally, we assess the efficacy of the TALP model on three widely adopted benchmark datasets and demonstrate the criticality of the TALP prior model. The experimental results demonstrate that our model surpasses the state-of-the-art models in terms of predictive accuracy and underscore the indispensability of the prior model in enhancing the overall performance of the model.

2   Related Work

2.1   Link Prediction Model Based on Bayesian Formula

Within the domain of link prediction, several models have been developed using the Bayesian formula. One such model is the knowledge graph embedding (KGE) model, which assumes that there exists a spatial structure between entities and relations in the knowledge graph. Another model is the TransE [9] model, which is a simple model based on the Bayesian formula that assumes that the embedding spaces of entities and relations can represent their respective features. Graph neural network-based models, such as GNN [10], also use the Bayesian formula to capture the feature information in the graph structure and transform it into a low-dimensional representation space. The TransT [11] model represents another Bayesian-based knowledge graph embedding model that incorporates type information to optimize link prediction accuracy. Although it achieves promising results, it wastes some useful type of information.

2.2   Models Using Type Information

There are several models that currently utilize type information to assist in link prediction tasks, albeit with slightly different methods. HAKE [12], proposed by Wang et al., represents the type information of each entity as a vector and incorporates it as an additional input feature to improve prediction accuracy. HKGE [13], proposed by Sun et al., uses relation embedding vectors for prediction and represents the type information of each entity as a vector, which is used as an additional input feature to improve prediction accuracy. HypER [14] improves prediction accuracy by representing the type information of each entity as an additional tensor. These models have made some use of type information, but there is still significant room for improvement.

3   TALP: Type Augmented Link Prediction Model

This section introduces a type-enhanced link prediction (TALP) framework. The framework consists of two parts: (a) an a priori model, which encodes type information as a priori probability and is used for final decision-making; and (b) a possibility model based on existing instance-level information. The proposed framework is based on a possibility model that leverages the available instance-level information. The likelihood model is implemented using a convolution-based approach, and two such models are briefly presented in this paper. The Bayesian formula is utilized to amalgamate the information from the prior and likelihood models to enhance the precision and dependability of link prediction. This methodology effectively capitalizes on the benefits of type information.

3.1 Type Information Coding

This section details the method of encoding type information as prior probability. In this paper, the knowledge graph is expressed as $G = \{E, R, T\}$, where E represents the entity set, R represents the relation set, and T represents the type se t [15]. First, we define the hierarchical type weight of each class entity $e \in E$ and relation $r \in R$ and calculate the prior probability by measuring the semantic similarity between the entity relation pairs (e_h, r) and (e_t, r) to encode the type information as the prior probability. This method fully leverages the benefits of type information, thereby enhancing the accuracy and reliability of link prediction.

3.2 Level-Based Type Weights

In most knowledge graphs, the types of entities are organized into a hierarchical structure, which reflects the abstraction level of the types. For instance, in Fig. 1, the hierarchical structure of the types {doctor, physicist, human} reflects the degree of abstraction of the types. Intuitively, more specific types, such as doctors, may be more representative in certain contexts (e.g., when a doctor treats a patient) than more abstract types, such as humans. Consequently, we exploit the hierarchical composition of types to allocate varying weights to types at distinct levels. Assuming an entity, denoted by e, its set of types is denoted as $T_e \in T$, T denotes the set encompassing all types within the knowledge graph. The possible subset of types in the hierarchy H is defined as $H = /t_1/t_2/.../t_k/.../t_K$, where $t_K \in T_e$. Here, K represents the comprehensive total count of levels present in the hierarchy, while t_k signifies the most detailed semantic type within the hierarchy, while t_1 represents the most overarching semantic type within the hierarchy. The allocation of weightage to types in the hierarchy is dependent on their hierarchical positions. By weighting the types in the hierarchy of entity e, we can capture the different levels of abstraction of the types and their relative importance for link prediction. The weight of the t_k type relative to its position in the entity e hierarchy H is defined as:

$$w_e^H(t_k) = \frac{ln(k + 1)}{\sum_{j=1}^{K} ln(j + 1)} \tag{1}$$

For a given entity, it is possible that there exist multiple hierarchical arrangements, each of which encompasses only a subset of the conceivable types. As an illustration, in Fig. 1, entity e = Albert Einstein can be categorized into three potential hierarchical arrangements based on its corresponding types.: H_1 =/human/physicist, H_2 =/human/Doctor, and H_3 =/human. Type human is contained in all three hierarchies. By Eq. 1, we have w_e^{Hi} (human)= 0.38, i = {1, 2}, w_e^{H3} (human)= 1. Regarding each type $t \in T_e$, its weight in the hierarchy is evaluated by means of the following calculation: $W_e(t) = min(w_e^{H1}(t), w_e^{H2}(t), ..., w_e^{HN}(t))$, whereas N denotes the aggregate count of hierarchies incorporating type t. In the above example, we can calculate that W_e (human)= 0.38,(Doctor)= 0.62,(physicist)= 0.62. The type set of the relationship is calculated by the following formula through the set of entities:

$$T_{r,head} = \cup_{e \in Head(r)} T_e \tag{2}$$

$$T_{r,tail} = \bigcup_{e\in\text{Tail}(r)} T_e \qquad (3)$$

In the aforementioned formula, the symbol Head(r) refers to the set of head entities associated with the relationship r, while Tail(r) denotes the set of tail entities associated with r. Subsequently, the weights of each type are computed as follows:

$$w_{r,\text{head}}(t) = \sum_{e\in\text{Head}(r)} w_e(t), \text{ for } t \in T_{r,\text{head}} \qquad (4)$$

$$w_{r,tail}(t) = \sum_{e\in\text{Tail}(r)} w_e(t), \text{ for } t \in T_{r,\text{tail}} \qquad (5)$$

3.3   Type Information Coding

The correlation between entities and relationships in a triplet (e_h, r, e_t) can be calculated using the following measurement:

$$s(e_h, r) = \frac{\sum_{t\in T_{r,\text{head}}\cap T_{e_h}} w_{r,\text{head}}(t)}{\sum_{t\in T_{r,\text{head}}} w_{r,\text{head}}(t)} \qquad (2)$$

$$s(e_t, r) = \frac{\sum_{t\in T_{r,\text{tail}}\cap T_{e_t}} w_{r,\text{tail}}(t)}{\sum_{t\in T_{r,\text{tail}}} w_{r,\text{tail}}(t)} \qquad (7)$$

where $T_{r,head} \cap T_{e_h} = \{T|T \in T_{r,head} \cap T \in T_{e_h}\}$, $T_{r,tail} \cap T_{e_t} = \{T|T \in T_{r,tail} \cap T \in T_{e_t}\}$, and the credibility of a predicted triple can be evaluated by utilizing the type similarity between the head entity and the tail entity. By incorporating this measure, we can calculate the probability of the predicted triple entity. For example, Albert Einstein was born in Germany. Albert Einstein is a head entity, and its type is {human, physicist, doctor}, while the type of tail entity Germany may be {country, European country}. There is no overlap between the two types, so it will have a negative impact on the final prediction results. Therefore, we can evaluate the probability of triplet entity prediction according to the type similarity between the entity and relation.

To improve the accuracy of prediction, we define a relationship X type weight matrices $W_{r,head}$ and $W_{r,tail}$ to ensure that different types obtain different weights under different relationships. To narrow the scope of prediction, we introduce a novel loss function to train the model and update the weight matrix, thereby enhancing the precision of the predicted results.

$$f(e_t, e_{true}) = \sum_{i=1}^{|T|} \left(W_e(t_i) - W_{e_{true}}(t_i)\right)^2 \qquad (8)$$

The right side of the formula represents the variance of all type weights in the type set of the predicted entity and the type set of the correct entity. We set a hyperparameter λ to filter our prior probabilities, where the initial value of λ is 0.01, which means that only the prior probabilities in the λ multiple of the prediction set size are considered valid, while other prediction results will be considered invalid. To avoid compromising the accuracy of the posterior probability, we assign a value of 0 to the invalid prediction results, thereby optimizing our screening outcomes Fig. 2.

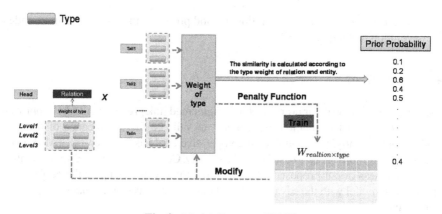

Fig. 2. Model diagram of TALP

3.4 Convolution-based Model

Based on the embedded convolution model, we use the better ConvE [16] model and InteractE [17] model.

ConvE model The ConvE model initially converts the representation of entity and relation vectors into two dimensions, conducts concatenation operations on them, and applies dropout operations on the encoded data. Subsequently, the model performs convolution and batch normalization and subsequently passes through a fully connected layer to generate the corresponding probabilities via softmax activation.

InteractE model Recent research has demonstrated that boosting the potential interaction between embeddings can effectively enhance the expressive capacity of the model. ConvE uses convolution in 2D reconstruction embedding. Although it uses a limited method, it also uses the same principle. InteractE extends the concept of capturing the interaction between entities and relational elements by using circular convolution. Compared with standard convolution, cyclic convolution can capture more feature interactions.

By using different convolution methods to increase the interaction between vectors, the final prediction result is improved, and then a prior model is used to standardize it to obtain higher prediction results.

3.5 Type Information Integeration

The last step of the model involves computing the posterior probability via the Bayesian formula, which combines the prior probability $p(e_t|\ \mathcal{T}(e_h, e_t, \mathcal{R}))$ calculated using type information and the likelihood probability $p(e_h, r|e_t)$ calculated using instance-level information. This enables the model to derive the ultimate result of the posterior probability $p(e_t|e_h, r, \mathcal{T}(e_h, e_t, \mathcal{R}))$.

$$p(e_t|e_h, r, \mathcal{T}(e_h, e_t, \mathcal{R})) \propto p(e_h, r|e_t)p(e_t|\ \mathcal{T}(e_h, e_t, \mathcal{R})) \tag{9}$$

The posterior probability $p(e_t|e_h, r, \mathcal{T}(e_h, e_t, \mathcal{R}))$, which is derived via the application of the Bayesian formula, encompasses both type information and instance-level information. The information integration method proposed in this paper is independent of convolution techniques.

4 Experiment

To assess the efficacy of our proposed type-augmented link prediction (TALP) approach, we initially conducted ablation experiments to validate the significance of the prior model. Subsequently, we conducted a comparative analysis between the TALP model and the baseline convolution-based models, namely, ConvE and InteractE, to substantiate the effectiveness of our proposed TALP method. Finally, we compare the performance of the TALP model with the advanced model that also uses type information to prove its superiority.

Table 1. Statistics of the dataset.

Dataset	#Rel	#Ent	#Types
FB15k-237	474	14505	663
YAGO26K-906	34	26078	226
DB111K-174	298	98336	242

4.1 Dataset

We use three different benchmark datasets - FB15k-237 [18], DB111K-174 [19] and YAGO26K-906 [20] to carry out the connection prediction task. FB15k-237 consists of.
    triples extracted from the FreeBase knowledge graph, while YAGO26K-906 and DB111K-174 are knowledge graphs that incorporate two forms of information, namely, KG and ontology KG, which are interconnected via type links. The statistical information of the three datasets can be found in Table 1.

4.2 Research on Ablation of Prior Model

To prove the effectiveness of our proposed hierarchical type weighting method and the importance of type information, we conducted ablation studies. Specifically, we compare the performance of our proposed model with uniform type weighting and the model based on hierarchical type weighting. We evaluated on the FB15k-237 dataset, and the results are shown in Table 2. It can be seen that the hierarchical weighting model proposed in this paper is superior to the uniform weighting model. These empirical results provide evidence for the effectiveness of our proposed hierarchical type weighting method.

    According to the experimental results in Table 2, this paper finds that the hierarchical prior model and the InteractE model have the best splicing effect, and their prediction performance and evaluation index scores are higher than those of the other two groups

Table 2. Averaged effectiveness of the hierarchy-based type weights.

Type weights	FB15k-237		
	MRR	Hits@1	Hits@10
Uniform	0.353	0.242	0.535
Hierarchy-based	**0.361**	**0.249**	**0.546**

of experiments. Specifically, the hierarchical prior model performs best on the MRR and Hits @ 10 indicators. This may be because there is no hierarchical type information directly taking a path as a type. In this case, the application of type information is not high, and the fault tolerance rate is relatively low, which easily produces errors. For example, when performing link prediction, the type of an entity may not play any role on the entire path because the underlying type is incorrect. If this happens, it will bring huge errors to the final result of link prediction, which will greatly affect the final prediction result. Therefore, the prior modulus of the hierarchical algorithm is adopted. The model has better prediction performance and higher evaluation index score than the prior model without the hierarchical algorithm, which also shows the effectiveness of the hierarchical algorithm in the prior model and provides a useful reference and idea for the improvement and optimization of the model. The hits @ 1 index is more difficult, and the improvement of the splicing group of the hierarchical prior model and the InteractE model is not obvious.

4.3 Comparison with Benchmark Models

The current study involved conducting ablation experiments on the FB15K-237 dataset with the aim of showcasing the efficacy of the proposed type hierarchy weighting method and the importance of type information. The hierarchical weighting model proposed in this study exhibited superior performance in comparison to other models. We conducted ablation experiments and validated the effectiveness of the proposed hierarchical weighting model on the FB15k-237 dataset. The comparative analysis revealed that the hierarchical weighting model proposed in this study outperformed other models, providing scientific evidence for our proposed approach.

By using type information without layering, we obtain two models: ConvE(t) and InteractE(t). By combining the prior models with the two convolution-based models, we obtained two TALP models: TALP-C and TALP-I. From Tables 3, 4, and 5, it can be seen that the TALP models achieved performance improvements over the corresponding baseline convolution-based models on all three benchmark datasets.

The prior model we proposed directly uses type information, so the errors introduced by type triples will not affect the TALP model. By combining type information, the TALP model always maintains comparable or superior performance to the original convolution model. Of particular significance, the TALP model exhibits superior overall performance compared to the convolution-based model, thereby attesting to the efficacy of our proposed decision-level integration process. In addition, the type-based training

Table 3. Evaluation of the TALP model on FB15k-237

Algorithm	FB15k-237		
	MRR	Hits@1	Hits@10
ConvE	0.325	0.237	0.501
InteractE	0.354	0.243	0.535
ConvE(t)	0.326	0.238	0.506
InteractE(t)	0.353	**0.250**	0.539
TALP-C	0.332	0.242	0.504
TALP-I	**0.361**	0.249	**0.546**

Table 4. Evaluation of the TALP model on DB111K-174

Algorithm	DB111K-174		
	MRR	Hits@1	Hits@10
ConvE	0.361	0.242	0.506
InteractE	0.368	0.256	0.524
ConvE(t)	0.363	0.244	**0.547**
InteractE(t)	0.362	0.252	0.532
TALP-C	0.362	0.243	0.506
TALP-I	**0.389**	**0.259**	0.538

Table 5. Evaluation of the TALP model on YAGO26K-906

Algorithm	YAGO26K-906		
	MRR	Hits@1	Hits@10
ConvE	0.291	0.202	0.478
InteractE	0.301	0.234	0.467
ConvE(t)	0.299	0.209	0.499
InteractE(t)	0.312	0.231	0.496
TALP-C	0.313	0.232	0.503
TALP-I	**0.319**	**0.251**	**0.514**

method proposed in this paper can alleviate some problems encountered by previous models.

4.4 Comparison of the TALP Model with Existing Model Methods

We compare the TALP model with the model that also references ontology information. On the FB15k-237 dataset, we compare the TALP model with HAKE, HKGE, HypER and other models based on type information. As depicted in Table 6, it can be seen from the table that the TALP model achieves good performance. Notably, the incorporation of type information resulted in a substantial improvement in Hits@10 scores, as the proposed TALP-I method ranked the triples between rank1-rank10. Consequently, TALP-I exhibited exceptional performance with significantly higher Hits@10 scores than the state-of-the-art (SoTA) methods across all three datasets.

Table 6. Comparison with models using type sets

Algorithm	FB15k-237		
	MRR	Hits@1	Hits@10
HAKE	0.346	0.250	0.542
HKGE	0.341	0.252	0.520
HypER	0.339	0.253	0.521
TALP-I(Ours)	**0.361**	0.249	**0.546**

The results demonstrate that the TALP method proposed in this study effectively enhances the incorporation of type information in the likelihood model. It can combine the type information on the basis of the likelihood model prediction results to avoid incorrect link prediction, thereby improving the link prediction results of the overall model. The model can not only effectively use type information but also overcome the conflict between type information and entity information. Under the premise of ensuring the excellent performance of the likelihood model itself, the TALP method uses type information as auxiliary information to help the overall model better predict the target entity so that the model can show better performance.

4.5 Determination Experiment of Hyperparameter γ

To determine the local optimal weight of the hyperparameter γ, two experiments are used in this paper. First, the influence of γ on the size of the dataset is observed by adjusting γ to narrow the range of γ. Second, the influence of γ on the final prediction results is observed by adjusting γ to determine the local optimal value of γ. These two aspects of experiments are important contents of this paper, which can better understand the role and influence of hyperparameter γ in the model. To ensure the accuracy of the experimental results, the experimental process is fully controlled and statistically analysed.

The analysis of Fig. 3 shows that as the hyperparameter γ continues to increase, the filtered data will also increase. When the hyperparameter γ exceeds a certain threshold, the types representing the low level will be filtered out, thus affecting the prediction effect of the model. Specifically, in the FB15k-237 dataset, the hyperparameter γ of 0.25 is a cut-off point, and more than 30% of the data will be filtered out if the value is exceeded; in the DB111K-174 dataset, the superparameter setting of 0.2 will filter out 40% of the type set; in the YAGO26K-906 dataset, when the hyperparameter γ is set to 0.2, most of the data will also be filtered out. To finally determine the value of the hyperparameter γ of each dataset, this paper further conducts experiments on the accuracy of prior probability prediction by hyperparameter adjustment.

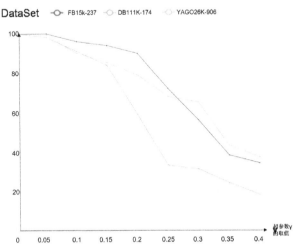

Fig. 3. Curves of the size of each dataset with hyperparameters

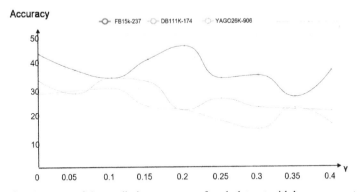

Fig. 4. The curve of the prediction accuracy of each dataset with hyperparameters

The analysis of Fig. 4 shows that after filtering out invalid or negative types, the prediction results of the prior probability model are improved. Specifically, on the FB15k-237 dataset, when the hyperparameter γ is set to 0.2, the prediction effect is the best, and this setting retains most of the type sets of the dataset, ensuring the validity and authenticity of the experimental results. On the DB111K-174 and YAGO26K-906 datasets, the experimental results show that when the hyperparameter γ is set to 0.1, the best effect can be achieved while retaining most type sets. This experimental result provides a basis for determining the optimal hyperparameter γ of each dataset.

When γ is set too small, it does not filter the type set, and there are still some redundant types in the dataset. In the previous work of TransT, the similarity between entities and relationships is calculated by the number of overlapping types. To better predict the prior probability, the similarity between entities and relationships is calculated by the weight of the overlapping type. If the low-level type is not filtered out, the numerator and denominator of the fraction are equivalent to adding the weight of the redundant type at the same time when the numerator and denominator of a fraction add a value at the same time or reduce a same value. The size of the score will change. When γ is set too large, most of the effective types will be lost. When only high-weight types participate in decision-making, it is easy to have large errors.

5 Conclusion

This paper proposes an effective type-aware link prediction (TALP) model, which builds a hierarchical feature model for computing type feature weights and makes type features specific to different relations through training, encoding weights as prior probabilities. Instance-level information is encoded as likelihood probabilities through convolution, and posterior probabilities of entity prediction are calculated by combining prior and likelihood probabilities using Bayes' formula. The model addresses the drawbacks of previous models that ignored most of the semantic information contained in knowledge graphs and the prior knowledge represented by semantic information. On the DB111K-174 and YAGO26K-906 datasets, the MRR increased by 5% and 6%, respectively. Additionally, this study presents a novel concept for utilizing semantic knowledge represented in the knowledge graph as prior knowledge in the link prediction model.

Acknowledgement. This work was supported by the National Key R&D Program of China under Grant No. 2020YFB1710200.

References

1. Huang, X., Zhang, J., Li, D., Li, P.: Knowledge graph embedding based question answering. In: Culpepper, J.S., Moffat, A., Bennett, P.N., Lerman, K. (eds.) Proceedings of the Twelfth ACM International Conference on Web Search and Data Mining, WSDM 2019, Melbourne, VIC, Australia, February 11–15, 2019, pp. 105– 113. ACM (2019)
2. Wang, X., He, X., Cao, Y., Liu, M., Chua, T.: KGAT: knowledge graph attention network for recommendation. In: Teredesai, A., Kumar, V., Li, Y., Rosales, R., Terzi, E., Karypis, G. (eds.) Proceedings of the 25th ACM SIGKDD International Conference on Knowledge Discovery & Data Mining, KDD 2019, Anchorage, AK, USA, August 4–8, 2019, pp. 950–958. ACM (2019)
3. Xu, H., Bao, J., Zhang, G.: Dynamic knowledge graph-based dialogue generation with improved adversarial meta-learning. CoRR abs/2004.08833 (2020)

4. Kapanipathi, P., et al.: Infusing knowledge into the textual entailment task using graph convolutional networks. In: The Thirty-Fourth AAAI Conference on Artificial Intelligence, AAAI 2020, The Thirty-Second Innovative Applications of Artificial Intelligence Conference, IAAI 2020, The Tenth AAAI Symposium on Educational Advances in Artificial Intelligence, EAAI 2020, New York, NY, USA, February 7–12, 2020, pp. 8074–8081. AAAI Press (2020)

5. Tang Z, Du L. A Knowledge Graph Embedding Model Based on Node-Relation Fusion Perception. In: 2020 The 9th International Conference on Networks, Communication and Computing, pp. 65–70 (2020)

6. Hao, J., Chen, M., Yu, W., Sun, Y., Wang, W.: Universal representation learning of knowledge bases by jointly embedding instances and ontological concepts. In: Teredesai, A., Kumar, V., Li, Y., Rosales, R., Terzi, E., Karypis, G. (eds.) Proceedings of the 25th ACM SIGKDD International Conference on Knowledge Discovery & Data Mining, KDD 2019, Anchorage, AK, USA, August 4–8, 2019, pp. 1709–1719. ACM (2019)

7. LeCun, Y., Bottou, L., Bengio, Y., et al.: Gradient-based learning applied to document recognition. Proc. IEEE **86**(11), 2278–2324 (1998)

8. West, R., Gabrilovich, E., Murphy, K., Sun, S., Gupta, R., Lin, D.: Knowledgebase completion via search-based question answering. In: Chung, C., Broder, A.Z.,Shim, K., Suel, T. (eds.) 23rd International Worldwide Web Conference, WWW'14, Seoul, Republic of Korea, April 7–11, 2014, pp. 515–526. ACM (2014)

9. Bordes, A., Usunier, N., García-Durán, A., Weston, J., Yakhnenko, O.: Translating embeddings for modelling multirelational data. In: NIPS (2013)

10. Zong, C., Strube, M.: Proceedings of the 53rd annual meeting of the association for computational linguistics and the 7th international joint conference on natural language processing (volume 2: Short papers). In: International Joint Conference on Natural Language Processing (2015)

11. Ma, S., Ding, J., Jia, W., Wang, K., Guo, M.: Transt: Type-based multiple embedding representations for knowledge graph completion. In: ECML/PKDD (2017)

12. Zhao, L., Akoglu, L.: Pairnorm: Tackling oversmoothing in gnns. vol. abs/1909.12223 (2019)

13. Balazevic, I., Allen, C., Hospedales, T.M.: Hypernetwork knowledge graph embeddings. In: ICANN (2021)

14. Leis, V., Kemper, A., Neumann, T.: The adaptive radix tree: Artful indexing for main-memory databases, pp. 38–49 (2013)

15. Cui Z, Kapanipathi P, Talamadupula K, et al.: Type-augmented relation prediction in knowledge graphs. In: Proceedings of the AAAI Conference on Artificial Intelligence vol. 35(8), pp. 7151–7159 (2021)

16. Dettmers, T., Minervini, P., Stenetorp, P., Riedel, S.: Convolutional 2d knowledge graph embeddings. In: McIlraith, S.A., Weinberger, K.Q. (eds.) Proceedings of the Thirty-Second AAAI Conference on Artificial Intelligence, (AAAI-18), the 30th innovative Applications of Artificial Intelligence (IAAI-18), and the 8th AAAI Symposium on Educational Advances in Artificial Intelligence (EAAI-18), New Orleans, Louisiana, USA, February 2–7, 2018, pp. 1811–1818. AAAI Press (2018)

17. Vashishth, S., Sanyal, S., Nitin, V., Agrawal, N., Talukdar, P.P. : Interacte: Improving convolution-based knowledge graph embeddings by increasing feature interactions. In: The Thirty-Fourth AAAI Conference on Artificial Intelligence, AAAI2020, The Thirty-Second Innovative Applications of Artificial Intelligence Conference, IAAI 2020, The Tenth AAAI Symposium on Educational Advances in Artificial Intelligence, EAAI 2020, New York, NY, USA, February 7–12, 2020, pp. 3009–3016. AAAI Press (2020)

18. Bordes, A., Usunier, N., García-Durán, A., Weston, J., Yakhnenko, O.: Translating embeddings for modelling multirelational data. In: Burges, C.J.C., Bottou, L., Ghahramani, Z., Weinberger, K.Q. (eds.) Advances in Neural Information Processing Systems 26: 27th Annual

Conference on Neural Information Processing Systems 2013. Proceedings of a meeting held December 5–8, 2013, Lake Tahoe,Nevada, United States, pp. 2787–2795 (2013)

19. Hao, J., Chen, M., Yu, W., et al.: Universal representation learning of knowledge bases by jointly embedding instances and ontological concepts. In: Proceedings of the 25th ACM SIGKDD International Conference on Knowledge Discovery & Data Mining, pp. 1709–1719 (2019)

20. Suchanek F M, Kasneci G, Weikum G. Yago: a core of semantic knowledge. In: Proceedings of the 16th international conference on Worldwide Web, pp. 697–706 (2007)

Multitask Graph Neural Network
for Knowledge Graph Link Prediction

Ye Wang[1(✉)], Jianhua Yang[1], Lijie Li[1(✉)], and Jian Yao[2]

[1] College of Computer Science and Technology, Harbin Engineering University, Harbin, China
{wangye2020,lilijie}@hrbeu.edu.cn
[2] TaoBao Software Co. Ltd, Hangzhou, China

Abstract. Predicting entities in knowledge graphs is a crucial research area, and convolutional neural networks (CNNs) have exhibited significant performance due to their ability to generate expressive feature embeddings. However, several existing methods in this field tend to disrupt entities and relational embeddings, disregarding the original translation characteristics in triples, leading to incomplete feature extraction. To address this issue and preserve the translation characteristics of triples, the present study introduces a novel representation technique, termed MultiGNN. The suggested approach uses a graph convolutional neural network for encoding and implements a parameter sharing technique. It employs a convolutional neural network and a translation model as decoders. The model's parameter space is expanded to effectively integrate translation characteristics into the convolutional neural network, which allows it to capture these characteristics and enhance the model's performance. The proposed method in this paper has demonstrated significant enhancements in several metrics on the public benchmark dataset when compared to the baseline method.

Keywords: Link Prediction · Multitask Learning · Graph Convolution Network

1 Introduction

Entities within the knowledge graph are interconnected through various relationships, forming a network structure of knowledge. For instance, Fig. 1 illustrates the relationship between 'Washington' and 'USA'. However, it is important to acknowledge that knowledge graphs only contain a subset of all possible facts, and many pieces of information may not be included in the current knowledge graph.

In Fig. 1, we can observe that there is a relationship between 'White House' and 'Washington' labelled as 'located', and a relationship between 'Washington' and 'USA' labelled as 'capital'. By using this information, we can deduce that the 'White House' entity is situated in the 'USA' entity. The relationships within the knowledge graph provide a helpful tool for exploring and comprehending the network of knowledge it represents. However, it is crucial to recognize that the graph only captures a snapshot of the complex and interconnected web of information that exists in the real world.

© The Author(s), under exclusive license to Springer Nature Singapore Pte Ltd. 2023
Z. Yu et al. (Eds.): ICPCSEE 2023, CCIS 1880, pp. 318–329, 2023.
https://doi.org/10.1007/978-981-99-5971-6_23

In many cases, the knowledge graph may lack certain pieces of information that are important for a particular task or analysis. In such instances, it is necessary to deduce or supplement the missing information. However, given the vast size of the current knowledge graph, it is often impractical to manually add new information. This would require significant labor and resources.

To address this issue, knowledge graph link prediction techniques can be utilized to predict missing information based on the existing triples in the graph. By analysing the relationships between entities and identifying patterns in the data, these techniques can make accurate predictions about missing information. This allows for the knowledge graph to continue to evolve and expand as new information becomes available.

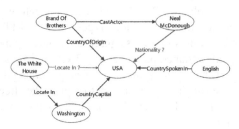

Fig. 1. Triple information

Numerous models have been proposed by experts and scholars in related fields to tackle the task of for link prediction. Link A link prediction model based on CNN can automatically extract features from the triplet of a knowledge graph, and generate feature embeddings with strong expression ability. It deforms the embedding before convolution, which leads to the destruction of the translation characteristics. We presents present a novel representation approach that at involves parameter sharing. The proposed representation approach utilizes the graph convolutional neural network as an encoder, with the addition of a translation model and a convolutional neural network as decoders. By extending the parameter space of the model, the translation properties are effectively integrated into the convolutional neural network, enabling it to capture the translation properties and enhance the model's performance. Our work has made makes several contributions, which can be summarized as follows:

- This paper propose proposes a model that utilizes the parameter space of the model to embed entities and relations. We demonstrate our model's effectiveness through experimental validation.
- This paper have has built a high-order relation modelling model within the framework of the GNN. It enables the exploration of translation characteristics within triples at a deeper level, while maintaining a clear and concise architecture.

2 Related Work

2.1 Non-Neural Based

TransE [1] regards the relationship as a transformation for entities. TransGate [2] adopts the idea of parameter sharing, capture and captures the internal relationship. TransR[3] assumes that entities may play different roles under different relationships. TuckER

[4] can model a very common interaction between each embedded item. DistMult[5], ComplEx [6] and SimplE [7] design special interaction mechanisms that limit entities and relationships to interact only between several entries, which greatly reduces the amount number of method parameters compared to previous methods.

2.2 Neural Network Based

ConvE [8] uses two-dimensional convolution kernels to capture deep-level feature interactions between entities and relationships. ConvR [9] takes the head entity embedding as the input and the relationship embedding as the convolution kernel. InteractE [10] enhances the expression ability of the model through feature arrangement, feature reorganization and cyclic convolution. R-GCN [11] uses a weight matrix specific to each relationship to discern the impact of various types. RR-GCN-PPV [12] uses random representations of random transformations aggregated from neighbors to construct node embeddings, retaining all untrained parameters. VR-GCN [13] uses a control variable-based algorithm to limit the sampling number of neighborneighboring nodes in each layer. KBGAT [14] uses an attention mechanism to capture the effects of neighborneighboring nodes. SACN [15] model models the correlation between adjacent entities with the same relationship type.

3 Model Description

3.1 Model Frame Detail

In this section, we aim to present a detailed account of our proposed approach, highlighting its key components. Figure 2 illustrates the overall architecture of our method, which we will explain in further detail. To effectively learn information from adjacent nodes and edges, we implemented an encoder-decoder structure.

Specifically, the encoder is responsible for capturing the neighbor information of the node, while the decoder applies this information to the link prediction task. This approach enables our model to effectively leverage the relationships between nodes and edges to make accurate predictions.

3.2 Encoder

The encoder in this model is based on the architecture of CompGCN [16]. It enables the information of each triple to propagate in two directions along the edges. Therefore, this paper adds an inverse relation r^{-1} to each relation r.

The encoder model utilizes an entity-relationship combination operation to embed the relationship into the entity embedding. The specific form of this operation is:

$$e_o = \varphi(e_o, e_r) \tag{1}$$

where $e_o, e_r \in R^d$, $\varphi(\cdot)$ is a combination for embedding relations into entity embedding.

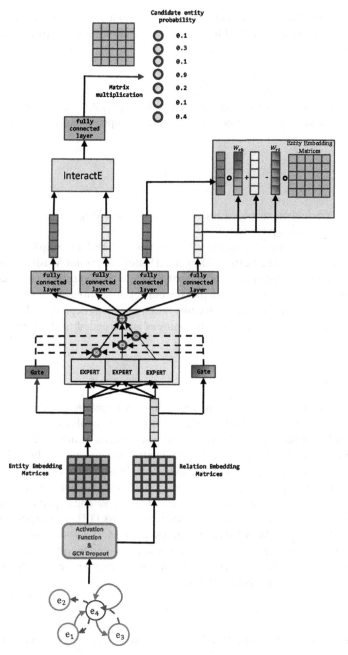

Fig. 2. The framework of the MultiGNN.

Building upon the ideas presented in the TransR model, the current study utilizes an interactive representation method for entities and relationships, which can be expressed as follows:

$$e_o = W_r e_o + e_o \tag{2}$$

W_r is a parameter matrix related to the relationship $e_r.e_v$ incorporates the structural information of its neighboring nodes, and the GCN update equation is formulated as follows:

$$e_v = f\left(\sum_{(o,r)N(v)} W_{dir}\varphi(e_o, e_r) \right) \tag{3}$$

The set of neighbors of node v is denoted as N(v), and $W_{dir(r)}$ are specific parameters for different relationship types.

As the triples are one-way ordered, the direction of the relationship between different nodes varies. For instance, the triple (the United States, the capital, Washington) indicates that Washington is the capital of the United States, whereas the reverse direction does not hold true. Hence, to precisely depict the impact of relationship types, it is essential to categorize the relationship of each entity based on its out-degree, in-degree, and self-circulation direction. Different weight matrices are assigned to different relationship types to achieve this.

3.3 Decoder

Based on the entity and relationship embeddings that contain neighbor node information, a convolution model and a translation model are simultaneously decoded using the multitask learning approach. This approach effectively integrates the translation characteristics into the CNN, enhancing the model's generalization ability [17].

An approach for link prediction is proposed and illustrated in Fig. 2. The input entity embedding and relationship embedding are initially fed into multiple expert modules, each consisting of a simple feedforward neural network. The experts are divided into two categories: shared experts and task-specific experts [18]. The experts in a shared learning system acquire knowledge of common parameters across multiple tasks, while task-specific experts gain expertise in the parameters unique to their respective tasks. The output layer of each task integrates the insights of both the shared and task-specific experts. The expert module is as follows:

$$y_i = f_i(x) \tag{4}$$

where i represents the i-th expert module, x is the input, and y_i is the i-th expert module's output.

The gated neural network is a technique that allows for the control of information flow in neural networks. It is commonly used by selectively allowing or blocking information at specific points in the network. The gate of the entity embedding vector is as follows:

$$g(x) = \sigma(Wx + b) \tag{5}$$

σ represents the softmax activation function, The weight matrix is denoted as W, while b represents the bias., and x represents the influence factor of the control gating network. The gating network of each task generates the distribution of n experts according to the input, and the ultimate output is calculated by taking the weighted sum of all the expert outputs.

To generate independent entity embedding vectors and relational embedding vectors for each task, the proposed method in this study uses both InteractE and a linear model as decoders for the input entity embedding vector and relational embedding. The formula for output is as follows:

$$y^k = F\left(\sum_{i=1}^{n} g(x)_i^k f_i(x)\right) \tag{6}$$

F is the fully connected function, the experts' number is n, and x are the given entity embedding vector and the relation embedding vector respectively. When $k = 1$, the output is related to the entity. When $k = 2$, it is related to the relation.

This paper employs two different decoders simultaneously: the InteractE and the linear model. The function of the linear model is to capture the translation characteristics between entities and relationships. Backpropagation is then used to propagate the specific links between entities and relationships to the shared layer, and the shared module is employed to expand the parameter space. The InteractE model is used to decode and apply the learned information to the link prediction tasks. This multitask learning approach allows the model to effectively integrate translation characteristics into the InteractE, thereby improving its generalization ability.

As shown in Fig. 3, InteractE first uses feature permutation to mix entity embeddings and relation embeddings, generating a randomly permuted sequence. Then, for each permutation sequence, cyclic convolution is utilized to interact the entity embeddings and relation embeddings, producing a tensor sequence that contains complex interaction information between the head entity and the relation. Throughout this process, InteractE captures the intricate interactions between the head entity and the relation through operations such as convolution and pooling, resulting in an embedding vector that incorporates information from both the head entity and the relation. This embedding vector can be matrix-multiplied with the entity set embedding matrix to obtain predicted scores for each candidate entity. Finally, these predicted scores are transformed into corresponding probabilities using the sigmoid function.

The specific operation process of InteractE can be found in formula 7:

$$p(s, r, o) = g(vec(f(\emptyset(P_t) * \omega))W)e_o \tag{7}$$

where $vec(\cdot)$ represents that the tensor is transformed into a vector, and $*$ represents deep convolution. ω represents the convolution filter. $\emptyset(\cdot)$ means feature remodeling. W is a weight matrix. P_t represents the mixed arrangement. f and g represent the ReLU function and sigmoid function, respectively. e_o represents the candidate tail entity.

Train the model to reduce loss values and update parameters through backpropagation:

$$loss_1 = -\frac{1}{N} \sum_{i=1}^{N} t_i \log p_i + (1 - t_i) \log(1 - p_i) \tag{8}$$

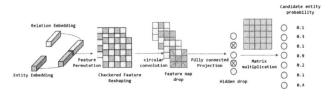

Fig. 3. InteractE decoding to obtain candidate entity probability.

If (e_s, e_r, e_o) appears in the dataset, the label is 1, otherwise the label is 0. p_i is the predicted score of the triple.

To enhance the interaction between entities and relationships, this paper use a linear model based on 1-N scoring.Inspired by the TransR model this paper creates a parameter embedding matrix W_{rh} and W_{rt} related to the relationship. The formula is as follows:

$$loss_2 = \sum_{(h,r,t)\in S} \sum_{(h',r,t')\in S'} \left[\gamma + d(W_{rh}h + r, W_{rt}t) - d(W_{rh}h' + r, W_{rt}t') \right] \quad (9)$$

where (h', r, t') is generated by randomly substituting either the head or tail entities in the original triple. γ is the setting interval, and the initial setting in this paper is 40.

In the back propagation process of a model, the convergence speed of different tasks may vary, resulting in different rates of loss reduction. As an example, the loss reduction of InteractE is approximately 20 times smaller than compared to that of a linear function that uses an interval loss function. If the direct addition method is employed, InteractE may have a negligible effect on the model's loss, and consequently, the learning effect may be diminished. To overcome this issue, Alex Kendall et al. suggested that the loss function of different tasks can be adjusted by setting appropriate parameters [19].The model's total loss can be expressed in formula 10:

$$L(\sigma_1, \sigma_2) = \frac{1}{2\sigma_1^2} loss_1 + \frac{1}{2\sigma_2^2} loss_2 + \log \sigma_1 + \log \sigma_2 \quad (10)$$

σ_1 and σ_2 are initially set as random parameters.

4   Experimental Analysis

4.1   Dataset

Table 1 provides the datasets' statistical information.

4.2   Training Protocol

In terms of embedding dimensions and other parameters, this paper follows the standards set by previous works such as CompGCN to ensure the soundness of the parameter settings.The model optimizer used is Adam, while GCN is set to a single layer. We set the batch size to 512 and the initial learning rate to 0.001. The final layer's entity and relationship embeddings have a dimension of 200.

Table 1. Dataset statistics.

Dataset	Entties	Relations	Train	Valid	Test
FB15K-237	14541	237	272115	17525	20466
WN18RR	40943	18	86835	3034	3134
UMLS	135	46	5216	652	661

4.3 Evaluation Protocol

We utilize three metrics to assess the model's performance: the mean reciprocal ranking (MRR), mean ranking (MR), and the proportion of correct entities among the top N entities (Hits @ N). The values of N used for Hits @ N are 1, 3, and 10.

4.4 Ablation Studies

By removing different module, the models can be roughly categorized into several types: (1) MultiGNN (W/I) Linear module as decoder;(2) MultiGNN (W/T) InteractE module as the decoder;(3) MultiGNN Linear module and InteractE module as decoder. The findings of this analysis are depicted in Fig. 4.

On the FB15k-237 dataset, the Hit @ 10 index of MultiGNN (W/I) is 1.7% higher than that of MultiGNN (W/T), the MR index is 61, and the MRR index is 5% higher. The reason for this could be that although MultiGNN (W/T) is capable of acquiring semantic information from triples, its model structure is relatively simple, and the learned entity and relation embeddings have limited expressive power. On the other hand, the MultiGNN (W/I) model expands the interaction area between entities and relationships by using operations such as cyclic convolution, thereby enhancing its expressive ability.

The MultiGNN (W/I) model achieved improvements in various performance indicators compared to the MultiGNN (W/I) model. Specifically, the hit @ 10 index increased by 2.8%, the MR index increased by 18, and the MRR index increased by 7.1% compared to MultiGNN (W/I). Additionally, compared to MultiGNN (W/T), the Hit @ 10 index increased by 5%, and the MRR index increased by 2%. However, MultiGNN (W/T) outperformed MultiGNN on the MR index. Because MultiGNN (W/T) only models the interaction information in triples, and the model obtains more information on the embedding of entities and relationships that have appeared in the training set. However, the model may suffer from overfitting issues due to its focus on interactive information, which may limit its ability to further improve performance.

By incorporating a shared layer, MultiGNN expands the parameter space, which can lead to better performance on indicators such as MRR. Overall, the convolutional neural network model can effectively capture shallow semantic information, while the translation model is better suited for learning deep information. This paper employs multitask learning to train these two tasks and achieve the best results. Through this method, we obtained better link prediction results than single-task learning.

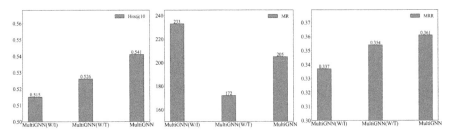

Fig. 4. Results of ablation experiments on the FB15K-237 dataset.

4.5 Analysis of Results

In this study, we conducted a comparative analysis between our proposed model and several existing link prediction methods. We used PyTorch to implement our model and compared our results with those reported in the original paper. Some models were not provided certain indicators in the original text. Due to limitations in equipment and time, these indicators were marked with a '−' in the tables presented in this chapter.

Table 2 and Table 3 provide an overview of the comparative analysis results between MultiGCN and other knowledge graph embedding methods that currently exist. Some indicators are supplementary experiments to compare the experimental results more intuitively.

MultiGCN outperformed CompGCN on multiple metrics, with improvements seen in the MRR index (up 1.6%), Hits @ 1 index (up 3%), Hits @ 3 index (up 1.5%), and Hits @ 10 index (up 1.4%). In comparison to InteractE, MultiGNN demonstrated superior performance on this dataset, with improvements observed in the MRR index (up by 0.007), Hits @ 1 index (up by 0.008), and Hits @ 10 index (up by 0.006).In summary, the proposed model demonstrated superior performance compared to both the CompGCN and InteractE models on the dataset. The model's ability to capture and handle complex interaction information allows for better performance in link prediction tasks.

The experimental results suggest that MultiGNN performs slightly worse than the InteractE model in terms of the MR index.However, the gap between the two evaluation indexes is small. This could be because the fact that the entity embedding of the Multi-GNN model not only contains its own information but also incorporates the message of adjacent entities connected through relationships. Entities with comparable graph structures and interconnected relationships can result in information leakage and duplication, posing a challenge in precisely identifying their associations. This situation could affect the MultiGNN model's ability to effectively express the internal information of the entity, thereby reducing the accuracy of the model prediction.

On the WN18RR dataset, MultiGNN outperforms the CompGCN and InteractE models in all evaluation indices, especially in the MR index. Compared to CompGCN, MultiGNN improves the MR index by 531, indicating that it can better capture the translation characteristics. Additionally, compared to InteractE, MultiGNN can more accurately simulate the complex interaction between entities and relationships, which is often more detailed and complex than the InteractE model.

Table 2. Comparative experimental results are obtained on the FB15k-237

Model	FB15k-237				
	MRR	MR	Hits@10	Hits@3	Hits@1
TransE	0.294	357	0.465	–	–
CompLEx	0.247	339	0.419	0.263	0.155
ConvE	0.325	244	0.501	0.356	0.273
ConvKB	0.243	311	0.421	0.371	0.155
InteractE	0.354	**172**	0.535	–	0.263
R-GCN	0.164	600	0.300	0.181	0.100
VR-GCN	0.248	–	0.432	0.272	0.159
RR-GCN-PPV	0.256	–	0.412	0.256	0.157
CompGCN	0.355	197	0.533	0.390	0.263
MultiGNN	**0.361**	205	**0.541**	**0.396**	**0.271**

Table 3. Comparative experimental results are obtained on the WN18RR

Model	WN18RR				
	MRR	MR	Hits@10	Hits@3	Hits@1
TransE	0.226	3384	0.501	–	–
CompLEx	0.44	5261	0.51	0.46	0.41
ConvE	0.43	4187	0.52	0.44	0.40
ConvKB	0.249	3324	0.524	0.417	0.436
InteractE	0.463	5202	0.528	–	0.43
R-GCN	0.123	6700	0.207	0.137	0.080
SACN	0.47	–	0.54	0.48	0.43
CompGCN	0.479	3533	0.546	0.494	**0.443**
MultiGNN	**0.481**	**3002**	**0.557**	**0.494**	0.442

SACN proposes a weighted graph convolutional network to address the issue of graph structure information being neglected and adopts the Conv-TransE model as a decoder to preserve the translation characteristics. However, due to the limitations of model parameters, it is difficult for SACN to fully retain the translation characteristics. In comparison, MultiGNN outperforms SACN in all metrics.

4.6 Parameter Experiment

Our objective in conducting experiments on the UMLS dataset was to identify the ideal number of shared experts for the two tasks. The findings presented in Table 4 demonstrate the following:

Table 4. Link prediction results of different numbers of shared experts on UMLS.

the number of shared experts	UMLS				
	MRR	MR	Hits @10	Hits @3	Hits @1
1	0.926	1.33	0.992	0.981	0.868
2	**0.932**	**1.26**	0.995	**0.983**	**0.881**
3	0.924	1.33	**0.997**	0.977	0.867
4	0.928	1.31	0.992	0.975	0.877

Based on the above experiments, it can be concluded that choosing two shared experts for decoding with two different decoders achieves relatively good results. The multitask learning process allows different experts to learn domain-specific knowledge. With two shared experts, each specific task can have enough parameter space without significantly affecting other tasks. The results demonstrate the influence of using multiple expert models in improving the performance of the model. Therefore, selecting an appropriate number of shared expert models is very important.

5 Conclusion

This paper proposes a method to help convolutional neural networks capture translation characteristics by expanding the parameter space. Nonetheless, there is scope for enhancing the comprehension of the interplay between entities and relationships. For our future endeavors, we intend to delve deeper into this aspect to more effectively incorporate the structural details of the knowledge graph.This will enable us to obtain more accurate and comprehensive representation of entities and relationships, leading to better performance in knowledge graph completion and related tasks.

Acknowledgements. This work was supported by the National Key R&D Program of China under Grant No. 2020YFB1710200.

References

1. Bordes, A., Usunier, N., Garcia-Duran, A., et al.: Translating embeddings for modeling multi-relational data. In: Advances in Neural Information Processing Systems, vol. 26 (2013)
2. Yuan, J., Gao, N,. Xiang, J.: TransGate: knowledge graph embedding with shared gate structure. In: Proceedings of the AAAI Conference On Artificial Intelligence. 2019, vol. 33(01), pp. 3100–3107 (2019)

3. Lin, Y., Liu, Z., Sun, M., et al.: Learning entity and relation embeddings for knowledge graph completion. In: Proceedings of the AAAI Conference on Artificial Intelligence, vol. 29(1) (2015)

4. Balažević, I., Allen, C., Hospedales, T M.: Tucker: Tensor factorization for knowledge graph completion. arXiv preprint arXiv:1901.09590 (2019)

5. Yang, B., Yih, W., He, X., et al.: Embedding entities and relations for learning and inference in knowledge bases. arXiv preprint arXiv:1412.6575 (2014)

6. Trouillon T, Welbl J, Riedel S, et al. Complex embeddings for simple link prediction. In: International Conference on Machine Learning. PMLR, pp. 2071–2080 (2016)

7. Kazemi, S.M., Poole, D.: Simple embedding for link prediction in knowledge graphs. In: Advances in Neural Information Processing Systems, vol. 31 (2018)

8. Dettmers, T., Minervini, P., Stenetorp, P., et al.: Convolutional 2D knowledge graph embeddings. In: Proceedings of the AAAI Conference on Artificial Intelligence, vol. 32(1) (2018)

9. Jiang, X., Wang, Q., Wang, B.: Adaptive convolution for multi-relational learning. In: Proceedings of the 2019 Conference of the North American Chapter of the Association for Computational Linguistics: Human Language Technologies, Volume 1 (Long and Short Papers), pp. 978–987 (2019)

10. Vashishth, S., Sanyal, S., Nitin, V., et al.: Interacte: Improving convolution-based knowledge graph embeddings by increasing feature interactions. In: Proceedings of the AAAI Conference on Artificial Intelligence, 34(03), pp. 3009–3016 (2020)

11. Schlichtkrull, M., Kipf, T.N., Bloem, P., et al.: Modeling relational data with graph convolutional networks. In: The Semantic Web: 15th International Conference, ESWC 2018, Heraklion, Crete, Greece, June 3–7, 2018, Proceedings 15. Springer International Publishing, pp. 593–607 (2018)

12. Degraeve, V., Vandewiele, G., Ongenae, F., et al.: R-GCN: the R could stand for random. arXiv preprint arXiv:2203.02424 (2022)

13. Chen, J., Zhu, J., Song, L.: Stochastic training of graph convolutional networks with variance reduction. arXiv preprint arXiv:1710.10568 (2017)

14. Nathani, D., Chauhan, J., Sharma, C., et al.: Learning attention-based embeddings for relation prediction in knowledge graphs. arXiv preprint arXiv:1906.01195 (2019)

15. Shang, C., Tang, Y., Huang, J., et al.: End-to-end structure-aware convolutional networks for knowledge base completion. In: Proceedings of the AAAI Conference on Artificial Intelligence, vol. 33(01), pp. 3060–3067 (2019)

16. Vashishth, S., Sanyal, S., Nitin, V., et al.: Composition-based multi-relational graph convolutional networks. arXiv preprint arXiv:1911.03082 (2019)

17. Ruder, S.: An overview of multi-task learning in deep neural networks. arXiv preprint arXiv: 1706.05098 (2017)

18. Ma, J., et al.: Modeling task relationships in multi-task learning with multi-gate mixture-of-experts. In: Proceedings of the 24th ACM SIGKDD International Conference on Knowledge Discovery & Data Mining (2018)

19. Kendall, A., Yarin, G., Roberto, C.: Multi-task learning using uncertainty to weigh losses for scene geometry and semantics. In: Proceedings of the IEEE Conference on Computer Vision and Pattern Recognition (2018)

A Short Text Classification Model Based on Chinese Part-of-Speech Information and Mutual Learning

Yihe Deng and Zuxu Dai[✉]

School of Mathematics and Physics, Wuhan Institute of Technology, Wuhan 430205, China
zxdai@wit.edu.cn

Abstract. Short text classification is one of the common tasks in natural language processing. Short text contains less information, and there is still much room for improvement in the performance of short text classification models. This paper proposes a new short text classification model ML-BERT based on the idea of mutual learning. ML-BERT includes a BERT that only uses word vector information and a BERT that fuses word information and part-of-speech information and introduces transmission flag to control the information transfer between the two BERTs to simulate the mutual learning process between the two models. Experimental results show that the ML-BERT model obtains a MAF1 score of 93.79% on the THUCNews dataset. Compared with the representative models Text-CNN, Text-RNN and BERT, the MAF1 score improves by 8.11%, 6.69% and 1.69%, respectively.

Keywords: Natural language processing · Neural network · Chinese short text classification · BERT · Mutual deep learning

1 Introduction

Automatic text classification is a research topic that closely combines pattern recognition and natural language processing (NLP). With the rapid development of the Internet today, quickly obtaining the required information from the massive amount of information on the Internet has become particularly important. When browsing massive amounts of information, the title information is an important basis for judging the relevance of the web page to the user's needs. However, as an important branch of short text classification, title classification has problems such as short text and lack of information. Therefore, the short text classification model still needs more in-depth research and exploration.

At the beginning of short text classification, text classification usually adopts a method based on machine learning. For example, Joachims [1] explored the application of support vector machines (SVMs) in text classification and analysed its particularity in text data learning and the reasons why it is suitable for text classification tasks. McCallum [2] compared two event models of the naive bayesian model, the multivariate Bernoulli event model and the multinomial event model, on classification effects on different text corpora. Sebastiani [3] summarised the application of machine learning in text classification and proposed the difficulties faced by text classification.

However, in recent years, with the improvement of computer computing power, deep learning algorithms have developed rapidly and have been widely used in the field of natural language processing. Yoon [4] proposed a Text-CNN model to apply convolutional neural networks [5] (CNN) to text classification tasks. Liu [6] proposed three different information sharing models, Text-RNN based on the Recurrent Neural Network [7] (RNN), and applied it to text multiclassification tasks. With the deepening of research, improved methods based on RNN networks have been proposed, including the long short-term memory network [8] (LSTM) and gated recurrent unit [9] (GRU). All are used to solve the gradient disappearance problem of RNNs. CNN and RNN are commonly used neural networks in NLP tasks. The difference is that the CNN model mainly mines local structural features, while RNN considers more connections between neurons. If CNN handles NLP tasks from a spatial perspective, RNN solves NLP tasks from a temporal perspective. To solve the limitations of the two architectures and combine their respective advantages, Zhou [10] proposed the C-LSTM model and applied it to the architecture of sentiment classification and question classification tasks.

Text classification based on deep learning is generally divided into two parts: pre-training models and nonpretraining models. In early text tasks, one-hot encoding and cooccurrence matrices were mainly used as word vector generation methods. However, both word embedding methods suffer from the curse of dimensionality. Mikolov [11] and others proposed a model based on a neural network to efficiently train word vectors: word2vec. Since then, an increasing number of domestic and foreign scholars have focused their attention on the research and discussion of pretraining models. In 2018, the Google team proposed the BERT [12] (bidirectional encoder representation from transformers) neural network model based on the transformer [13] structure. The BERT model can not only serve as a pretraining model but also participate in downstream tasks as a classification model. Since then, combining BERT with Chinese text tasks has been a hot topic for scholars.

In the field of deep learning, knowledge distillation is an effective technique to improve the performance of small models. It was first proposed by Hinton [14] and applied to the CNN network. During the training process of the model, a large-scale teacher network is trained first. After the teacher network is trained, the predicted value of the teacher network for the dataset will be passed to the student network to assist the student network in calculation. In 2018, Zhang [15] proposed a concept that is different from the distillation model but related: deep mutual learning (DML). The usual pattern of traditional distillation methods is to first train a large teacher network and then transfer the knowledge to a small, untrained student network. Mutual learning, on the other hand, starts with a set of untrained student networks and lets the models learn simultaneously to solve tasks together. Compared with the traditional "teacher-student" mode, this method is more similar to the "student–student" mode. The two student networks learn from each other during the training process; that is, they will be teacher or student during each training.

In the DML model, the two models are trained at the same time, and information is exchanged after each round of training to participate in the loss calculation. During the loss calculation, the cross-entropy function is used to calculate HardLoss, and the KL divergence is used to calculate SoftLoss:

$$L_{\Theta 1} = L_{c1}(q_1, p_1) + KL(p_2 \| p_1),$$ (1)

$$L_{\Theta 2} = L_{c2}(q_2, p_2) + KL(p_1 \| p_2)$$ (2)

Among them, $L_{\Theta 1}$ and $L_{\Theta 2}$ represent the two student networks. q_1 and q_2 represent the true label distribution of the two models, usually $q_1 = q_2$. p_1 and p_2 represent prediction results of the two models.

From the definition of the loss function, it can be seen that the loss calculation of a single model is related to the result of another model in the same round. This method is not a one-way transfer between teacher and student under static conditions but a process of cooperative learning between two students. On this basis, Rajamanoharan [16] proposed a multi-task mutual learning model that simultaneously learns discriminative features from multiple branches and applies them to the task of vehicle recognition. Xue [17] combined deep adversarial networks with mutual learning methods, proposed a new deep adversarial mutual learning method involving two sets of features, and applied it to sentiment classification. Bhat [18] adopted the idea of mutual learning and proposed Distill-on-the-Go to improve the representation learning effect of small models by using online distillation.

To solve the problems of less information in short texts and poor correlation between Chinese characters and characters. A new short text classification model ML-BERT is proposed based on the idea of mutual learning. The main contributions are as follows:

(1) Improved mutual learning methods are used to enrich the information content of the model and achieve information transfer between models to improve the performance of the model.
(2) Improved mutual learning methods are used to enrich the information content of the model and achieve information transfer between models to improve the performance of the model.
(3) Propose a new short text classification model ML-BERT and conduct comparative experiments with the base model to verify the performance of the model.

2 Method

2.1 ML-BERT Structure

This paper combines BERT with the idea of mutual learning, proposes a new classification model ML-BERT, and applies it to the task of classifying Chinese news titles. ML-BERT includes a BERT that only uses word vector information and a BERT that fuses word information and part-of-speech information. During the training process, the two student networks will cross-learn and transfer information. Different from the traditional MDL model that continuously transmits information between models, this paper introduces transmission flag (TF) as the signal for whether the "student–student" model transmits information during the training process of the two models. During the training process, the fitting results on the training set are used as the basis for whether two students transmit information to each other, and the verification results on the test set are used as the basis for which side is the guiding party. The model flowchart is shown in Fig. 1, and the two networks are cross-trained during the training process.

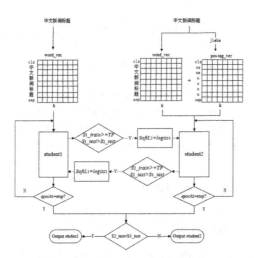

Fig. 1. Model Flowchart

In Fig. 1, student1 and student2 represent two student networks, and TF represents transmission flag. If the fitting effect of the student1 network on the training set exceeds the Transmission Flag set by the evaluation index, it is considered that the student1 network has the ability to guide another student network. Then, the test set results of the student1 network are compared with the test set results of the student2 network. If the test set results of the student1 network are higher than the test set results of the student2 network, it is considered that the student1 network will serve as the teacher to guide the student2 network in the next epoch. That is, the probability distribution obtained by the studet1 network in this calculation will be used as a soft label to guide the Student2 network to perform loss calculation. The judgment method of the student2 network is the same. In this way, the ability of the model as a guide can be guaranteed to be better than that of the model it guides to learn.

2.2 BERT Student Network

BERT [12] is a neural network model proposed by Google that is pretrained through a masked language model and uses a deep bidirectional transformer component to build the entire model. BERT can not only be used as a pretraining model but can also handle sequence-level tasks and token-level tasks in NLP. As a sequence-level task, title classification only needs to add an additional output layer to BERT to achieve the classification effect.

Text information is a kind of sequence information that computers cannot understand. Before the title information is input to the BERT network, the characters in the sentences participating in the training must be converted into the form of word vectors (word_vec). The label vectors corresponding to [cls] and [sep] are added at the head and the end of the sentence, respectively. Assume that the sentence vector composed of word vectors input into the model is:

$$Input\_BERT = [w_{[CLS]}^T, w_1^T, w_2^T, ..., w_{len}^T, w_{[SEP]}^T]^T, \tag{3}$$

$$w_{[CLS]}, w_{[SEP]}, w_n \in R^{1 \times k}, n = 1, ..., len.$$

where *len* represents the length of the sentence, $w_1, w_2,..., w_{len}$ represents the word vector corresponding to the characters in the sentence, and the word vectors are all *k*-dimensional row vectors. After the sentence passes through the BERT student network, it outputs an output vector *Output* with the same dimension as the input vector *Input_BERT*:

$$Output = BERT(Input\_BERT) = [o_{[CLS]}^T, o_1^T, o_2^T, ..., o_{len}^T, o_{[SEP]}^T]^T, \tag{4}$$

$$o_{[CLS]}, o_{[SEP]}, o_n \in R^{1 \times k}, n = 1, ..., len.$$

where $o_1, o_2,..., o_{len}$ represent the word vector output after operation.

2.3 BERT_{pos-tag} Student Network

The BERT_{pos-tag} student network adds part-of-speech information to assist in judgment based on the BERT student network. BERT_{pos-tag} converts the sentences participating in the training into the form of word vectors and part-of-speech vectors and accumulates the two vectors. The word vectors are generated in the same way as described above. The part-of-speech vector is mainly used to obtain the part-of-speech tag (pos-tag) after the sentence is segmented by the jieba tokenizer [19]. Then, the part-of-speech tag is expressed in the form of a part-of-speech vector (pos-tag_vec). After the word vector and part-of-speech vector are accumulated, it is the input of the BERT_{pos-tag} student network. The addition of the part-of-speech vector solves the problem of lack of association between words. The part-of-speech vector conversion based on jieba is shown in Fig. 2.

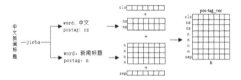

Fig. 2. Pos-tag to Pos-tag_vec

Let *Input_word* and *Input_pos* be the sentence vector composed of word vectors and the part-of-speech sentence vector composed of part-of-speech vectors in the

BERTpos-tag student network, respectively; then, the final accumulated input vector is $Input\_BERT_{pos\text{-}tag}$. The calculation method is as follows:

$$Input\_word = [w_{[CLS]}^T, w_1^T, w_2^T, ..., w_{len}^T, w_{[SEP]}^T]^T, \tag{5}$$

$$Input\_pos = [p_{[CLS]}^T, p_1^T, p_2^T, ..., p_{len}^T, p_{[SEP]}^T]^T, \tag{6}$$

$$Input\_BERT_{pos-tag} = Input\_word + Input\_pos, \tag{7}$$

$$w_{[CLS]}, w_{[SEP]}, w_n \in R^{1 \times k}, p_{[CLS]}, p_{[SEP]}, p_n \in R^{1 \times k}, \ n = 1, ..., len.$$

where len represents the length of the sentence, w_1, w_2..., w_{len} represents the word vector corresponding to the characters in the sentence, p_1, p_2..., p_{len} represents the part-of-speech vector corresponding to the characters in the sentence, and both the word vector and the part-of-speech vector are k-dimensional row vectors. The rest of the calculation methods are the same as the BERT student network.

2.4 Classification Prediction

This article uses the average pooling method to obtain the sentence vector and uses the softmax function as the output layer of the multiclassification problem. Compared with only extracting the CLS vector as the sentence vector in BERT, the average pooling method can retain the information in the sentence to a greater extent. As shown in the following formula:

$$Pool\_vec = MeanPool(Output), \tag{8}$$

$$Pool\_vec \in R^{1 \times k}.$$

Input the result obtained after pooling into the fully connected layer for linear transformation:

$$Linear\_vec = Pool\_vec \cdot W_{Linear\_vec}^T + bias_{Linear\_vec}, \tag{9}$$

$$Linear\_vec \in R^{1 \times k}, W_{Linear\_vec} \in R^{k \times k}, bias_{Linear\_vec} \in R^{1 \times k}.$$

Then, the *tanh* function is used to activate the neurons in the linearly transformed *Linear_vec* to obtain the sentence vector *Sentence_vec*. The calculation method is as follows:

$$Tanh(x) = \frac{e^x - e^{-x}}{e^x + e^{-x}}, \tag{10}$$

$$Sentence\_vec = Tanh(Linear\_vec), \tag{11}$$

$$Sentence\_vec \in R^{1 \times k}.$$

Input the obtained sentence vector into the fully connected layer again, and convert the sentence vector whose dimension is a k-dimensional row vector into an l-dimensional row vector. The result after the operation is *Logits*:

$$Logits = Sentence\_vec \cdot W_{Logits}^T + bias_{Logits}, \tag{12}$$

$$Logits \in R^{1 \times l}, W \in R^{l \times k}, bias_{Logits} \in R^{1 \times k}.$$

where l represents the number of label categories and k represents the word vector dimension. For multiclassification tasks, the obtained *Logits* are a logistic distribution. After the *Logits* are normalised using the *softmax* function, a probability distribution S of length l can be obtained, which is used as the basis for classification. The calculation method of the *softmax* activation function of a certain category probability is as follows:

$$s_j = softmax(j) == \frac{e^j}{\sum\limits_{i=1}^{l} e^i}. \tag{13}$$

Then, the probability distribution obtained through model operation is:

$$S = [s_1, s_2, ..., s_l]. \tag{14}$$

where S represents the obtained probability distribution, and s_1, s_2...s_l represents the probability of each class.

2.5 Loss Function

In this paper, KL divergence [18] is used as a method to measure the matching degree between hard labels and predicted values when calculating *HardLoss*. The KL divergence can be expressed in the following form:

$$KL(P\|Q) = \sum\limits_{i=1}^{l} P(x_i) log \frac{P(x_i)}{Q(x_i)}. \tag{15}$$

Let *HardL* be the actual probability distribution of the label, that is, the hard label; S is the probability distribution calculated by the passed model, and S_i represents the probability that the input is category i. According to formula (15), the *HardLoss* loss calculation can be expressed as:

$$HardLoss = KL(HardL\|S) = \sum\limits_{i=1}^{l} HardL_i \cdot log \frac{HardL_i}{S_i}. \tag{16}$$

In this paper, JS divergence [20] is adopted as a method to measure the matching degree between soft labels and predicted values when calculating SoftLoss. JS divergence is a variant of KL divergence, and its advantage is that it solves the asymmetric problem of KL divergence. The symmetric nature of JS divergence makes it more accurate in judging the similarity, and it is more in line with the symmetric idea in the ML-BERT model.

$$JS(P\|Q) = \frac{1}{2}KL\left(P\|\frac{P+Q}{2}\right) + \frac{1}{2}KL\left(Q\|\frac{P+Q}{2}\right). \tag{17}$$

Let $M = \frac{SoftL+S}{2}$, where $SoftL$ is the probability distribution of the soft label and S is the probability distribution calculated by the passed model. According to formula (17), the loss calculation of the $SoftLoss$ part can be expressed as:

$$SoftLoss = JS(SoftL\|S) = \frac{1}{2}KL(SoftL\|M) + \frac{1}{2}KL(S\|M). \tag{18}$$

Therefore, the complete loss function is expressed as:

$$Student_n Loss = KL(HardL, S_n) + JS(SoftL_n, S_n), \quad n = 1, 2. \tag{19}$$

where n represents student network 1 or student network 2 (i.e., BERT student network or $\text{BERT}_{\text{pos-tag}}$ student network). The two student networks are the same dataset, so the $HardL$ of the same data is consistent. Through the above method, each model can not only use the predicted value and the real label to calculate the supervised loss through the KL divergence but also use the predicted value and the soft label to calculate the same amount of probability estimation through the JS divergence.

2.6 Optimisation

Traditional mutual learning algorithms compute the predicted probability distributions of the two models at each training iteration and update the soft labels of the two networks based on the predicted probability distributions of the other model. However, transferring the probability distribution to another model when the initial model underfits may lead to learning the wrong probability distribution. Moreover, there will gradually be a gap between the two student networks during the training process, and if the probability distribution of the poorer network is updated with soft labels for the better network, the network may be in trouble. Therefore, this article chooses to use the transmission flag and compare the test set results to solve the above problems. The optimisation details are summarised in Algorithm 1.

Algorithm1 Optimised Mutual Deep Learning Algorithm

Initialise: Models $\Theta_{1,t}$ and $\Theta_{2,t}$, train result x_1 and x_2, test result y_1 and y_2, soft label *SoftL*$_1$ and *SoftL*$_2$, logistic distribution S_1 and S_2, transmission flag *TF*.

Repeat:

    1. $t = t + 1$

    2.   if $x_2 >= TF$ and $y_2 > y_1$: update *SoftL*$_1 = S_2$

    3.   if *SoftL*$_1$ is not None:

    4. $Loss_{\Theta_1} = KL(HardL, S_1) + JS(SoftL_1, S_1)$

    5. $\Theta_{1,t+1} = AdamW(\Theta_{1,t})$

    6.   else:

    7. $Loss_{\Theta_1} = KL(HardL, S_1)$

    8. $\Theta_{1,t+1} = AdamW(\Theta_{1,t})$

    9.   updata S_1, x_1 and y_1

   10. if $x_1 >= TF$ and $y_1 > y_2$: update *SoftL*$_2 = S_1$

   11. if *SoftL*$_2$ is not None:

   12. $Loss_{\Theta_2} = KL(HardL, S_2) + JS(SoftL_2, S_2)$

   13. $\Theta_{2,t+1} = AdamW(\Theta_{2,t})$

   14. else:

   15. $Loss_{\Theta_2} = KL(HardL, S_2)$

   16. $\Theta_{2,t+1} = AdamW(\Theta_{2,t})$

   17. updata S_2, x_2 and y_2

2.7 Evaluation Indicators

Macro-average accuracy (MAA), macro-average precision (MAP), macro-average recall (MAR) and macro-average F1-score (MAF1) are used as evaluation indicators for text classification models. Among them, macro-average means the arithmetic average of accuracy, precision, recall and F1-score. The calculation methods of the four evaluation indicators are as follows:

$$MAA = \frac{1}{l} \sum_{i=1}^{l} Accuracy_i = \frac{1}{l} \sum_{i=1}^{l} \frac{TP_i + TN_i}{TP_i + TN_i + FN_i + FP_i}, \tag{20}$$

$$MAP = \frac{1}{l} \sum_{i=1}^{l} Precision_i = \frac{1}{l} \sum_{i=1}^{l} \frac{TP_i}{TP_i + FP_i}, \tag{21}$$

$$MAR = \frac{1}{l} \sum_{i=1}^{l} Recall_i = \frac{1}{l} \sum_{i=1}^{l} \frac{TP_i}{TP_i + FN_i}, \tag{22}$$

$$MAF1 = \frac{1}{l} \sum_{i=1}^{l} F1_i = \frac{1}{l} \sum_{i=1}^{l} \frac{Precision_i \times Recall_i \times 2}{Precision_i + Recall_i}. \tag{23}$$

where l represents the number of label categories. TP and TN belong to the correct judgment. TP means that the positive is predicted as positive, and TN means that the negative is predicted as negative. FN and FP are judgment errors. FN means predicting the positive as negative, and FP means predicting the negative as positive. For multi-classification models, it is positive if the predicted label is the same as the actual label and negative if the predicted label is different from the actual label.

3 Experiment and Results Analysis

3.1 Dataset

The THUCNews dataset is a Chinese dataset for text classification created by Tsinghua University Natural Language Processing Laboratory (THUNLP). This article screened 100,000 pieces of news data, covering ten categories including Finance, Property, Stocks, Education, Science, Society, Politics, Sports, Game, and Entertainment. As shown in Table 1, 10,000 pieces of data are selected for each category and are divided into a training set and a test set at a ratio of 8:2.

In this experiment, the PyTorch framework is used to encode the relevant models, and a GPU (NVIDIA GeForce RTX 2080 SUPER) is used to train and debug the model on the Windows 10 system. The specific parameters of the model are shown in Table 2.

Table 1. Dataset

Category	Train set	Test set	Total
Finance	8000	2000	10000
Property	8000	2000	10000
Stocks	8000	2000	10000
Education	8000	2000	10000
Science	8000	2000	10000
Society	8000	2000	10000
Politics	8000	2000	10000
Sports	8000	2000	10000
Game	8000	2000	10000
Entertainment	8000	2000	10000

Table 2. Parameter Settings

Parameter	Value
Epoch	50
Batch size	32
Transformer block	12
Head number	12
Word embedding	768
Dropout	0.2
Learning rate	1e−5
Transmission Flag	0.8

3.2 Results

The comparison models are as follows:

"Text-CNN" [4]: A neural network model for text tasks proposed by Yoon K based on the original convolutional neural network.

"Text-RNN" [6]: A neural network model for text tasks proposed by Liu based on recurrent neural networks.

"BERT" [12]: A neural network model proposed by the Google Kenton team based on Transformer.

In this paper, the abovementioned MAA, MAP, MAR, and MAF1 are used as the evaluation indicators of the model, and the model results are shown in Table 3. According to the experimental results in Table 3, the ML-BER model constructed in this paper achieves the best results in the classification experiments on the THUCNews dataset, and its MAA, MAP, MAR, and MAF1 are 93.80%, 93.79%, 93.80% and 93.79%, respectively.

Table 3. Text classification results of each model

Dataset	MAA	MAP	MAR	MAF1
Text-CNN	85.66%	85.80%	85.66%	85.68%
Text-RNN	87.06%	87.17%	87.06%	87.10%
BERT	92.15%	92.16%	92.15%	92.10%
ML-BERT	**93.80%**	**93.79%**	**93.80%**	**93.79%**

Table 3 shows that the indicators of the Text-CNN model are the lowest among all experimental models, and MAF1 is only 85.68%. The reason may be that the Text-CNN model cannot adapt well to different situations when dealing with texts of different lengths. If the length of the input text does not match the size of the convolution kernel, padding or truncation is needed, which may result in loss or redundancy of information. The Text-RNN model performs better than the Text-CNN model, with an MAF1 of

87.10%. However, because short texts usually contain fewer features, the discrimination ability of Text-RNN is relatively poor. The BERT model is stacked by multiple Transformer encoders, which can generate multilayer complex representations, learn deeper semantic information and have higher expressive capabilities. Therefore, the test scores are higher than Text-CNN and Text-CNN, reaching 92.10%, but 1.69% lower than the ML-BERT model.

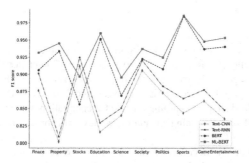

Fig. 3. F1-source for each type

To more intuitively compare the classification effect of the model on each category, this paper uses a line chart, as shown in Fig. 3. It can be seen from Fig. 3 that ML-BERT has the highest F1 source in all categories except the stocks category. Text-CNN and Text-RNN have classification capabilities in the stocks category but are weaker than the ML-BERT model in other categories. The classification ability of the BERT model is slightly weaker than that of ML-BERT in each category.

4 Conclusion

This article proposes a new short text classification model, ML BERT, based on the idea of mutual learning. This model includes a BERT that only uses word vector information and a BERT that integrates word information and part of speech information as the student network and uses transmission flags to control information transmission between student networks to simulate the mutual learning process between two students. Finally, KL divergence is used as the matching degree measurement method of HardLoss, and JS divergence is used as the matching degree method of SoftLoss to build the loss function. From the experimental results, it can be seen that compared with the BERT model, the ML-BERT model introduces part of speech information, and the two models dynamically transfer information during the learning process, making up for the problem of less text information in short text classification. Compared with text CNN, text RNN, and BERT, MAF1 scores increased by 8.11%, 6.69%, and 1.69%, respectively. From a single category perspective, ML BERT achieved the best performance in all categories except for the stock category. ML BERT can enable the model to obtain more external information and express the correlation between Chinese characters through part of speech information. This article uses two BERT models of the same scale to learn from each other. Compared with the "teacher student" model that requires large models to guide small model training, the improved mutual learning method in this article is more flexible, allowing the two models to dynamically switch between guidance and guidance during the learning process based on the evaluation indicators of the test set. However, the model still has the problem of long training time. In future work, we can try to use distributed thinking to optimise the calculation methods of the model.

References

1. Joachims, T.: Text categorization with support vector machines: learning with many relevant features. In: Nédellec, C., Rouveirol, C. (eds.) European Conference on Machine Learning, pp. 137–142. Springer, Heidelberg (1998). https://doi.org/10.1007/BFb0026683
2. McCallum, A., Nigam, K.: A comparison of event models for Naive Bayes text classification. In: AAAI-1998 Workshop on Learning for Text Categorization, vol. 752(1), pp. 41–48 (1998)
3. Sebastiani, F.: Machine learning in automated text categorization. ACM Comput. Surv. (CSUR) 34(1), 1–47 (2002)
4. Yoon, K.: Convolutional neural networks for sentence classification. In: Proceedings of the 2014 Conference on Empirical Methods in Natural Language Processing (EMNLP), pp. 1746–1751 (2014)
5. Bouvrie, J.: Notes on convolutional neural networks (2006)
6. Liu, P., Qiu, X., Huang, X.: Recurrent neural network for text classification with multi-task learning. arXiv preprint arXiv:1605.05101 (2016)
7. Elman, J.L.: Finding structure in time. Cogn. Sci. 14(2), 179–211 (1990)
8. Hochreiter, S., Schmidhuber, J.: Long short-term memory. Neural Comput. 9(8), 1735–1780 (1997)
9. Cho, K., Van Merriënboer, B., Gulcehre, C., et al.: Learning phrase representations using RNN encoder-decoder for statistical machine translation. arXiv preprint arXiv:1406.1078 (2014)
10. Zhou, C., Sun, C., Liu, Z., et al.: A C-LSTM neural network for text classification. arXiv preprint arXiv:1511.08630 (2015)
11. Mikolov, T., Chen, K., Corrado, G., et al.: Efficient estimation of word representations in vector space. arXiv preprint arXiv:1301.3781 (2013)
12. Devlin, J., Chang, M.-W., Lee, K., Toutanova, K.: BERT: pre-training of deep bidirectional transformers for language understanding. In: Proceedings of NAACL-HLT, pp. 4171–4186 (2019)
13. Vaswani, A., Shazeer, N., Parmar, N., et al.: Attention is all you need. In: Advances in Neural Information Processing Systems, vol. 30 (2017)
14. Hinton, G., Vinyals, O., Dean, J.: Distilling the knowledge in a neural network. arXiv preprint arXiv:1503.02531, vol. 2(7) (2015)
15. Zhang, Y., Xiang, T., Hospedales, T.M., et al.: Deep mutual learning. In: Proceedings of the IEEE Conference on Computer Vision and Pattern Recognition, pp. 4320–4328 (2018)
16. Rajamanoharan, G., Kanacı, A., Li, M., et al.: Multi-task mutual learning for vehicle re-identification. In: IEEE Conference on Computer Vision and Pattern Recognition (CVPR) (2019)
17. Xue, Q., Zhang, W., Zha, H.: Improving domain-adapted sentiment classification by deep adversarial mutual learning. Proc. AAAI Conf. Artif. Intell. 34(05), 9362–9369 (2020)
18. Bhat, P., Arani, E., Zonooz, B.: Distill on the go: online knowledge distillation in self-supervised learning. In: Proceedings of the IEEE/CVF Conference on Computer Vision and Pattern Recognition, pp. 2678–2687 (2021)

19. Sun, J.: Jieba Chinese word segmentation tool (2012)
20. Higuchi, H., Suzuki, S., Shouno, H.: Measuring shift-invariance of convolutional neural network with a probability-incorporated metric. In: Mantoro, T., Lee, M., Ayu, M.A., Wong, K.W., Hidayanto, A.N. (eds.) ICONIP 2021. CCIS, vol. 1516, pp. 719–728. Springer, Cham (2021). https://doi.org/10.1007/978-3-030-92307-5_84

The Analysis of Phase Synchronisation in the Uniform Scale-Free Hypernetwork

Juan Du[1,2], Xiujuan Ma[1,2(✉)], Fuxiang Ma[1], Bin Zhou[1,2], and Wenqian Yu[1,2]

[1] College of Computer, Qinghai Normal University, Xining 810016, Qinghai, China
qhnumaxiujuan@163.com
[2] The State Key Laboratory of Tibetan Intelligent Information Processing and Application, Xining 810008, Qinghai, China

Abstract. Many phenomena in realistic complex systems can be explained by the synchronisation behavior of complex systems, such as cricket chirping in unison. The synchronisation behavior occurring on a hypernetwork can be used to explain the swarming behavior occurring on a multivariate interacting system, such as the synchronised forwarding of group messages. There is a lack of results related to phase synchronization of hypernetwork in the existing studies on the synchronization behavior of hypernetworks. To address this problem, this paper investigates the node-based and hyperedge-based phase synchronisation of a scale-free hypernetwork using the Kuramoto model with the order parameter r as the synchronisation degree indicator. The comparative analysis reveals that the phase synchronisation of the scale-free hypernetwork is related to the uniformity k of the hypernetwork but not to the number of nodes and hyperedges, and the phase synchronisation based on hyperedges is more likely to occur than that based on nodes as the coupling strength increases. In addition, the degree of phase synchronisation of scale-free hypernetworks is related to the number of *new_nodes* of newly added nodes when the hyperedge grows during the construction of the hypernetwork, which shows that the smaller the *new_nodes* is, the better the degree of synchronisation of the hypernetwork is.

Keywords: k-uniform · scale-free hypernetwork · phase synchronisation

1 Introduction

Complex systems exist widely in various scenarios of daily life, such as power networks [1], communication networks [2], citation networks [3], etc. In recent years, the dynamic properties of complex systems have become a research hotspot as research on complex systems has intensified. Synchronisation, as a class of collective behaviors prevalent in nature, is widely found in various types of complex systems, such as cricket chirping in unison [4] and firely flickering [5]. Synchronisation is also one of the most fundamental and important dynamical properties of complex systems [6–8]. Currently, among the methods to study the synchronisation phenomenon of complex systems, phase synchronisation has important research value and significance [6]. Winfree [9] proposed a method

© The Author(s), under exclusive license to Springer Nature Singapore Pte Ltd. 2023
Z. Yu et al. (Eds.): ICPCSEE 2023, CCIS 1880, pp. 344–363, 2023.
https://doi.org/10.1007/978-981-99-5971-6_25

to study the synchronisation behavior of oscillators with the help of mathematical theory, in which the system consists of many weakly coupled limit ring oscillators coupled together, while the phase between oscillators is influenced by a certain function. Based on this, Kuramoto [10] further studied and proposed using the phase equation to represent the synchronisation behavior of a system coupled by a finite number of oscillators, the Kuramoto model. This model is a mathematical model to describe the synchronisation behavior of the oscillators in the system by defining the order parameter r to represent the phase correlation between the oscillators and the degree of synchronisation. As research progresses, an increasing number of researchers have successfully studied the phase synchronisation of various types of complex systems using the Kuramoto model, and the different characteristics of the phase synchronisation dynamics of different complex systems are given analytically [11–14].

However, with the rapid development of human society, the interaction between nodes of many complex systems occurs not only between any two nodes, but also between multiple nodes, so the coupling relationship between nodes becomes more complex and multiple [15–17]. Complex networks based on ordinary graphs can only represent the coupling relationship between two nodes and cannot clearly express the interactions between multiple nodes. Therefore, to express the coupling relationships among multiple nodes more efficiently and clearly, hypergraph-based hypernetwork structures were developed [18–20]. Modelling many real systems using hypernetwork theory can more clearly describe the complex interactions between nodes and exhibit better applicability than complex networks [21–23]. Research on the modelling of hypernetworks has become increasingly mature, but research on the nature of the synchronous dynamics of hypernetworks is still in its infancy due to the complexity of the hypernetwork structure. Daniel Irving et al. [24] studied reducing the dimensionality of the problem by simultaneous block diagonalisation of the matrix and gave a general framework for the stability of synchronous solutions of dynamic hypernetworks. Md Sayeed et al. [25] proved that the intralayer synchronous state in a dynamic multilayer hypernetwork is an invariant solution, and the coherent state stability under the associated time-averaged network structure was obtained using the transformation stability basis condition. They also obtained the condition for the stability of the synchronisation without considering the time-averaged network structure by diagonalising the coupling matrix in blocks simultaneously. Sarbendu Rakshit et al. [26] studied the interlayer synchronisation of stochastic multilayer hypernetwork under two types of connections and obtained invariance and stability conditions for the interlayer synchronisation manifold using the principal stability function method and a sufficient condition. Sorrentino et al. [27] constructed dynamic hypernetworks coupled by two or more different networks and used the principal stability function method to obtain the approximate form of the principal stability function only when the network satisfies certain conditions and generalised the stability results to the general hypernetwork case by constructing a neural hypernetwork. Zhao et al. [28] introduced a joint degree to construct an evolutionary hypernetwork model, studied the synchronisation problem of hypernetworks for the first time and obtained several simple and effective synchronisation criteria. Sarbendu Rakshit et al. [29] obtained stability conditions for the synchronisation of a time-varying multilayer neuronal hypernetwork by the principal stability function method and found that the hypernetwork layerwise and interlayerwise synchronisation could be enhanced by changing the intralayer connection probability. The intra- and interlayer synchronisation was found to be enhanced by varying the intralayer connectivity probability. However, the existing studies on the synchronisation dynamics of hypernetworks still lack results related to the phase synchronisation problem of hypernetworks. The study of the phase synchronisation of hypernetworks is a very important reference for further understanding the synchronisation behavior of hypernetworks.

To this end, this paper describes the nodes in the Barabasi-Albert (BA) scale-free hypernetwork as Kuramoto phase oscillators using the Kuramoto model and compares the values of the node-based and hyperedge-based order parameters r to study and analyse the node-based and hyperedge-based phase synchronisation characteristics of the scale-free hypernetwork.

2 Related Knowledge

2.1 Hypernetwork Based on Hypergraph

Hypernetworks are a class of hypergraph-based network models that can be used to describe multivariate interactions between nodes in complex systems. Berge et al. [30, 31] first introduced the concept of hypergraphs, defined as follows:

If the binary relation $H = (V,E)$ satisfies the condition:

$$\emptyset \neq e_i \in P(V), \quad i \in \{1, 2, \cdots, m\}$$

$$\bigcup_{i=1}^{m} e_i = V$$

where $V = \{v_1, v_2, \ldots, v_n\}$ and $E = \{e_1, e_2, \ldots, e_m\}$; then H is a hypergraph. The elements in set V are called the nodes or vertices of the hypergraph; the elements in set E are called the hyperedges of the hypergraph. The hyperdegree of node v_i in the hypergraph is defined as the number of hyperedges containing node v_i, denoted as $d_H(v_i)$. The node degree of node v_i is defined as the number of connected edges of node v_i, denoted as $d(v_i)$. The common degree of node v_i within the hyperedge e_i is denoted as $k_{ei}(v_i)$.

The 2-section graph [32] of the hypergraph $H = (V,E)$ is denoted as $[H]_2$, where the set of nodes of H is denoted as the set of nodes of $[H]_2$, and if two different nodes appear within the same hyperedge, a connected edge is formed between these two nodes. Therefore, all nodes within each hyperedge in $[H]_2$ are fully connected with each other. The adjacency matrix of the 2-section graph shows the adjacency between nodes in the corresponding hypergraph.

The line graph [32] of the hypergraph $H = (V,E)$ is denoted as $L(H) = (V',E')$, and if there are no heavy edges in the hypergraph, then $V': = E$; also, if any two hyperedges in the hypergraph satisfy, then nodes ej and ek are adjacent to each other in $L(H)$ through a common edge. The adjacency matrix of the line graph shows the adjacency of the corresponding hypergraph to the hyperedges in the hypergraph. Figure 1 depicts a hypernetwork consisting of 14 nodes with 6 hyperedges, where (a) shows the topology of a hypernetwork structure, (b) shows the 2-section graph network corresponding to the hypernetwork shown in (a), and (c) shows the line graph network corresponding to this hypernetwork.

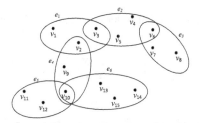

(a) A hypernetwork containing 14 nodes and 6 hyperedges

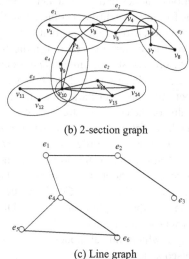

(b) 2-section graph

(c) Line graph

Fig. 1. Hypernetwork and its 2-section graph, line graph

2.2 Phase Synchronisation Model for Hypernetwork

The Kuramoto model [33] is used to describe a class of synchronisation phenomena with a relatively weak degree of synchronisation. In this model, only the phase difference of each oscillator needs to be considered, and if the phase difference of any two coupled oscillators is constant, then these two coupled oscillators are said to reach phase synchronisation.

Researchers have used the Kuramoto model to describe the phase synchronisation of a network consisting of N nodes coupled to each other [12]. Let us denote the phase of node i and the phase variation of node i as shown in Eq. (1):

$$\dot{\theta}_i = \omega_i + \frac{C}{N} \sum_{i=1}^{N} A_{ij} \sin(\theta_j - \theta_i), \quad i = 1, \dots N \tag{1}$$

where C denotes the coupling strength, and the default coupling strength is equal among all nodes; ω_i denotes the natural frequency of i, usually $\{\omega_i | i = 1, 2, ..., N\}$ obeys the Gaussian distribution of single-peaked symmetry $g(\omega)$, and takes a range of values $(-\frac{\pi}{2}, \frac{\pi}{2})$. A_{ij} denotes the adjacency matrix of the network; if node i is adjacent to node j, then $A_{ij} = 1$; otherwise, $A_{ij} = 0$.

For a general N, define the global order covariates of the network at moment t as shown in Eq. (2):

$$r(t)e^{i\Psi(t)} = \frac{1}{N} \sum_{j=1}^{N} e^{i\theta_j(t)} \tag{2}$$

where $^{\Psi(t)}$ denotes the average phase of all nodes at time t. $r(t)$ is used to measure the degree of phase synchronisation of nodes in the whole network at time t. It indicates the ratio of the number of nodes forming synchronous clusters in the network to the total number of network nodes. $r(t)$ closer to 1 indicates that the degree of phase synchronisation of nodes in the network is higher, and $r(t) = 0$ indicates that the nodes in the network are in a completely disordered state at time t. The order parameter $r(t)$ is used to describe the degree of phase synchronisation of the nodes in the network. Figure 2 represents the relationship between the order parameter $r(t)$ and the coupling strength C, where C_0 is the critical value of the coupling strength. When $C > C_0$, the degree of phase synchronisation among the nodes in the network increases as the coupling strength increases. Figure 3 shows the variation in the order parameter $r(t)$ with time. When $C < C_0$, the order parameter $r(t) = 0$, at which time the nodes in the network all move at their respective initial natural frequencies without any coupling, and the phases of the nodes are in a completely disordered state. When $C > C_0$, the value of the order parameter $r(t)$ rises rapidly with the change in the coupling strength, and the phase of the nodes in the network is in the local synchronisation state. When the coupling strength $C \to \infty$, the order parameter $r(t) \to 1$, at which time all nodes in the network are considered to be in a completely synchronised state.

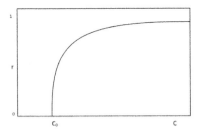

Fig. 2. Variation curve of the sequence parameter $r(t)$ with the coupling strength C

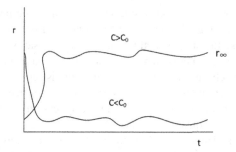

Fig. 3. Variation curve of the order parameter $r(t)$ with the moment t

3 Uniform Scale-Free Hypernetwork Phase Synchronisation Analysis Method

The phase synchronisation of a hypernetwork is similar to that of a normal network but more complex. In this paper, we propose two types of synchronisation that occur on hypernetworks and the analysis methods by combining the different phase synchronisation methods of interpersonal hypernetworks and message group hypernetworks.

3.1 Node-Based Phase Synchronisation Analysis Method

In an interpersonal hypernetwork, nodes denote different peoples, hyperedges denote different social relationships, and the number of nodes in a hyperedge is denoted as the total number of peoples in that social relationship. If a person is in two social circles at the same time, there is an adjacency relationship between these two social circles. We can represent the synchronous behavior of nodes in the network in terms of people in the network infected with similar diseases. Suppose that in an interpersonal hypernetwork, as shown in Fig. 4(a), if some people in the hypernetwork are infected with the same disease, it is considered that this part of the network has achieved local synchronisation; if all people are infected with the same disease, it is considered that the hypernetwork has achieved global synchronisation. Here, the number of people infected with the disease can be used to indicate the degree of synchronisation of the network. For example, a people 1 is infected with a disease for some reason. At this time, the phase of people 1 has changed, without taking any preventive measures. After a certain time of evolution, as shown in Fig. 5, the four people in the social relationship Family 1 will be infected with the same disease. At this time, these four people are infected with the same disease and form a small synchronisation cluster, that is, reach local synchronisation. Then, at the next evolution time, 2 people in the social relationship with colleagues and 4 people in the social relationship with classmates are also infected with the same disease, at which time 6 people newly infected with the virus are added to the synchronisation cluster. Subsequently, after a certain time of evolution, all people in the interpersonal hypernetwork are infected with the same disease, all people in the interpersonal hypernetwork have reached synchronisation. This phenomenon is similar to the phase synchronisation of nodes in a network, where the phases between nodes interact through neighboring relationships to eventually form the phase synchronisation of the whole network. In this paper, this phase synchronisation in the hypernetwork through the mutual influence between nodes is called node-based phase synchronisation.

When the phase synchronisation of the hypernetwork is performed through the nodes, it is equivalent to the fact that phase synchronisation is performed between the nodes of its 2-section graph network. In this case, the phase synchronisation between the nodes of the 2-section graph is equivalent to the phase synchronisation between the nodes of the normal network. Therefore, when analysing the node phase synchronisation of the hypernetwork, this paper refers to the 2-section graph analysis method proposed in the literature [34] to convert the hypernetwork into the form of its corresponding 2-section graph network, such as Fig. 4(b), which is the 2-section graph network of the interpersonal relationship hypernetwork in Fig. 4(a). Then the Kuramoto model of the ordinary network is applied to the 2-section graph network of the hypernetwork, and the characteristics of the node-based phase synchronisation of the hypernetwork are obtained by analysing the phase synchronisation of its 2-section graph network.

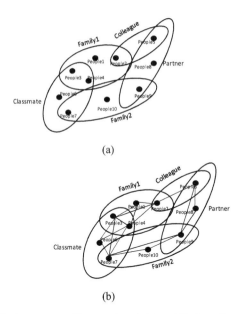

(a)

(b)

Fig. 4. An interpersonal hypernetwork (a) and its 2-section graph (b)

3.2 Hyperedge-Based Phase Synchronisation Analysis Method

The phase synchronisation that occurs in a message group hypernetwork is different from the node-based phase synchronisation in an interpersonal relationship hypernetwork. In a message group hypernetwork, a node represents a group member and a hyperedge represents a message group composed of multiple group members. If a group member

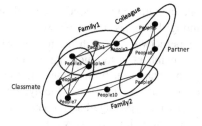

The first evolutionary moment

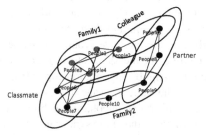

The second evolutionary moment

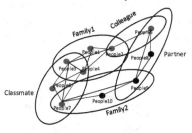

The third evolutionary moment

Fig. 5. Interpersonal hypernetwork phase change process

is present in two message clusters at the same time, there is an adjacency between these two message clusters in the hypernetwork. In a message group hypernetwork as shown in Fig. 6(a), when a group member receives a message, all members in its message group will receive the same message, and then all group members are considered to have reached local synchronisation; when all members in the whole message group hypernetwork receive the same message, they are considered to have reached global synchronisation. For example, in the message group hypernetwork shown in Fig. 6(a), when a group member Q_1 of Message group 1 sends a message to the group, every member of Message group 1 will receive the message, and all members of Message group 1 form a local synchronisation, and other message groups adjacent to Message group 1 through Q_1 will also receive the message through forwarding by group members. Then the message will be forwarded in the whole message group hypernetwork, which eventually leads to the message group hypernetwork all receiving the message, and the network reaches global synchronisation at this time. In this case, a message is initially sent by a node in a hyperedge, and then the hyperedge in which the node is located changes. All the nodes contained in the hyperedge receive the message to achieve local phase synchronisation among the nodes in the hyperedge. After that, the message is forwarded in the whole hypernetwork through the neighboring relationship between hyperedges, and finally all members in the message group hypernetwork receive the message, thus achieving global phase synchronisation. Since this phase synchronisation is triggered by the adjacency relationship between the hyperedges, this paper calls the phase synchronisation in this case hyperedge-based phase synchronisation.

For phase synchronisation through a hypernetwork, this paper analyses the characteristics of hypernetwork-based phase synchronisation by using the line graph network of the hypernetwork. In a message group hypernetwork, a message is initially sent by a group member in a message group, and then the message is received by the whole message group, at which time the message group is considered to have undergone a phase change. At the same time, the message is forwarded to other message groups and other hyperedges through the neighboring relationship between hyperedges, and when the other hyperedges receive the same message, the message groups are considered to be globally synchronised. At this point, the phase synchronisation based on hyperedges in the hypernetwork is equivalent to the phase synchronisation in its line graph network. Then, the Kuramoto model of the ordinary network is applied to the line graph network of the hypernetwork, and the characteristics of the phase synchronisation based on hyperedges in the hypernetwork are obtained by analysing the phase synchronisation of its line graph network. Figure 6(b) shows the line graph network of the message group hypernetwork of Fig. 6(a).

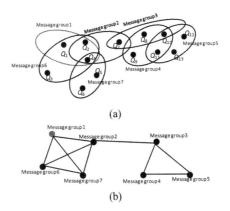

(a)

(b)

Fig. 6. A message group hypernetwork (a) and its line graph (b)

4   Simulation Analysis of the Phase Synchronisation of a Uniform Scale-Free Hypernetwork

In this paper, the node-based phase synchronisation and the hyperedge-based phase synchronisation of the scale-free hypernetwork are analysed separately using the scale-free hypernetwork constructed in the literature [35] as a model. To ensure the validity and reliability of the experimental results, the experimental results are averaged over 50 results.

4.1   Node-Based Phase Synchronisation

In analysing the node-based phase synchronisation, this paper analyses the effects of the uniformity k of the hypernetwork, the number of nodes N, and the number of *new_nodes* added at each time step during the construction of the hypernetwork model on the degree of phase synchronisation of the scale-free hypernetwork and obtains the characteristics of the phase synchronisation of the scale-free hypernetwork.

Figure 7 shows the variation curves of the order parameter r with the coupling strength C for k uniformity scale-free hypernetworks with the same number of nodes N at $k = 3$, 4 and 5. The results in Fig. 7 show that in node-based phase synchronisation, the degree of phase synchronisation is related to the uniformity k of the hypernetwork, and the greater the uniformity k is, the better the degree of phase synchronisation of the hypernetwork. The results in Fig. 7 further show that when the coupling strength C is at a small value, there is no significant difference in the value of the order parameter r at different values of k. However, when the coupling strength C increases to a certain value, the values of the order parameter r show obvious differences at different values of k. Combined with the data in Table 1, it can be seen that when the coupling strength $C > 1.8$, the larger the uniformity k is at the same coupling strength the larger the value of the order parameter r is. It can be seen that in the node-based phase synchronisation when the coupling strength C is at a small value, most of the nodes in the network are in disorder, the percentage of nodes in the synchronisation cluster is smaller, and therefore the value of the order parameter r is also smaller. At this time, the effect of uniformity k on the value of the order parameter r is not significant. However, as the coupling strength C increases, the degree of phase synchronisation of the hypernetwork is closely related to its uniformity k. When k is larger, the value of the order parameter r is larger, the nodes in the synchronisation cluster are more occupied, and the degree of synchronisation of the hypernetwork is better at this time.

In the theoretical analysis of the Kuramoto model, the value of the order parameter r is always 0 when the coupling strength C less than the coupling strength threshold C_0. However, in the experiments of this paper, it can be seen from the simulation data given in Table 1 that the value of the order parameter r is not equal to 0 but a very small value when the coupling strength $C = 0$.

It is clear from the analysis that the nodes in each hyperedge in the 2-section graph analysis method are connected in a fully connected manner, which increases the possibility of each node being connected to any node. This leads to the possibility that a very small fraction of nodes in the network may be connected to nodes with the same phase at the time of initial connection, at which point that small fraction will form a small synchronous cluster even without any effect of coupling strength. Thus the value of the order parameter r will appear as a very small value even without any coupling strength C.

In addition, this paper analyses the relationship between node-based phase synchronisation and the number of newly added nodes *new_node* during the evolution of the hypernetwork with $k = 4$, $k = 6$ and $k = 8$. Figure 8 shows the variation curves of the

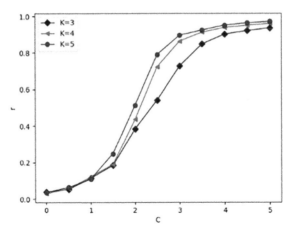

Fig. 7. Curves of the variation in the order parameter r with the coupling strength C for a k-uniform scale-free hypernetwork at $N = 500$ ($k = 3, 4, 5$)

Table 1. The values of the order parameter r for different uniformities k ($k = 3, 4, 5$) and different coupling strengths C

Uniformity	The values of the order parameter r for different coupling strengths C							
	$C = 0$	$C = 0.5$	$C = 1$	$C = 1.5$	$C = 2.5$	$C = 3.5$	$C = 4.5$	$C = 5$
$k = 3$	0.039	0.054	0.114	0.184	0.542	0.848	0.922	0.936
$k = 4$	0.030	0.056	0.123	0.189	0.724	0.913	0.950	0.960
$k = 5$	0.036	0.064	0.109	0.248	0.791	0.924	0.963	0.971

value of the order parameter r with the coupling strength C for different *new_nodes*. Analysis of the results in Fig. 8 shows that the degree of synchronisation of the scale-free hypernetwork decreases with the increase in *new_node*, regardless of the uniformity k of the hypernetwork. The results in Fig. 8 further show that the value of the order parameter r is the largest when *new_node* $= 1$, the optimal degree of synchronisation of the scale-free hypernetwork; the value of the order parameter r is the smallest when *new_node* $= k - 1$, the worst degree of synchronisation of the scale-free hypernetwork. In this paper, the reasons for the correlation between the degree of phase synchronisation and *new_node* of the scale-free hypernetwork are analysed as follows.

Figure 9 shows the hypernetwork hyperdegree distribution plots constructed at different *new_nodes* for $k = 8$. The results in Fig. 9 show that: the smaller the *new_node* is, the steeper the hyperdegree distribution curve is, the larger the maximum value of the hyperdegree is, and the stronger the scale-free characteristic of the hypernetwork is. The analysis shows that when the uniformity k is certain, the smaller the number of newly added nodes *new_node* in the evolution of the hypernetwork, the larger the maximum value of the hyperdegree will be, the scale-free characteristic of the hypernetwork will be stronger at this time, and the connection between nodes will be tighter. Therefore, the

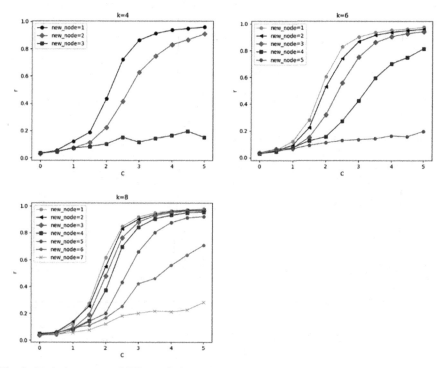

Fig. 8. Variation curves of different timing order parameters r of *new_node* with coupling strength C (a : $k = 4$, b : $k = 6$, c : $k = 8$)

degree of synchronisation is also relatively good. As the number of *new_node* increases, the maximum value of hyperdegree in the hypernetwork decreases, and the scale-free characteristic of the hypernetwork decreases, so the degree of synchronisation of the hypernetwork is relatively poor.

In addition, the effect of the node size of the scale-free hypernetwork on the degree of phase synchronisation of the hypernetwork is analysed in this paper with $k = 3$, $k = 4$ and $k = 5$. Figure 10 shows the variation of the value in the order parameter r with the coupling strength C of the k-uniform scale-free hypernetwork at different node sizes N. From the results in Fig. 10., it can be seen that when the uniformity k of the hypernetwork is certain, the variation curves of the values of the order parameter r with the coupling strength C for three different node sizes do not differ much, which indicates that the node size N of the hypernetwork has less influence on the values of the order parameter r, and the node size of the scale-free hypernetwork has less influence on the degree of phase synchronisation of the hypernetwork.

4.2   Hyperedge-Based Phase Synchronisation

In analysing the phase synchronisation based on hyperedges, this paper verifies the effects of uniformity k, the number of hyperedges, and the number of *new_nodes* added

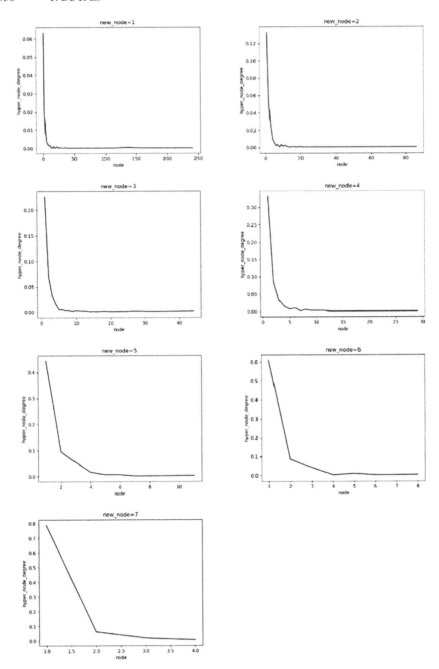

Fig. 9. The distribution of hyperdegree of hypernetwork under different *new_nodes*

at each time step during the construction of the hypernetwork model on the degree of phase synchronisation of scale-free hypernetworks.

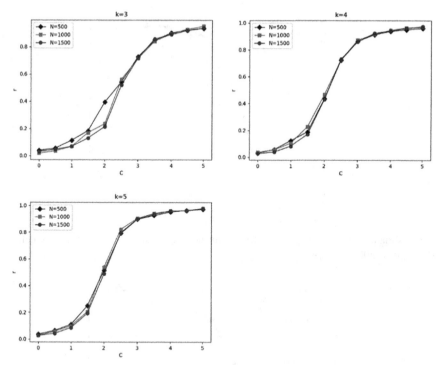

Fig. 10. Variation curves of the order parameter r with the coupling strength C (a : $k = 3$, b : $k = 4$, c : $k = 5$)

Figure 11 shows the curves of the variation in the value of the order parameter r with the coupling strength C when the number of hyperedges $= 500$ and the uniformity k varies. The results in Fig. 11 show that there is no significant difference in the value of the order parameter r when the number of hyperedges is certain and the uniformity k is different. The uniformity k has less influence on the degree of phase synchronisation based on hyperedges when the number of hyperedges is certain. The analysis shows that the phase synchronisation of a scale-free hypernetwork based on a hyperedge is obtained by analysing the corresponding line graph network of the hypernetwork, and in the process of converting the hypernetwork into a line graph network, only the neighboring relationship between hyperedge and hyperedge is considered without considering the number of nodes inside the hyperedge and the neighboring relationship between nodes inside the hyperedge, so the uniformity k of the hypernetwork has little effect on the degree of hypernetwork based on hyperedge synchronisation.

Figure 12 shows the variation curves of the value of the order parameter r with the coupling strength C for different numbers of hyperedges of the scale-free hypernetwork. From the results in Fig. 12, it can be seen that there is no significant difference in the variation curves of the value of the order parameter r with the coupling strength C for three different hyperedge sizes. The number of hyperedges of the scale-free hypernetwork has a small effect on the degree of phase synchronisation of the hypernetwork based on the hyperedge.

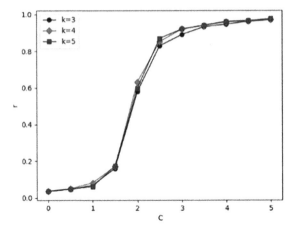

Fig. 11. Variation curve of the order parameter r with the coupling strength C (*hyperedge* $= 500$, $k = 3, 4, 5$)

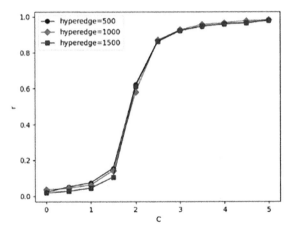

Fig. 12. Variation curve of the order parameter r with the coupling strength C (*hyperedge* $= 500, 1000, 1500$)

Figure 13 shows the variation curves of the values of different order parameters r of *new_node* with the coupling strength C for $k = 6$. From the results in Fig. 13, it can be seen that when the coupling strength $C < 1$, *new_node* has a small effect on the degree of phase synchronisation of the hypernetwork based on hyperedges, but as the coupling strength increases, the smaller the *new_node* is, the better the degree of phase synchronisation of the hypernetwork based on hyperedges.

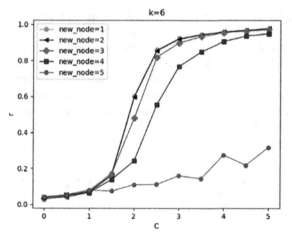

Fig. 13. Variation curves of different timing order parameters r of *new_node* with coupling strength C

5 Comparison of Node-Based Phase Synchronisation and Hyperedge-Based Phase Synchronisation

To obtain the similarities and differences between node-based phase synchronisation and hyperedge-based phase synchronisation in scale-free hypernetworks, this paper compares the degree of node-based phase synchronisation with that of hyperedge-based phase synchronisation at the same scale. Figure 14 shows the variation curves of the values of the order parameters r with the coupling strength C for $N = 500$ and hyperedge $= 500$. From the results in Fig. 14, it can be seen that when the coupling strength C is small, the degree of node-based phase synchronisation of the scale-free hypernetwork is better than that of the hyperedge-based phase synchronisation, and it is easier to achieve synchronisation between nodes in the scale-free hypernetwork. However, as the coupling strength C increases, hyperedge-based phase synchronisation again outperforms node-based phase synchronisation, and it is easier to achieve phase synchronisation between hyperedges when the coupling strength is increased. Combined with the data in Table 2, it can be seen that the value of the node-based order parameter r is higher than that of the hyperedge-based order parameter r when the coupling strength $C < 1.6$. This indicates that node-based phase synchronisation is better than hyperedge-based phase synchronisation when the coupling strength is small. However, when the coupling strength $C \geq 1.6$, the value of the hyperedge-based order parameter r grows faster and significantly higher than that of the node-based order parameter r. This further indicates that the degree of phase synchronisation based on the hyperedge is better than that based on the node when the coupling strength is larger for a scale-free hypernetwork.

This paper further analyses the reason why the degree of phase synchronisation based on hyperedges of the hypernetwork outperforms the degree of phase synchronisation based on nodes at larger coupling strengths. In this paper, when analysing the degree of phase synchronisation based on hyperedges, it is obtained by analysing the phase synchronisation of the nodes in the line graph corresponding to the hypernetwork. In the process of mapping the hypernetwork into the line graph, the individual characteristics of the nodes inside the hyperedge are ignored, and the unification of the nodes inside the hyperedge is achieved. In contrast, when analyzing the degree of node-based phase synchronisation, it is obtained by analyzing the node phase synchronisation of the 2-section graph network corresponding to the hypernetwork, and the characteristics of each node need to be synchronised in the synchronisation process, which makes the hyperedge-based phase synchronisation superior to the node-based phase synchronisation under the same coupling strength.

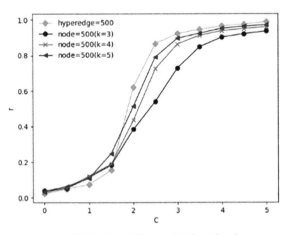

Fig. 14. Variation curves of node-based and hyperedge-based order parameter r values with coupling strength C

Table 2. The variation in the order parameter r with the coupling strength C for a hypernetwork of the same scale

Scale	The values of the order parameter r for different coupling strengths C							
	$C = 0$	$C = 0.5$	$C = 1$	$C = 1.5$	$C = 2.5$	$C = 3.5$	$C = 4.5$	$C = 5$
hyperedge	0.022	0.050	0.075	0.155	0.866	0.946	0.972	0.988
$k = 3$	0.039	0.054	0.114	0.184	0.542	0.848	0.922	0.936
$k = 4$	0.030	0.056	0.123	0.189	0.724	0.913	0.950	0.960
$k = 5$	0.036	0.064	0.109	0.248	0.791	0.924	0.963	0.971

6 Conclusion

In this paper, the node-based and hyperedge-based phase synchronisation of scale-free hypernetworks is studied analytically based on the Kuramoto model. In the node-based phase synchronisation, the uniformity of the hypernetwork has a greater influence on the node-based phase synchronisation of the hypernetwork; the greater the uniformity k of the hypernetwork with the increase of the coupling strength C, the better the node-based phase synchronisation of the hypernetwork. At the same time, the number of *new_nodes* added in the scale-free hypernetwork model construction process also has a certain influence on the node-based phase synchronisation of the scale-free hypernetwork; the larger the *new_node* is, the worse the node-based phase synchronisation of the hypernetwork, and the node-based phase synchronisation of the hypernetwork is optimal when *new_node* = 1. When *new_node* = $k - 1$, the node-based phase synchronisation of the hypernetwork is the worst when *new_node* = 1. In addition, the degree of phase synchronisation of scale-free hypernetworks based on hyperedges is independent of the uniformity k of the hypernetwork. However, the *new_node* in the evolution of the hypernetwork also has some influence on the hyperedge-based phase synchronisation. When the uniformity k of the hypernetwork is certain, the degree of hyperedge-based phase synchronisation of the scale-free hypernetwork is related to the *new_node*, and the larger the *new_node* is, the worse the degree of hyperedge-based phase synchronisation of the hypernetwork is. The comprehensive analysis shows that in the node-based and hyperedge-based phase synchronisation, the degree of synchronisation of the scale-free network is independent of the number of nodes and hyperedges, and when the coupling strength C is small, the node-based phase synchronisation degree is better than the hyperedge-based phase synchronisation degree, but as the coupling strength C increases, the hyperedge-based phase synchronisation degree is better than the node-based phase synchronisation degree again.

References

1. Ren, H.P., Gao, Y., Huo, L., et al.: Frequency stability in modern power network from complex network viewpoint. Physica A **545**, 123558 (2020)
2. Zhou, B., Ma, X.J., Ma, F.X., Gao, S.J.: Robustness analysis of random hyper-networks based on the internal structure of hyper-edges. AIMS Math. **8**(2), 4814–4829 (2023). https://doi.org/10.3934/math.2023239
3. Cai, W.X., Liang, F.F., Wang, Y.C., et al.: An innovative approach for constructing a shipping index based on dynamic weighted complex networks. Physica A **578**, 126101 (2021)
4. Walker, T.J.: Acoustic synchrony: two mechanisms in the snowy tree cricket. Science **166**, 891–894 (1969)
5. Buck, J.: Synchronous rhythmic flashing of fireflies. II. Q. Rev. Biol. **63**, 265–289 (1988)
6. Pikovsky, A., Rosenblum, M., Kurths, J.: Synchronisation: A Universal Concept in Nonlinear Science. Cambridge University Press, Cambridge (2001)
7. Rogge, J.A., Aeyels, D.: Existence of partial entrainment and stability of phase locking behavior of coupled oscillators. Prog. Theoret. Phys. **112**, 921–942 (2004)
8. Strogatz, S.H.: From Kuramoto to Crawford: exploring the onset of synchronisation in populations of coupled oscillators. Physica D. **143**, 1–20 (2000)
9. Winfree, A.T.: The Geometry of Biological Time. Springer, Cham (1980). https://doi.org/10.1007/978-3-662-22492-2
10. Kuramoto, Y.: Chemical Oscillations, Waves, and Turbulence. Springer, Cham (1984). https://doi.org/10.1007/978-3-642-69689-3
11. Wang, X.F., Chen, G.: Synchronisation in scale-free dynamical networks: robustness and fragility. IEEE Trans. Circ. Syst. Part I: Fundam. Theory Appl. **49**, 54–62 (2002)

12. Rodrigues, F.A., Peron, T., Ji, P., et al.: The Kuramoto model in complex networks. Phys. Rep. **610**, 1–98 (2016)

13. Lehnert, J.: Synchronisation in complex networks. In: Controlling Synchronization Patterns in Complex Networks. Springer Theses. Springer, Cham (2016). https://doi.org/10.1007/978-3-319-25115-8_3

14. Yamir, M., Amallo, F.P.: Synchronisation of Kuramoto oscillators in scale-free networks. Europhys. Lett. **68**, 603 (2004)

15. Chen, G.R.: Exploring Higher-Order Topologies of Complex Networks and Applications. Chinese Institute of Command and Control, Beijing (2021)

16. Federico, B., Giulia, C., Iacopo, I., et al.: Networks beyond pairwise interactions: structure and dynamics. Phys. Rep. **874**, 1–92 (2020)

17. Sinan, G.A., Cliff, J., Carlos, O.M., et al.: Hypernetwork science via high-order hypergraph walks. EPJ Data Sci. **9**(1), 519–535 (2020)

18. Wang, J.W., Rong, L.L., Deng, Q.H., et al.: Evolving hypernetwork model. Eur. Phys. J. B **77**(4), 493–498 (2010)

19. Estrada, E., Rodriguez-Velazquez, J.A.: Subgraph centrality and clustering in complex hypernetworks. Physica A **364**, 581–594 (2006)

20. Suo, Q., Guo, J.L.: The structure and dynamics of hypernetworks. Syst. Eng. Theory Pract. **37**(03), 720–734 (2017)

21. Hu, F., Zhao, H.X., He, J.B., et al.: An evolving model for hypergraph-structure-based scientific collaboration networks. Acta Physica Sinica **62**(19), 547–554 (2013)

22. Hu, F., Liu, M., Zhao, J., et al.: Analysis and application of the topological properties of protein complex hypernetworks. Complex Syst. Complexity Sci. **15**(04), 31–38 (2018)

23. Luo, H.X., Zhao, H.X., Xiao, Y.Z., et al.: Topology characteristics and robustness analysis of bus hypernetwork based on hypergraph. J. Southwest Univ. Nat. Sci. Ed. **43**(10), 181–191 (2021)

24. Irving, D., Sorrentino, F.: Synchronisation of dynamical hypernetworks: dimensionality reduction through simultaneous block-diagonalization of matrices. Phys. Rev. E Stat. Nonlinear Soft Matter Phys. **86** (2012)

25. Anwar, S., Rakshit, S., Ghosh, D., et al.: Stability of intralayer synchronisation in dynamic multilayer hypernetwork with generic coupling functions. Phys. Rev. E **105** (2021)

26. Rakshit, S., Bera, B.K., Ghosh, D.: Invariance and stability conditions of interlayer synchronisation manifold. Phys. Rev. E **101**(1), 012308 (2020)

27. Sorrentino, F.: Synchronisation of hypernetworks of coupled dynamical systems. New J. Phys. **14**(3), 33035–33058(24) (2012)

28. Wu, Z.Y., Duan, J.Q., Fu, X.C.: Synchronisation of an evolving complex hypernetwork. Appl. Math. Model. **38**(11–12), 2961–2968 (2014)

29. Rakshit, S., Bera, B.K., Ghosh, D.: Synchronisation in a temporal multiplex neuronal hypernetwork. Phys. Rev. E **98**, 032305 (2018)

30. Berge, C., Minieka, E.: Graph and Hypergraph North Holland, pp. 389–413. North-Holland Publishing Company, Amsterdam (1973)

31. Berge, C., Sterboul, F.: Equipartite colorings in graphs and hypergraphs. J. Comb. Theory Ser. B, **22**(2), 97–113 (1977)

32. Bretto, A.: Hypergraph Theory. Mathematical Engineering. Springer, New York (2013). https://doi.org/10.1007/978-3-319-00080-0

33. Kuramoto, Y.: Self-entrainment of a population of coupled non-linear oscillators. In: Araki, H. (eds.) International Symposium on Mathematical Problems in Theoretical Physics. LNP, vol. 39, pp. 420–422. Springer, Heidelberg (1975). https://doi.org/10.1007/BFb0013365

34. Ma, X.J., Zhao, H.X.: Cascading failure analysis in hypernetwork based on the hypergraph. Acta Phys. Sin. **65**(08), 374–383 (2016)
35. Hu, F.: Research on the Structure, Model and Application of Complex Hypernetwork. Shaanxi Normal University, Wuhan (2014)

LGHAE: Local and Global Hyper-relation Aggregation Embedding for Link Prediction

Peikai Yuan[1], Zhenheng Qi[2(✉)], Hui Sun[2], and Chao Liu[2]

[1] College of Computer Science and Technology, Harbin Engineering University, Harbin, China
yuanpeikai@hrbeu.edu.cn
[2] Institute of Systems Engineering, Academy of Military Sciences, Beijing 100089, China
qizhenheng1234@163.com

Abstract. The Knowledge Graph (KGs) have profoundly impacted many research fields. However, there is a problem of low data integrity in KGs. The binary-relational knowledge graph is more common in KGs but is limited by less information. It often has less content to use when predicting missing entities (relations). The hyper-relational knowledge graph is another form of KGs, which introduces much additional information (qualifiers) based on the main triple. The hyper-relational knowledge graph can effectively improve the accuracy of predicting missing entities (relations). The existing hyper-relational link prediction methods only consider the overall perspective when dealing with qualifiers and calculate the score function by combining the qualifiers with the main triple. However, these methods overlook the inherent characteristics of entities and relations. This paper proposes a novel Local and Global Hyper-relation Aggregation Embedding for Link Prediction (LGHAE). LGHAE can capture the semantic features of hyper-relational data from local and global perspectives. To fully utilize local and global features, Hyper-InteractE, as a new decoder, is designed to predict missing entities to fully utilize local and global features. We validated the feasibility of LGHAE by comparing it with state-of-the-art models on public datasets.

Keywords: Knowledge Graph · Hyper-relation · Link Prediction · Knowledge Graph Completion

1 Introduction

With the vigorous development of artificial intelligence, the critical problems encountered in the subdivision research fields of KGs, such as knowledge extraction, knowledge question and answer, knowledge representation fusion, and knowledge reasoning, have been solved and broken through. There is a problem of incomplete information in KGs, such as in FreeBase. Over 70% of individual entities do not have birthplace information, and over 99% of individuals lack racial information. Over 65% of individual entities in DBpedia do not have birthplace information, and over 55% of researchers lack research field information [1]. The integrity constraints of KGs also limit the development of research in many fields, forcing researchers to supplement missing information in KGs through various technical means. This technology is generally referred to as link prediction technology.

© The Author(s), under exclusive license to Springer Nature Singapore Pte Ltd. 2023
Z. Yu et al. (Eds.): ICPCSEE 2023, CCIS 1880, pp. 364–378, 2023.
https://doi.org/10.1007/978-981-99-5971-6_26

The KG is represented by (h, r, t), where h and t represent entities, and r represents relation from h to t. This structure is called the binary-relational knowledge graph in this paper. Figure 1 (Albert Einstein, educated at, University of Zurich) indicates that Albert Einstein is educated at the University of Zurich. According to different data forms, link prediction tasks can be divided into binary-relational link prediction tasks and hyper-relational link prediction tasks. The process of binary-relational link prediction is symbolized as $(h, r, ?)$, $(h, ?, t)$, and $(?, r, t)$.

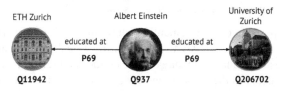

Fig. 1. An example of a binary-relational fact

Restricted by the fact that the triple in the binary-relational knowledge graph carries less information and expression ability is limited, the effect is not ideal when predicting missing entities or relations. Therefore, the type of data stored in the general knowledge graph is not a simple triple structure, but a hyper-relational data storage method, also called n-ary relational data.

The hyper-relational knowledge graph adds additional information based on the binary-relational knowledge graph. This additional information is also known as qualifiers [2]. The hyper-relational knowledge graph includes a main triple (h, r, t) and a set of qualifiers $(q_{k_1}, q_{v_1}, \ldots, q_{k_i}, q_{v_i}, \ldots, q_{k_n}, q_{v_n})$. Among them, q_k and q_v represent additional relation and entity information, respectively. As shown in Fig. 2, when performing prediction (Albert Einstein, Doctorate educated at, ?), The hyper-relational knowledge graph can use the degree and major information in qualifiers to find the corresponding tail entity (University of Zurich). The hyper-relational link prediction can predict all entities and relations in hyper-relational data, which performs better than the binary-relational knowledge graph.

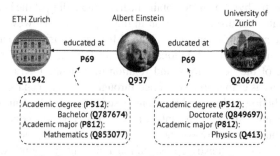

Fig. 2. An example of a hyper-relational fact

In previous methods for hyper-relational link prediction tasks, processing hyper-relational data often takes a holistic approach by concatenating or combining qualifiers with the main triple. These methods do not pay attention to the inherent characteristics of entities and relations in hyper-relational data and lack consideration of various aspects of hyper-relational data. Therefore, the Local and Global Hyper-relation Aggregation Embedding for Link Prediction (LGHAE) model introduces local and global concepts, learning the semantic features of hyper-relational data from two different perspectives. The local entity and relation features are extracted from hyper-relational data through local feature aggregation operations. Then the association between the local entity feature and local relation feature is learned through global feature aggregation operation. Therefore, the contributions of this paper are as follows:

- LGHAE is proposed to capture semantic features of hyper-relational data from local and global perspectives. LGHAE divides hyper-relational data into entities and relations, calculates the semantic features of the entities and relations from a local perspective, and calculates the association between entity features and relation features from a global perspective.
- To fully utilize the captured local and global features, this paper also designed and implemented a Hyper-InteractE decoder for predicting missing entities in hyper-relational data.
- We have verified the excellent performance of LGHAE.

2   Related Work

2.1   Binary-Relational Models

Based on various research methods, link prediction tasks in KGs can be categorized into three main areas: geometric models, tensor decomposition models, and deep learning models.

Geometric models mainly focus on studying the association between entity and relation embeddings. TransE [3] predicts tail entities by calculating distance, adding the head entity and relation embeddings, and subtracting the tail entity embeddings to represent the final triple score. StransE [4] adds two independent matrices generated by relation to TransE, where the head and tail entities are multiplied by the independent matrices, and the distance is calculated. TransH [5] embeds and projects the head entity and tail entity onto the hyperplane to obtain the new embedding of the head entity and tail entity and calculates the distance fraction in the hyperplane. However, TransH cannot handle triples under multiple semantics. TransR [6] enables triplets to adapt to multiple semantic environments through relational space and entity mapping. CrossE [7] utilizes cross-embedding of the head entity and relation to enhance interactivity. RotatE [8] represents relations as embeddings obtained through rotation in space.

The link prediction entities and relations based on tensor decomposition are transformed into semantic space representations. In recent years, tensor decomposition has also been popular among researchers. RESCAL [9] decomposes high-dimensional embedding into third-order tensor representation, reducing the complexity of data. However, as the dimensions of RESCAL increase, the number of training parameters also increases, which can easily cause overfitting. Therefore, DistMult [10] uses a diagonal matrix representation relation matrix, which reduces the number of parameters in training. However, due to the structural characteristics of DistMult, it can only deal with the symmetric relations in KGs and performs poorly in the face of asymmetric relations. Complex [11] decomposes embedding into genuine and complex parts, alleviating the problems in DistMult. Complex uses a complex form to represent entity and relation embedding. Complex solves the problem of asymmetric relations in DistMult. TuckER [12] decomposes the original tensor of a triple into the product of a core tensor and three hyperplanes using Tucker decomposition.

Deep learning link prediction tasks learn embedded features between entities and relations through commonly used neural network methods. ConvE [13] reconstructs and concatenates entities and relations into a two-dimensional matrix, utilizing convolution and fully connected operations to extract interaction features between entities and relations. ConvR [14] embeds entities and relations into vectors representing different dimensions, and uses head entities as matrices, and relations as convolutional filters to extract interactive features between entities and relations. InteractE [15] enhances the interactivity between entities and relations by randomly shuffling them, generating multiple random feature sequences and circular convolutions. ConvKB [16] concatenates entity and relation embedding features, utilizes convolution operations to extract the overall features of triplets, and uses a straightforward structure to predict triple scores. SCAN [17] uses WGCN as the encoder, a Graph Convolutional Network (GCN) with weights to pretrain entities. However, WGCN only pretrains entities and ignores pretraining on relations. CompGCN [18] introduces the concept of direction based on WGCN and utilizes aggregation operations to integrate relations into entity calculations. In the CompGCN, entities, and relations are added to pretraining. RAGAT [19], based on CompGCN, weights neighbor information with different weights onto the central node and achieves good results.

2.2 Hyper-relational Models

Similar to binary-relational link prediction methods, hyper-relational link prediction methods can also be divided into three fields: geometric models, tensor decomposition models, and deep learning models.

m-transH [20] is improved from TransH, m-transH the interaction between entities in hyper-relational data and captures the relation between entities based on TransH. However, some entities and relations attributes are permanently lost due to the need to transform data in m-TransH. RAE [21] introduces a multilayer awareness mechanism to improve the efficiency of m-transH. Using the hyper-relational data to learn the pairing relation of each triple further, RAE can predict one or more missing entities in hyper-relational data.

Fatemi et al. [22] proposed HSimple and Hype models based on Simple [23]. HSimple obtains entity embedding based on the position value of a given entity in the corresponding relation. Hype learns position embedding based on HSimple, which separates position embedding from entity embedding and is used to transform and obtain entity embedding based on the entity's position in the relation. The number of entities and relations in hyper-relational data limits the core tensor of Tucker decomposition. Due to the large number of entities and relations in each statement of hyper-relational dataset, the core tensor is too large, and the number of parameters is too large, which is prone to noise problems. GETD [24] utilizes tensor ring decomposition to decompose the core tensor in TuckER, reducing the high-dimensional core tensor to two-dimensional, reducing the number of computational parameters and complexity. RAM [25] learns the role features between entities and relations and uses pattern matrices to determine the degree of association between roles and entities.

NaLP [26] correlation between each triple with qualifiers extracts the features of entities and relations in hyper-relational data through convolutional operations. tNaLP$^+$ [27] proposes a new negative sampling method based on NaLP and introduces type information by means of unsupervised learning. HINGE [28] utilizes Convolutional Neural Network (CNN) to extract the features of hyper-relational data. However, the training and testing time of HINGE is too long, and the mixed matrix composed of hyper-relational data has poor interactivity. StarE [29] draws inspiration from the pretraining idea in CompGCN, using the Graph Neural Network (GNN) in the form of message passing to aggregate qualifiers with the main triple for pretraining of entities and relations embedding. StarE utilizes Transformer [30] as a decoder to predict missing entities. Hy-Transformer [31] utilizes Layer Normalization (LN) to improve the inefficient training of StarE.

3   Model Description

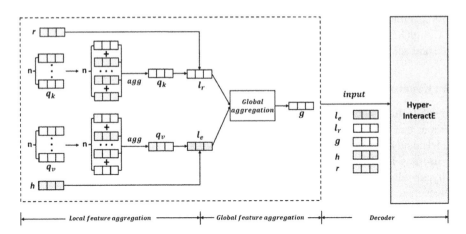

Fig. 3. The structure of LGHAE

This section briefly introduces Local and Global Hyper-relation Aggregation Embedding for Link Prediction (LGHAE), a hyper-relational link prediction model. The structure of LGHAE is shown in Fig. 3. LGHAE introduces local and global concepts, learning the semantic features of hyper-relational data from two different perspectives. Local entity features and local relation features are extracted from hyper-relational data through local feature aggregation operations. Then the association between local entity features and local relation features is learned through global feature aggregation operations. In addition, the LGHAE model has designed a new decoder, Hyper-InteractE, based on

InteractE and introduced local and global features to predict missing entities based on the main triple.

Our model consists of three parts: 1. Local feature aggregation. This process mainly obtains local entity and relation features of hyper-relational data. 2. Global feature aggregation. Calculate the correlation between the local entity and relation features. 3. Hyper-InteractE. Hyper-InteractE is proposed in this paper, which inputs local entity features, local relation features, global features, and the main triple information into learning hyper-relational data features. In the following, we present the details of these modules.

3.1 Local Feature Aggregation

The number of qualifiers is often uncertain. For example, in the WikiPeople dataset, some statements contain seven pairs of qualifiers. In comparison, others contain two pairs of qualifiers, but more statements still need qualifiers, which poses a considerable challenge to model training.

In some models, such as GETD and RAM, more standardized datasets, such as WikiPeople-3, WikiPeople-4, JF17K-3, and JF17K-4, are used for training. The training data of these datasets have a uniform amount of qualifiers. However, this approach is not conducive to promotion and use because the number of qualifiers per statement in existing knowledge bases differs. There are also some methods, such as HINGE, that uses batch training to train the model based on the dataset's logarithmic batch training of qualifiers. However, the experimental results of this method could be better.

LGHAE aggregates qualifiers into a unified format to facilitate model training and accelerate model training speed to address the varying qualifiers in hyper-relational data. However, some statements may have too many qualifiers, which may cause the aggregated qualifiers to be embedded in the vector value and interfere with the training and testing of the model. To reduce the impact of qualifier embedding vectors, a normalization operation was adopted for the additional information embedding vectors during the aggregation process. The process of aggregating additional entities is represented by formula (1), and the process of aggregating additional relations is shown in Eq. (2), where q_v represents the additional entity information embedding vector after aggregation and q_k represents the additional relation embedding vector after aggregation, n represents the logarithm of qualifiers.

$$q_v = \frac{1}{n} \sum_{i=1}^{n} q_{v_i} \tag{1}$$

$$q_k = \frac{1}{n} \sum_{i=1}^{n} q_{k_i} \tag{2}$$

Aggregated additional entity information q_v and aggregated additional relation information q_k are fused with head entity h and relation r. The process of local entity feature fusion is shown in Eq. (3), and the process of local relation feature fusion is shown in Eq. (4), where α is a hyperparameter, which can freely set the ratio between qualifiers and head entity and relation, controlling the size of qualifiers flowing into the main triple.

l_e and l_r represent the local entity feature embedding vector and local relation feature embedding vector, respectively, and $\odot$ represents the Hadamard product.

$$l_e = \alpha \odot q_v + (1 - \alpha) \odot h \tag{3}$$

$$l_r = \alpha \odot q_k + (1 - \alpha) \odot r \tag{4}$$

3.2   Global Feature Aggregation

l_e and l_r contain entity and relation features in hyper-relational data. However, these are only two local parts of the hyper-relational data and do not establish the overall connection of the hyper-relational data. The LGHAE model learns the association between l_e and l_r through global feature embedding g.

In this paper, the attention mechanism is used to fuse l_e and l_r. Dynamically calculate the ratio between l_e and l_r through a learnable vector λ. Specifically, l_e and l_r are multiplied by λ to calculate weight scores a_e and a_r. The calculation process of a_e is shown in Eq. (5), and the calculation process of a_r is shown in Eq. (6), where *LeakyReLU* represents a nonlinear activation function.

$$a_e = \frac{exp\left(LeakyReLU\left(l_e\lambda^T\right)\right)}{\sum_{l\in\{l_e,l_r\}} exp\left(LeakyReLU\left(l\lambda^T\right)\right)} \tag{5}$$

$$a_r = \frac{exp\left(LeakyReLU\left(l_r\lambda^T\right)\right)}{\sum_{l\in\{l_e,l_r\}} exp\left(LeakyReLU\left(l\lambda^T\right)\right)} \tag{6}$$

After calculating a_e and a_r, they are combined with l_e and l_r to calculate the global feature embedding vector g. The calculation process is shown in (7).

$$g = a_e \odot l_e + a_r \odot l_r \tag{7}$$

3.3   Hyper-InteractE

The overall structure of Hyper-InteractE is shown in Fig. 4. Hyper-InteractE is improved from InteractE [15], which alleviates the problem of insufficient interaction between head entity h and relation r in ConvE. InteractE utilizes chequered feature reshaping, feature permutation, and circular convolution to enhance the interactivity between the head entity and relation. Chequered feature reshaping randomly disrupts the head entity embedding vector and relation embedding vector into a two-dimensional mixed matrix. Feature permutation generates multiple randomly disordered interleaved mixed matrices, with each mixed matrix having different data from other mixed matrices; In the past, when performing matrix convolution operations, the data at the edge of the mixed matrix often lacked sufficient neighbor information, and the interactivity at the edge was poor compared to the center during convolution operations. Circular convolution uses the data in the mixed matrix to fill the edge, improving the edge's data interactivity.

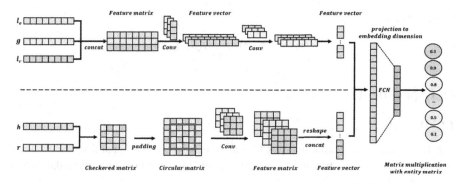

Fig. 4. Overview of Hyper-InteractE

Hyper-InteractE is divided into two parts. The first part mainly processes the obtained local and global features, concatenating the local entity feature embedding vector l_e, local relation feature embedding vector l_r, and global feature embedding vector g into a two-dimensional matrix. Using a set of 3×1 convolutional filters to fuse local and global features, the feature vector contains the local and global feature information of hyper-relational data. The extracted feature vector is learned using a set of 1×3 convolutional filters to combine the extracted information into a new feature vector f_1. Calculating f_1 is shown in Eq. (8).

where *concat* represents the concatenation operation, ω_1 represents a 3×1 convolutional filter, ω_2 represents a 1×3 convolution filter, and *vec* represents the features extracted by vectorization.

$$f_1 = vec(concat(l_e, g, l_r)\omega_1\omega_2) \tag{8}$$

The second part of Hyper-InteractE mainly addresses the head entity embedding vector h and relation embedding vector r, which are processed using the InteractE model. h and r are interleaved and combined into a mixed matrix, filling the edge parts with data from the mixed matrix. The features of the mixed matrix are extracted through a set of 9×9 convolutional filters, and the features are reshaped and concatenated into feature vectors f_2. The calculation process of f_2 is shown in Eq. (9), where ϕ represents the combination of h and r interlaced into a mixed matrix, $\oplus$ represents the circular filling operation, ω_3 represents a 9×9 convolutional filter, and σ represents the *Relu* activation function.

$$f_2 = vec(\sigma(\oplus(\phi(h, r))\omega_3)) \tag{9}$$

After splicing the feature vectors f_1 and f_2 of the two parts, the feature vector is converted into a vector with the same size as the entity embedding through a transpose matrix. It is multiplied by the entity embedding to calculate the prediction score of all tail entities. The prediction score is mapped to the range of [0,1] through the *sigmoid* activation function. The entity with the highest prediction score is selected as the tail entity of the hyper-relational data. The calculation function of Hyper-InteractE is shown in Formula (10). ψ represents the *sigmoid* activation function, and W represents the transpose matrix.

$$\varphi(h, r, t) = \psi\big(\sigma\big(concat(\boldsymbol{f}_1, \boldsymbol{f}_2)\boldsymbol{W}\big)\boldsymbol{t}\big) \tag{10}$$

4 Experimental Analysis

4.1 Datasets

The JF17K and WikiPeople datasets [26] are widely used in comparative experiments of hyper-relational link prediction models. Detailed information on the two datasets is shown in Table 1.

Table 1. Datasets

Dataset	Statements	Qualifiers (%)	Train	Valid	Test	Entities	Relations
WikiPeople	382,229	44,315(11.6%)	305,725	38,223	38,281	47,765	193
JF17K	100,947	46,320(45.9%)	76,379	–	24,568	28,645	501

– WikiPeople: The WikiPeople dataset is extracted from Wikidata. It contains a large amount of information related to humans. The WikiPeople dataset represents all data as n-ary facts, each containing a main triple and several qualifiers. However, qualifiers in the WikiPeople dataset contain much digital information, such as time and date. This information is often used as interference data in prediction tasks and is often ignored by researchers [28] or merged and processed in some special ways [32]. In this paper, we used the method of [29] to process digital information.
– JF17K: The JF17K dataset data is extracted from Freebase, and the maximum logarithm of qualifiers in the JF17K dataset is four pairs. The JF17K dataset has much hyper-relational data compared to the WikiPeople dataset. Statements containing qualifiers account for 45.9%.

4.2 Baselines

In this experiment, we compare LGHAE with current state-of-the-art methods, including m-TransH [20], RAE [21], NaLP [26], tNaLP$^+$ [27], HINGE [28], StarE [29], and Hy-Transformer [31].

– m-TransH [20]: The m-TransH model is improved from TransH. As a binary-relational link prediction model, the TransH model abstracts the relation into a hyperplane and calculates the head and tail entity on the hyperplane formed by the relation through entity mapping. m-TransH requires data transformation, resulting in the permanent loss of some entity and relation attributes.
– RAE [21]: The RAE introduces a multilayer awareness mechanism to improve the efficiency of m-TransH. However, due to the performance of the TransH model, the RAE model cannot achieve good experimental results.

- NaLP [26]: To reduce the problem of structural information loss caused by complex transformations of hyper-relational data, NaLP treats entities and relations as key-value pairs, uses one-dimensional convolution to capture the structural features of key-value pairs, and outputs predicted scores through a Fully Connected Neural Network (FCNN).
- tNaLP$^+$ [27]: tNaLP$^+$ is improved from NaLP. tNaLP introduces type information and calculates the compatibility between entities and relations through unsupervised learning. NaLP$^+$ improved the negative sample sampling scheme commonly used in binary-relational link prediction and designed a more plausible negative sampling mechanism for hyper-relational data. The tNaLP$^+$ model combines the advantages of tNaLP and NaLP$^+$. However, tNaLP$^+$ only considers from a global perspective and ignores the local characteristics of the correlation coefficients.
- HINGE [28]: HINGE processes hyper-relational data through CNN, which is improved from ConvE in binary-relational link prediction. HINGE embeds the head entity, relation, and tail entity into a concatenated representation. It uses convolution operations to extract features between the main triple and the overall features between the main triple and each pair of qualifiers. However, the training and testing time of HINGE is too long, and the mixed matrix composed of hyper-relational arrays has poor interactivity.
- StarE [29]: StarE draws on the pretraining entity and relation embedding in CompGCN, using message passing to aggregate qualifiers with the main triple through GNN, achieving pretraining of entity and relation embedding matrices.
- Hy-Transformer [31]: Hy-Transformer uses LN and dropout operations to improve the inefficient training speed of the GNN in StarE. Hy-Transformer reduces the model training time compared to the StarE model without sacrificing prediction accuracy. In addition, the Hy-Transformer model has added directed qualifiers to assist in training tasks, improving the model's prediction accuracy.

4.3 Training and Evaluation

LGHAE uses the Adam optimizer, with a batch size of 128, an entity and relation embedding size of 200, and a learning rate of {0.001, 0.0001}. LGHAE first uses a 0.001 learning rate to accelerate the training process and then continues training with a 0.0001 learning rate when the key evaluation metrics (MRR) no longer improve. LGHAE also uses the label smoothing method [13] to process labels and sets it to 0.1. Hyper-InteractE uses 32 convolutional filters of sizes 3×1 and 1×3 to extract local and global features, 96 convolutional filters of size 9×9 to extract head entity and relation features from the main triple, and the dropout used in the model is set to {0.2, 0.5}. LGHAE uses binary cross entropy as the loss function (11), where y_i is the label of the main triple i and x_i is the prediction score. The experimental code is published in https://github.com/yuanpeikai/LGHAE.

$$\mathcal{L} = -\frac{1}{N} \sum_{i=1}^{N} (y_i \cdot \log(x_i) + (1 - y_i) \cdot \log(1 - x_i)) \tag{11}$$

To better compare with baselines, LGHAE uses the Mean Reciprocal Rank (MRR), Hit@1, and Hit@10 as evaluation metrics, which are widely used in link prediction tasks [10]. In the link prediction task, because the tail entities obtained from only the head entities and relations are not unique, there is only one tail entity in each test statement. Inspired by StarE and other models predicting missing entities, it is necessary to filter out other correct tail entity prediction scores from the test data, sort the filtered candidate entity set, and evaluate the model's performance using MRR, Hit@1, and Hit@10 evaluation metrics.

4.4 Results and Discussion

The experimental results of LGHAE on the WikiPeople and JF17K datasets are shown in Table 2. All evaluation metrics of LGHAE on the JF17K dataset exceeded the baseline. Compared to StarE, LGHAE improved by approximately 1.74% MRR, 1.81% Hit@1, and 2.48% Hit@10 points. In the WikiPeople dataset, LGHAE achieved 3.05% MRR and 8.79% Hit@1 points improvement compared to StarE.

Table 2. Experimental results on JF17K and WikiPeople

Methods	JF17K			WikiPeople		
	MRR	Hit@1	Hit@10	MRR	Hit@1	Hit@10
m-TransH	0.206	0.206	0.463	0.063	0.063	0.300
RAE	0.215	0.215	0.469	0.059	0.059	0.306
NaLP	0.386	0.317	0.517	0.338	0.272	0.466
HINGE	0.449	0.361	0.624	0.476	0.415	0.585
tNaLP+	0.449	0.370	0.598	0.339	0.269	0.473
Hy-Transformer	0.582	0.501	0.742	0.501	0.426	0.634
StarE	0.574	0.496	0.725	0.491	0.398	**0.648**
LGHAE	**0.584**	**0.505**	**0.743**	**0.506**	**0.433**	0.632

Although the Hit@10 point is lower than StarE, it is superior to StarE in other evaluation metrics. The low evaluation metrics of LGHAE in Hit@10 point may be because: 1. The entity and relation features in qualifiers need to be more effectively utilized. In the LGHAE local feature aggregation operation, only qualifiers are aggregated together, ignoring the possibility that qualifiers may contain data unrelated to the prediction, which may interfere with the model's prediction. Therefore, reducing the interference of low correlation data in qualifiers on prediction tasks can improve the model's predictive performance. 2. This may also be because qualifiers in the dataset are too sparse, with only 2.6% of the hyper-relational data in the WikiPeople dataset containing qualifiers. Therefore, LGHAE can only aggregate to fewer qualifiers during aggregation. The entities and relations in StarE are embedded in the GNN, which learns more features with qualifiers.

Although LGHAE performed poorly on the Hit@10 point, it exceeded all baselines in MRR and Hit@1 points, proving that LGHAE has a more accurate hit rate and performs better in hyper-relational link prediction tasks.

4.5 Ablation Study

This section conducts an ablation study on the JF17K dataset to verify the effectiveness of various modules of LGHAE. The experiment removes local features, global features, head entities, and relations in the main triple while ensuring that other conditions remain unchanged. The experimental results are shown in Table 3.

Verify the role of local features. We removed the local entity feature embedding l_e and local relation feature embedding l_r in Hyper-InteractE to verify the performance of LGHAE without local features. At this point, the model experienced significant performance degradation, indicating that local entity features and local relation features played an essential role in LGHAE.

Verify the role of global features. To verify the impact of without global features on LGHAE. At this point, the performance of LGHAE also decreased, but the decrease was less significant than removing the influence of local information. Global feature embedding g effectively combines local entity feature embedding l_e and local relation feature embedding l_r. The information in the global embedding g only includes some of the features in the hyper-relational data. However, the features in the hyper-relational data are more in the local information. The global features contain the association between the entity and relation parts in the hyper-relational data. Therefore, the experimental effect of removing global features is less evident than removing local information, but global features still occupy an important position in the LGHAE.

Verify the overall effect of local and global features. Remove local and global features to explore the impact on LGHAE, where Hyper-InteractE degenerates into InteractE. After the absence of local and global features, the experimental effect of LGHAE showed a severe decrease, with a decrease of approximately 16.78% in MRR compared to the original experimental effect, Hit@1 decreased by approximately 19.6%, and Hit@10 decreased by approximately 14.26%. The experimental results validate the importance of local and global features in LGHAE.

Verify the role of the head entity and relation of the main triple. Remove the input of the head entity h and relation r in Hyper-InteractE and only rely on the global feature embedding g, local entity feature embedding l_e, and local relation feature embedding l_r as inputs. The performance of LGHAE significantly decreased on the JF17K dataset, and our analysis suggests that qualifiers are only used to assist in the main triple prediction. Therefore, we should focus on the main triple in prediction tasks.

In summary, local features, global features, and Hyper-InteractE can effectively improve the predictive performance of LGHAE.

Table 3. Ablation study on JF17K

Method	MRR	Hit@1	Hit@10
LGHAE	**0.584**	**0.505**	**0.743**
w/o local feature	0.495	0.418	0.642
w/o global feature	0.580	0.501	0.736
w/o local and global feature	0.486	0.406	0.637
w/o head entity and relation	0.579	0.498	0.737

5 Conclusion

With the continuous development of artificial intelligence technology, knowledge graphs are increasingly receiving widespread attention. This paper proposes that LGHAE handles hyper-relational data from multiple perspectives, which can capture the semantic features of hyper-relational data from both local and global perspectives. LGHAE divides hyper-relational data into entity and relation parts, calculates the semantic features of entities and relations from a local perspective, and calculates the connections between local entity features and local relation features from a global perspective. In addition, to better utilize the captured local and global features, LGHAE has also designed a Hyper-InteractE decoder. Finally, the reliability and feasibility of LGHAE are verified through experiments. In future work, we plan to address the sparse data problem in LGHAE and improve the model's predictive performance.

References

1. Krompaß, D., Baier, S., Tresp, V.: Type-constrained representation learning in knowledge graphs. In: Arenas, M., et al. (eds.) The Semantic Web-ISWC 2015: 14th International Semantic Web Conference, Bethlehem, PA, USA, 11–15 October 2015, Proceedings, ISWC 2015, Part I 14, vol. 9366, pp. 640–655. Springer, Cham (2015). https://doi.org/10.1007/978-3-319-25007-6_37
2. Möller, C., Lehmann, J., Usbeck, R.: Survey on English entity linking on Wikidata: datasets and approaches. Semant. Web **13**, 925 (2022)
3. Bordes, A., Usunier, N., Garcia-Duran, A., et al.: Translating embeddings for modeling multi-relational data. In: Advances in Neural Information Processing Systems, vol. 26 (2013)
4. Nguyen, D.Q., Sirts, K., Qu, L., et al.: STransE: a novel embedding model of entities and relationships in knowledge bases. In: Proceedings of NAACL-HLT, pp. 460–466 (2016)
5. Wang, Z., Zhang, J., Feng, J., et al.: Knowledge graph embedding by translating on hyperplanes. Proc. AAAI Conf. Artif. Intell. **28**(1) (2014)
6. Lin, Y., Liu, Z., Sun, M., et al.: Learning entity and relation embeddings for knowledge graph completion. Proc. AAAI Conf. Artif. Intell. **29**(1) (2015)
7. Zhang, W., Paudel, B., Zhang, W., et al.: Interaction embeddings for prediction and explanation in knowledge graphs. In: Proceedings of the Twelfth ACM International Conference on Web Search and Data Mining, pp. 96–104 (2019)
8. Sun, Z., Deng, Z.H., Nie, J.Y., et al.: RotatE: knowledge graph embedding by relational rotation in complex space. In: International Conference on Learning Representations (2019)

9. Nickel, M., Tresp, V., Kriegel, H.P.: A three-way model for collective learning on multi-relational data. ICML **11**(10.5555), 3104482–3104584 (2011)
10. Yang, B., Yih, S.W., He, X., et al.: Embedding entities and relations for learning and inference in knowledge bases. In: Proceedings of the International Conference on Learning Representations (ICLR) (2015)
11. Trouillon, T., Dance, C.R., Gaussier, É., et al.: Knowledge graph completion via complex tensor factorization. J. Mach. Learn. Res. **18**, 1–38 (2017)
12. Balažević, I., Allen, C., Hospedales, T.: TuckER: tensor factorization for knowledge graph completion. In: Proceedings of the 2019 Conference on Empirical Methods in Natural Language Processing and the 9th International Joint Conference on Natural Language Processing (EMNLP-IJCNLP), pp. 5185–5194 (2019)
13. Dettmers, T., Minervini, P., Stenetorp, P., et al.: Convolutional 2D knowledge graph embeddings. Proc. AAAI Conf. Artif. Intell. **32**(1) (2018)
14. Jiang, X., Wang, Q., Wang, B.: Adaptive convolution for multi-relational learning. In: Proceedings of the 2019 Conference of the North American Chapter of the Association for Computational Linguistics: Human Language Technologies, vol. 1 (Long and Short Papers), pp. 978–987 (2019)
15. Vashishth, S., Sanyal, S., Nitin, V., et al.: InteractE: improving convolution-based knowledge graph embeddings by increasing feature interactions. Proc. AAAI Conf. Artif. Intell. **34**(03), 3009–3016 (2020)
16. Nguyen, T.D., Nguyen, D.Q., Phung, D.: A novel embedding model for knowledge base completion based on Convolutional Neural Network. In: Proceedings of the 2018 Conference of the North American Chapter of the Association for Computational Linguistics: Human Language Technologies, vol. 2 (Short Papers), pp. 327–333 (2018)
17. Shang, C., Tang, Y., Huang, J., et al.: End-to-end structure-aware convolutional networks for knowledge base completion. Proc. AAAI Conf. Artif. Intell. **33**(01), 3060–3067 (2019)
18. Vashishth, S., Sanyal, S., Nitin, V., et al.: Composition-based multi-relational graph convolutional networks. In: International Conference on Learning Representations (2020)
19. Liu, X., Tan, H., Chen, Q., et al.: RAGAT: relation aware graph attention network for knowledge graph completion. IEEE Access **9**, 20840–20849 (2021)
20. Wen, J., Li, J., Mao, Y., et al.: On the representation and embedding of knowledge bases beyond binary relations. In: Proceedings of the Twenty-Fifth International Joint Conference on Artificial Intelligence, pp. 1300–1307 (2016)
21. Zhang, R., Li, J., Mei, J., et al.: Scalable instance reconstruction in knowledge bases via relatedness affiliated embedding. In: Proceedings of the 2018 World Wide Web Conference, pp. 1185–1194 (2018)
22. Fatemi, B., Taslakian, P., Vazquez, D., et al.: Knowledge hypergraphs: prediction beyond binary relations. In: Proceedings of the Twenty-Ninth International Conference on International Joint Conferences on Artificial Intelligence, pp. 2191–2197 (2021)
23. Kazemi, S.M., Poole, D.: Simple embedding for link prediction in knowledge graphs. In: Advances in Neural Information Processing Systems, vol. 31 (2018)
24. Liu, Y., Yao, Q., Li, Y.: Generalizing tensor decomposition for n-ary relational knowledge bases. In: Proceedings of the Web Conference, pp. 1104–1114 (2020)
25. Liu, Y., Yao, Q., Li, Y.: Role-aware modeling for n-ary relational knowledge bases. In: Proceedings of the Web Conference, pp. 2660–2671 (2021)
26. Guan, S., Jin, X., Wang, Y., et al.: Link prediction on n-ary relational data. In: The World Wide Web Conference, pp. 583–593 (2019)
27. Guan, S., Jin, X., Guo, J., et al.: Link prediction on n-ary relational data based on relatedness evaluation. IEEE Trans. Knowl. Data Eng. **35**(1), 672–685 (2021)
28. Rosso, P., Yang, D., Cudré-Mauroux, P.: Beyond triplets: hyper-relational knowledge graph embedding for link prediction. In: Proceedings of the Web Conference, pp. 1885–1896 (2020)

29. Galkin, M., Trivedi, P., Maheshwari, G., et al.: Message passing for hyper-relational knowledge graphs. In: Proceedings of the 2020 Conference on Empirical Methods in Natural Language Processing (EMNLP), pp. 7346–7359 (2020)
30. Vaswani, A., Shazeer, N., Parmar, N., et al.: Attention is all you need. In: Advances in Neural Information Processing Systems, vol. 30 (2017)
31. Yu, D., Yang, Y.: Improving hyper-relational knowledge graph completion. arXiv preprint arXiv:2104.08167 (2021)
32. Kristiadi, A., Khan, M.A., Lukovnikov, D., et al.: Incorporating literals into knowledge graph embeddings. In: Ghidini, C., et al. (eds.) The Semantic Web–ISWC 2019: 18th International Semantic Web Conference, Auckland, New Zealand, 26–30 October 2019, Proceedings, Part I 18, vol. 11778, pp. 347–363. Springer, Cham (2019). https://doi.org/10.1007/978-3-030-30793-6_20

Efficient *s*-Core Community Search on Attributed Graphs

Yuesheng Fu[1](✉) and Ruilu Sun[2]

[1] The 722th Research Institute of China Shipbuilding Corporation Limited, Wuhan 430079, China
scorecs@163.com
[2] School of Computer Science, Fudan University, Shanghai 200438, China
rlsun22@m.fudan.edu.cn

Abstract. With the advantage of high personalization in many applications, community search on attributed graphs has received increasing attention. The communities found in an attributed graph, called attributed communities, show inherent community structure and attribute cohesion. However, most of the traditional community search algorithms only consider the existence of query attributes in the resulting communities, which ignores the importance of attribute quantities. In this paper, we study the attributed community search problem and formulate this problem as finding the tightest connected subgraph, named the *s*-core attributed community, that meets the given query condition. We introduce an efficient algorithm using local search and attribute inspection techniques to search the communities. Additionally, pruning techniques that exploit community structure and attribute information are proposed to prevent unnecessary community construction and attribute inspection. Finally, we conduct extensive experiments on real-world datasets. The experimental results verified the pruning strategy's effectiveness and the algorithm's efficiency.

Keywords: Community Search · Attributed Graph · Keyword Search

1 Introduction

The internet has become an important model for representing complex relationships in the real world. communities in a network are subgraphs composed of densely interconnected nodes, which are the most widely studied structural features for understanding latent relationships a network may contain. as a result, community search, especially attribute-driven community search, has gained increasing attention [1–3].

Currently, a variety of community search models are proposed, such as quasiclique [4], densest subgraph [5], k-clique [6], k-truss [7, 8], and k-core [9, 10]. However, these models only consider searching for compactly structured communities in simple graphs. In real-world networks, nodes often possess some attributes, and such networks are referred to as attributed graphs [11]. Community search on attributed graphs is defined as attributed community search. Given an attributed graph and a set of query conditions, attributed community search aims to find communities that include nodes with attributes similar to those of the given query conditions [12]. In other words, it seeks communities with both structure and attribute cohesion [13].

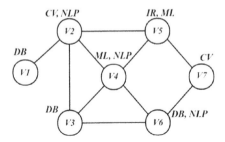

Fig. 1. Example of Attributed Graph

Existing attributed community search can be mainly separated into two categories. The first is homogeneous attributed community search [12, 14, 15], where given a query, the aim is to find communities with high similarity in terms of attributes to the query. The second is personalized attributed community search [16–18], where a set of attributes is given, and it searches for communities that contain all of these attributes. This type of problem is often referred to as keyword search within communities. However, these approaches have a common limitation: they only consider the presence of query attributes in the resulting communities while disregarding the quantity of attributes. To address this issue, Wang et al. [19] proposed the OACS algorithm and CSHA algorithm for community search based on local search and attribute checking techniques, aiming to support keyword search that considers the quantity of community attributes. In this paper, we refer to this problem as the s-core attributed community search problem.

As shown in Fig. 1, the seven nodes represent seven experts and their respective domains of expertise, while the edges represent historical collaborations between the experts. Now we assume that a knowledge graph project is planned, which requires the formation of an expert team. The team should consist of at least one expert in DB, at least two experts in NLP, and at least one expert in ML. However, due to the limited capacity of each expert, even if an expert excels in multiple areas, they can only contribute to one specific domain within a project. Additionally, to control the budget, the team size should not exceed five members. This paper formulates the above requirements as a combined query condition in the form of $Q = \{\{(DB, 1), (NLP, 2), (ML, 1)\}, 5\}$, transforming the problem of forming an expert team into searching for the s-core attributed community in the graph that satisfies the combined query condition. The s-core attributed community has the following requirements: (1) The community should contain a sufficient number of experts specializing in the required domains. (2) The size of the community should not exceed the specified total number limit. (3) The community structure should be as compact as possible. The compactness of the community structure is measured using the minimum degree among the nodes in the community. (4) Among the communities that satisfy the above conditions, smaller communities are preferred to minimize costs. For the given combined query condition, the optimal s-core attributed community that can be found is subgraph $H_1 = \{v_2, v_3, v_4, v_5\}$ in Fig. 1. Currently, there are several algorithms available for performing *s*-core attributed community search. Nevertheless, current algorithms inadequately leverage the attribute information within the attributed graph, resulting in unnecessary attribute checking and the potential for enhancing search efficiency.

In summary, we tackle the problem of s-core attributed community search, proposing a bitmap-based method to represent the node attribute for attribute graphs and developing a set of node pruning strategies without any false negatives and an efficient search algorithm based on these representations. This algorithm is called BMPA (Pruning Algorithm based on BitMap Match). In addition, the algorithm incorporates a series of optimization strategies based on the properties of the *s*-core attributed community to further accelerate the search algorithm. Experimental results demonstrate that the proposed algorithm can effectively and efficiently search for the s-core attributed community.

2 Preliminaries

In this section, we provide a formal definition of the problem and introduce the basic ideas of existing efficient search algorithms.

2.1 Problem Definition

Wang et al. [19] provided a comprehensive definition for *s*-core attributed community search under combination query conditions. To ensure the rigor of the technical presentation in the following sections, we formally introduce important concepts related to the core attribute community search in this part.

Definition 1 (Undirected Attributed Graph): An undirected graph where each node is associated with one or more attributes is referred to as an undirected attributed graph, denoted as $G = (V, E, L)$, where V represents the set of nodes, E represents the set of edges, and L denotes the attribute mapping. For each $v \in V$, we have $L(v) = \{A_1, A_2, \ldots, A_m\}$, where A_i indicates the specific attribute value.

Later in this paper, $|G|$ is used to represent the number of nodes in graph G, $deg_{G_{min}}(v)$ represents the minimum degree of the nodes in graph G, and $V_{A_i}(G)$ denotes the set of nodes that possess attribute A_i.

Definition 2 (Combined Query Condition): Combined query condition specifies the attributes and quantity restrictions for the nodes included in the target community. It is formally represented as $Q = \{W_Q, s\}$, which consists of a set of attribute pairs W_Q and an integer s. The attribute pair set is denoted as

$$W_Q = \{(A_1, count_1), (A_2, count_2), \ldots, (A_m, count_m)\}$$

where each attribute pair contains an attribute A_i and an integer $count_i$, representing that the target community should contain at least $count_i$ nodes with attribute A_i. The integer s indicates the maximum number of members in the target community.

The *s*-core community model [19] requires at most one attribute selected for each node in the target community. Therefore, the concept of attribute mapping is introduced as follows.

Definition 3 (Attribute Mapping): Given an attribute graph $G = (V, E, L)$, attribute mapping requires each node to select only one attribute, resulting in a new attribute graph denoted as $G\prime = AM(G)$.

Definition 4 (s-core Attributed Community): Given an attributed graph G and a combined query condition Q, an s-core attributed community is a subgraph H of the graph G, satisfying the following constraints:

- Connectivity: The subgraph H is connected.
- Multiattribute Compactness: The subgraph H has an attribute mapping $H' = AM(H)$ such that $|H'| \leq s$ and $V_{A_i}(H') \geq count_i$.
- Structural Compactness: Among all subgraphs satisfying conditions (1) and (2), the s-core attributed community H is the subgraph with the largest $deg_{H_{min}}(V)$.

Definition 5 (s-core Attributed Community Search): Given an attributed graph G and combined query condition Q, s-core attributed community search returns the optimal s-core attributed community, such that there is no s-core attributed community in graph G where $|H'| < |H|$.

2.2 Related Search Algorithm

For the problem of s-core attributed community search, the OACS algorithm and the optimized CSHA algorithm [19] propose solution strategies based on local search and attribute checks. The algorithm we propose in this paper follows the basic framework of these algorithms. Therefore, we introduce their basic ideas in this subsection.

OACS Algorithm. The OACS algorithm can be understood as a brute-force algorithm for solving the s-core attribute community problem, consisting of two stages: community search and attribute check. The basic idea is as follows:

- Community Search: Starting from each node in graph G, the algorithm performs a depth-first search to explore all possible communities C.
- Attribute Check: For each possible community C, the algorithm uses a maximum matching algorithm to determine whether the community satisfies the attribute constraints specified in the set of attribute pairs W_Q. If the conditions are all met, the current optimal community $C_{optimal}$ is updated.

After all possible communities have been checked, $C_{optimal}$ represents the final result found by the OACS algorithm.

CSHA Algorithm. The attribute check in OACS requires constructing a bipartite graph and conducting a maximum matching algorithm, which are quite time-consuming. However, the OACS algorithm traverses a large number of unnecessary communities and performs attribute checks, resulting in poor query efficiency. Therefore, the CSHA algorithm introduces various pruning strategies to reduce the number of communities that need to be checked. The following outlines the basic flow of the CSHA algorithm.

Algorithm 1: *CSHA Algorithm*

Data: $G = (V, E, L), Q = \{W_Q, s\}$

Result: $C_{optimal}$

1   Construct the set of starting nodes $StartID$;

2   for each $v \in StartID$ do

3   if $CoreNumber(v) < C_{optimal}.d$ then

4   break;

5   end if

6       $C_{optimal} \leftarrow Search(G, Q, v, C \cup \{v\}, C_{optimal})$;

7       $G \leftarrow G - \{v\}$;

8   end for

9   return $C_{optimal}$;

Algorithm 1 presents the basic logic of CSHA. Firstly, the CSHA algorithm incorporates attribute conditions, node degrees, and other information to filter some nodes when choosing starting nodes, reducing unnecessary searches (line 1). For each starting node in $StartID$, the algorithm calls the search algorithm and obtains the current optimal community $C_{optimal}$ (lines 2–8). Here, $C_{optimal}.d$ represents the minimum degree of the current optimal community, and the algorithm terminates prematurely if the core number of a new node is less than $C_{optimal}.d$ (lines 3–5). This search algorithm provides the search logic on the attributed graph, including community search logic based on depth-first search and attribute check logic. It also introduces several pruning strategies to reduce unnecessary nodes. The specific details and correctness of the algorithm can be found in the paper of Wang [19].

3   BMPA Search Algorithm

The multiple pruning strategies provided by the CSHA algorithm are all related to the degree of nodes but do not fully utilize the attribute information of nodes in the graph. Additionally, several pruning strategies can still be further optimized. Therefore, this section thoroughly provides the ideas and formal descriptions of multiple pruning strategies based on the CSHA algorithm. We call the proposed search algorithm BMPA.

3.1   Optimization Strategies

In this section, we introduce the basic ideas and specific details of various pruning strategies one by one, providing corresponding definitions and descriptions.

Pruning Based on Attribute Bitmap. The existing methods do not fully utilize the attribute information in the graph. Therefore, we propose the attribute bitmap, representing node attributes and topological information as binary data, for pruning nodes and accelerating the algorithm. Here, we begin by providing the definition of an attribute bitmap.

Definition 6 (Attribute Bitmap of Subgraph): Given an attribute graph $G = (V, E, L)$, where the set of all attribute sets is listed as $A = \{A_1, A_2, \ldots, A_n\}$, the attribute bitmap of the subgraph H is a binary string of length n, denoted as $B(H) = b_1 b_2 \ldots b_n$, such that $b_i = 1$ if and only if H contains attribute A_i.

Definition 7 (k-Hop Subgraph of Node): Given an attribute graph $G = (V, E, L)$, for a node $v \in V$, the k-hop subgraph of the node v is defined as the set of nodes whose shortest distance from node is less than or equal to k, denoted as $J_k(v) = \{u | Dis_{min}(u, v) \le k, u \in V\}$.

Definition 8 (k-Hop Attribute Bitmap of Node): Given an attribute graph G, the k-hop attribute bitmap of a node v is the attribute bitmap of the k-hop subgraph of node v, denoted as $B(J_k(v))$, or simply $B_k(v)$.

Now, we utilize the attribute bitmap to determine whether the node v can be added to the current community C during the search procedure. The specific strategy is as follows: First, initialize a binary string of length n. Then, for each attribute A_i in the combined query condition Q, count the number c_i of A_i in the community C and compare c_i with the required number specified in Q. If c_i is greater than or equal to the required number, set the i-th position of the binary string to 0; otherwise, set it to 1. The obtained binary string is denoted as $B(Q - C)$, where if the i-th bit was set as 1, it indicates that the current community C lacks attribute A_i. Due to the maximum community size limit, at most $(s - C.size)$ additional nodes can be added to the current community C. Therefore, if the $(s - C.size - 1)$-hop attribute bitmap of node v does not contain the attributes that C lacks, i.e., $B(Q - C) \& B_{s-C.size-1}(v) = 0$, node v will not be added to C.

Example. Assume the attribute graph in Fig. 1 and the combined query condition $Q = \{\{(CV, 1), (NLP, 1), (IR, 1)\}, 5\}$. Assuming that the current searched community is $C = \{v_2, v_3, v_4\}$, we now consider whether to add node v_6 to the community. The set of all attributes in the graph is listed as $A = \{A_1 = DB, A_2 = CV, A_3 = NLP.A_4 = ML, A_5 = IR\}$, consisting of 5 attributes. Using the concept of attribute bitmaps, we initialize a binary string of length 5. The attributes and their numbers in C are $\{(DB, 1), (CV, 1), (NLP, 2), (ML, 1)\}$, which lacks the IR attribute compared to the required attribute constraints in the combined query condition. Therefore, we set the 5th position of the binary string to 1 and the other positions to 0, resulting in $B(Q - C) = 00001$. Since the maximum community size is 5, $C.size = 3$, at most 2 more nodes can be added into C. We then examine the 1-hop attribute bitmap of node v_6, which is denoted as $J_1(v_6) = \{v_3, v_4, v_6, v_7\}$, containing the attributes $\{DB, CV, ML, NLP\}$, so the 1-hop attribute bitmap of node v_6 is $B_1(v_6) = 11110$. Then we have that $B(Q - C) \& B_1(v_6) = 0$. Therefore, node v_6 will not be added to the current community. This avoids unnecessary attribute checks and reduces the number of search iterations.

Pruning Based on Complete Graph. If there exists a $(s - 1)$-order complete subgraph in the attribute graph G that satisfies the combined query condition, then the $(s - 1)$-order complete subgraph is the optimal s-core attributed community. Therefore, if a $(s - 1)$-order complete subgraph satisfying the query condition is found in the graph, the search algorithm will stop and return the complete subgraph as result.

Pruning Based on Total Attribute Count. Here we consider the total number of query attributes, *CountN*. Since the attribute mapping restricts each node in the community to select at most one attribute, for a community C that satisfies the combination query condition, the size of the set of nodes satisfying the attribute condition must be greater than or equal to *CountN*. Therefore, we can dynamically maintain the set S of nodes that satisfy the attribute condition, and if $S.size$ is less than *CountN*, there is no need to perform attribute check on the community C.

Pruning Based on Single Attribute Count. For the specific attribute A_i in query, if the number of nodes in a community containing attribute A_i is less than $Count_i$, then the community does not satisfy the combined query condition, and there is no need to perform attribute check.

Algorithm 2: *BMPA-Search Algorithm*

Data: $G = (V, E, L), Q = \{W_Q, s\}, v, C, C_{optimal}$

Result: $C_{optimal}$

1   if $C_{optimal}.d = s - 1$ then

2   return $C_{optimal}$;

3   end if

4   if $ExpectedCore(C).d < C_{optimal}.d$ or $C.size > s$ then

5   return $C_{optimal}$;

6   end if

7   if $(C.d = C_{optimal}.d$ and $C.size < C_{optimal}.size)$ or $C.d > C_{optimal}.d$ then

8   if C satisfies the attribute constraints then

9           $C_{optimal} \leftarrow C$;

10   end if

11   end

12   for each $u \in N(v)$ do

13   if $(CoreNumber(u) = C_{optimal}.d$ and $C.size < C_{optimal}.size)$ or $CoreNumber(u) > C_{optimal}.d$ then

14   if $B(Q - C) \& B_{s-C.size-1}(u) \neq 0$ then

15              $C_{optimal} \leftarrow BMPA - Search(G, Q, u, C \cup \{u\})$;

16   end if

17   end if

18     $C_{optimal} \leftarrow BMPA - Search(G, Q, u, C, \{u\})$;

19   return $C_{optimal}$;

3.2 BMPA Search Algorithm

In this section, we will combine the pruning strategies presented in the previous section to provide the search algorithm, which can directly replace the *search* logic in the CSHA algorithm.

Algorithm 2 presents the search logic of the BMPA. Building upon the CSHA search algorithm, this algorithm adds a condition for early stopping of the search (complete graph check in line 1). Furthermore, it further tightens the criteria for adding nodes by adjusting the judgment from node degree to node core number (line 13). Last, it introduces attribute bitmaps to reduce the inclusion of unnecessary nodes (line 14), significantly reducing the search space.

4 Experiment

This section presents comprehensive experiments conducted on four real datasets to evaluate three algorithms: OACS, CSHA [19], and the proposed BMPA algorithm. Section 4.1 introduces the hardware environment and datasets we used. Section 4.2 presents the specific experimental results, demonstrating the efficiency of the BMPA algorithm proposed in this paper.

4.1 Experimental Setup

This section provides an overview of the experimental setup. First, we describe the hardware environment used in the experiments. Then, we present the specific information and generation methods of the four datasets used in the experiments.

Hardware Environment. All programs in this paper are implemented in C++. The hardware environment for the experiments is 11th Gen Intel(R) Core (TM) i5-11320H @ 3.20 GHz, 16.0 GB RAM.

Datasets. Four real datasets of different sizes are used for experimental research in this paper. Table 1 summarizes the statistical data of the four attributed graphs. The DBLP dataset is used in this paper, and four different-sized expert collaboration networks are constructed based on different sampling ratios. DBLP is a well-known document integration database system in the field of computer science, that includes papers published in international journals and conferences over the past few decades. Each node represents an author, and the attribute labels associated with each node represent the fields in which the author is familiar. For example, if author A has published papers in CVPR and KDD, their labels would be DM and CV. If two authors have coauthored a paper, there would be an edge between two nodes.

4.2 Query Efficiency

In this subsection, the running time of the existing OACS, CSHA, and the proposed BMPA algorithms are tested on four different-sized DBLP datasets by varying three parameters in the combined query conditions, validating the efficiency of the BMPA algorithm.

Table 1. Datasets

Attributed Graph	#Nodes	#Edges
DBLP-1	890	1495
DBLP-2	3517	7096
DBLP-3	20431	49140
DBLP-4	88527	329198

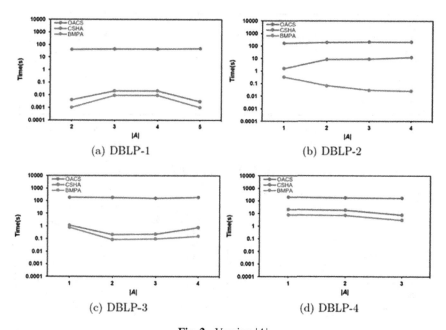

(a) DBLP-1   (b) DBLP-2

(c) DBLP-3   (d) DBLP-4

Fig. 2. Varying |*A*|

Varying the Number of Query Attribute Types. While keeping the other two parameters in the combined query conditions fixed, we varied the number of query attribute
types |*A*| from 1 to 5. We tested different values on the four attributed graphs. For each
value of |*A*|, we randomly generated 10 queries that satisfy the conditions and tested
them on the three algorithms, recording the average running time. The results are shown
in Fig. 2. The OACS algorithm performs the worst, taking approximately 100 to 1000 s
to find results. The optimized CSHA algorithm is better than OACS by 1 to 2 orders of
magnitude. The proposed BMPA algorithm outperforms the CSHA algorithm, improving performance by 70% to 80%. This is because the BMPA algorithm avoids many
unnecessary community searches and attribute checks. The running time of all three
algorithms increases with the increase of |*A*|. However, the complexity of the expert
collaboration network structure and the variation in the filter ratio for different queries
may lead to a decrease in running time with an increasing in |*A*| in certain cases.

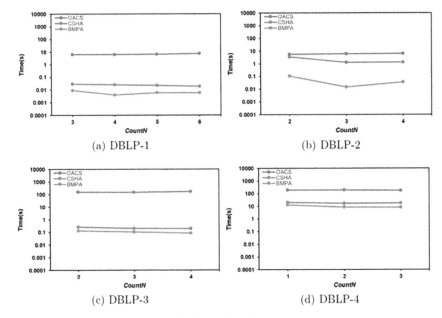

Fig. 3. Varying *CountN*

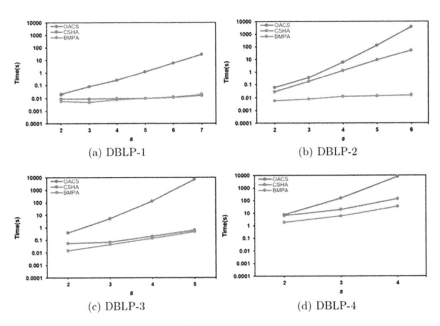

Fig. 4. Varying *s*

Varying the Total Number of Query Attributes. While keeping the other conditions constant, we tested the running time of the OACS, CSHA, and BMPA algorithms for different values of *CountN*, as shown in Fig. 3. In each test, OACS found the optimal result in approximately 10 to 100 s. The CSHA algorithm reduced the running time by 1 to 2 orders of magnitude. The BMPA algorithm further improved efficiency, being 50% to 80% faster than CSHA. The running time of all three algorithms basically increases with increasing *CountN*.

Varying the Community Size Limit. We tested the performance of the three algorithms under different community size limits *s* on the four attribute graphs, as shown in Fig. 4. As *s* increases, the running time of the three algorithms increases significantly. Compared to the experiments with varying |A| and *CountN*, the running time of the algorithms increases more rapidly. When *s* is large, the OACS algorithm requires 1000 to 10000 s of running time. The CSHA algorithm reduces the running time to approximately 100 s. The optimization strategies of the proposed BMPA algorithm effectively avoid unnecessary community searches, reducing the running time to less than 10 s.

5 Conclusion

In this paper, we proposed an attribute-driven community model called *s*-core and approached the problem of searching for s-core attributed communities. Unlike traditional community search problems, we considered the number of attributes in the target community. We introduced an optimization algorithm called BMPA to further improve the performance of the CSHA algorithm, leveraging the structural information and attribute features of the network. Finally, we conducted extensive experiments on real-world datasets to evaluate our proposed algorithm. The experimental results demonstrate the effectiveness and efficiency of our proposed model and algorithm. There are many directions for future work. In this paper, we ensured the tightness of the community structure by maximizing the minimum degree of the community. To make the community structure even tighter, upper constraints on the distance between nodes containing query attributes could be introduced. In addition to constraints on attribute quantity, further research can be conducted on other relationships between attributes, such as attribute diversity measurement and attribute distance measurement, to handle other real-world application scenarios.

References

1. Fang, Y., Cheng, R., Chen, Y., Luo, S., Hu, J.: Effective and efficient attributed community search. VLDB J. **26**(6), 803–828 (2017)
2. Xu, T., Lu, Z., Zhu, Y.: Efficient triangle-connected truss community search in dynamic graphs. Proc. VLDB Endow. **16**(3), 519–531 (2022)
3. Jiang, Y., Fang, Y., Ma, C., Cao, X., Li, C.: Effective community search over large star-schema heterogeneous information networks. Proc. VLDB Endow. **15**(11), 2307–2320 (2022)
4. Cui, W., Xiao, Y., Wang, H., Lu, Y., Wang, W.: Online search of overlapping communities. In: Proceedings of the 2013 ACM SIGMOD International Conference on Management of Data, pp. 277–288 (2013)
5. Yubao, W., Jin, R., Li, J., Zhang, X.: Robust local community detection: on free rider effect and its elimination. Proc. VLDB Endow. **8**(7), 798–809 (2015)

6. Yuan, L., Qin, L., Lin, X., Chang, L., Zhang, W.: Diversified top-k clique search. VLDB J. **25**(2), 171–196 (2016)
7. Huang, X., Cheng, H., Qin, L., Tian, W., Yu, J.X.: Querying k-truss community in large and dynamic graphs. In: Proceedings of the 2014 ACM SIGMOD International Conference on Management of Data, pp. 1311–1322 (2014)
8. Huang, X., Lakshmanan, L.V.S., Yu, J.X., Cheng, H.: Approximate closest community search in networks (2015). arXiv preprint arXiv:1505.05956
9. Barbieri, N., Bonchi, F., Galimberti, E., Gullo, F.: Efficient and effective community search. Data Min. Knowl. Disc. **29**(5), 1406–1433 (2015)
10. Li, R.H., Qin, L., Yu, J.X., Mao, R.: Influential community search in large networks. Proc. VLDB Endow. **8**(5), 509–520 (2015)
11. Zhou, Y., Cheng, H., Yu, J.X.: Graph clustering based on structural/attribute similarities. Proc. VLDB Endow. **2**(1), 718–729 (2009)
12. Huang, X., Lakshmanan, L.V.S.: Attribute-driven community search. Proc. VLDB Endow. **10**(9), 949–960 (2017)
13. Liu, Q., Zhu, Y., Zhao, M., Huang, X., Xu, J., Gao, Y.: Vac: vertex-centric attributed community search. In: 2020 IEEE 36th International Conference on Data Engineering (ICDE), pp. 937–948. IEEE (2020)
14. Fang, Y., et al.: On spatial-aware community search. IEEE Trans. Knowl. Data Eng. **31**(4), 783–798 (2018)
15. Huang, X., Cheng, H., Yu, J.X.: Dense community detection in multivalued attributed networks. Inf. Sci. **314**, 77–99 (2015)
16. Chen, L., Liu, C., Liao, K., Li, J., Zhou, R.: Contextual community search over large social networks. In: 2019 IEEE 35th International Conference on Data Engineering (ICDE), pp. 88–99. IEEE (2019)
17. Li, G., Ooi, B.C., Feng, J., Wang, J., Zhou, L.: Ease: an effective 3-in-1 keyword search method for unstructured, semistructured and structured data. In: Proceedings of the 2008 ACM SIGMOD International Conference on Management of Data, pp. 903–914 (2008)
18. Zhu, Y., Zhang, Q., Qin, L., Chang, L., Yu, J.X.: Querying cohesive subgraphs by keywords. In: 2018 IEEE 34th International Conference on Data Engineering (ICDE), pp. 1324–1327. IEEE (2018)
19. Wang, L.: On search algorithms for optimal s-core attributed communities with composite query conditions, Publication No. S21/201/18212010033. Master dissertation, Fudan University. PQDT Open (2021)

Education Using Big Data, Intelligent Computing or Data Mining, etc.

A Study on English Classroom Learning Anxiety of Private Vocational College Students

Han Wu[✉]

Sanya Aviation and Tourism College, Sanya 572000, Hainan, China
516095898@qq.com

Abstract. Private higher vocational colleges in Hainan play an important role in higher education. As Hainan vigorously develops free trade port construction, it is in urgent need of a number of professional and compound talents who know foreign languages and applied skills. In line with the characteristics of serving regional development, Hainan private vocational colleges need to further strengthen the cultivation of students' English ability in future teaching to better serve the construction of Hainan free trade port. As we all know, there are many factors that affect students' English learning. In addition to cognitive ability, emotional factors are also among the most important factors. Among the emotional factors, anxiety, as a negative factor, hinders students' English learning to some extent. In this paper, 282 non-English major students from Sanya Aviation and Tourism College were selected as the research objects, and SPSS22.0 was used to analyse the collected data through a questionnaire survey to determine the current English classroom learning anxiety of private vocational college students and whether there is a difference in the level of English classroom anxiety among students of different genders, regions and majors. Finally, some relevant suggestions have been put forward to relieve students' English classroom learning anxiety, providing a valuable reference for teachers' future English classroom teaching.

Keywords: Private vocational college · English classroom learning anxiety · Suggestions

1 Introduction

Private colleges and universities are important parts of higher vocational education and are also important in promoting the development of China's higher education. On Hainan Island, among 8 undergraduate universities, there are 3 private undergraduates. Among 13 higher vocational colleges, there are 6 private vocational colleges. The number of private colleges accounted for nearly half the number of colleges and universities in Hainan, which play an important role in higher education. In June 2021, The State Council issued the overall plan for the construction of Hainan Free Trade Port, which stated that Hainan would be built into an international tourism and consumption center in the future and a high-level free trade port with strong international influence. Accordingly, we know that the construction of the Hainan free trade port needs to be in line with international standards. Therefore, it is necessary to cultivate a large number of talents with international vision and proficiency in English.

© The Author(s), under exclusive license to Springer Nature Singapore Pte Ltd. 2023
Z. Yu et al. (Eds.): ICPCSEE 2023, CCIS 1880, pp. 393–407, 2023.
https://doi.org/10.1007/978-981-99-5971-6_28

Privately- run vocational colleges, as the backbone of higher education in Hainan, in line with the characteristics that higher education serves the local economy, will surely take on the responsibility of training professional and skilled talent with strong foreign language ability for the construction of FTP in the future. In reality, private vocational colleges can only recruit students who are usually enrolled in the fifth batch of college entrance examination. These students generally have poor grades and low quality, so it is a very challenging job to cultivate their English application ability at college. In private vocational college students' English learning classes, there are many factors affecting their English learning, such as emotional factors, motivational factors, discourse factors and cognitive ability. Among these factors, emotional factors play a crucial role. Positive and optimistic emotional factors can enhance students' interest in English learning, while negative factors hinder students' English learning. As one of the five emotional factors affecting language learning, anxiety ranks first, and as a negative emotional factor, it plays a certain role in hindering students' English learning.

In view of the important impact of anxiety on English class learning, since the 1970s, many foreign scholars have carried out studies on the impact of anxiety on language learning and obtained very valuable results. Horwitz et al., Jakobovits, and Maclintyre have clearly indicated that language anxiety was one of the important factors affecting students' foreign language learning in their research results. However, domestic research on language learning anxiety started late in the early 21st century. From 2007 to 2012, studies on students' learning anxiety experienced swift and violent development. The object of these studies covered all disciplines, including primary school, middle school, university, etc., and the research achievements were rich. However, studies on private higher vocational college students' English learning anxiety were relatively few. Based on the above reality, this paper takes 282 students from Sanya Aviation and Tourism College as the research object, collects data in the form of a questionnaire, uses SPSS 22.0 to analyse the data, and then clarifies the English class learning anxiety level of private vocational college students and whether there are significant differences in anxiety levels in English class learning among students of different genders, regions and majors. On the basis of data analysis, some suggestions will be given to alleviate the English class anxiety of higher vocational college students.

2 English Learning Status of Private Vocational College Students

There are various sources of students in private vocational colleges. Most of them are students enrolled in the fifth batch of college entrance examination, while there are quite a number of adult education students and single enrollment students. These students have a weak English foundation, and some even have no English foundation before entering college. In many years of English learning, these students have not formed good learning habits, and their learning initiative is poor. Moreover, China is not an English-speaking country, so students lack the environment for English learning. Meanwhile, the negative comments given by teachers, parents and their classmates lead to their exclusion or aversion to learning, especially English learning. With the popularity of smartphones and campus mobile networks, students are easily attracted by their smartphones in class, and the phenomenon of phubbing in class is serious. Figure 1 below shows the results of the Practical English Test for Colleges Level A from Sanya Aviation and Tourism College in December 2021 to further clarify the current English level of private vocational college students. According to Fig. 1, among the 1631 students who took the exam, the number of students with 60–89 points was 295, among which only 5 students had 80–90 points, 21 students had more than 70 points and 269 students had 60–69 points. It can be seen that among the qualified students, the scores were generally not high, and most of them were 60–69 points. The overall pass rate for the test was only 18%. Among the students who failed, 33% had a score of less than 40. These students have a poor English foundation. In future exams, it will be difficult for them to pass the Level-A test if they do not change their learning attitude and find the appropriate learning strategies. Students with scores of 40–59 accounted for 49%. If they can change their learning attitude in time and use appropriate learning methods and strategies in future learning, they are expected to pass the level-A test. The statistical analysis shows that students in private vocational colleges have low English levels and poor English learning initiative, fail to develop good learning habits, lack appropriate English learning strategies and lack self-confidence in English learning.

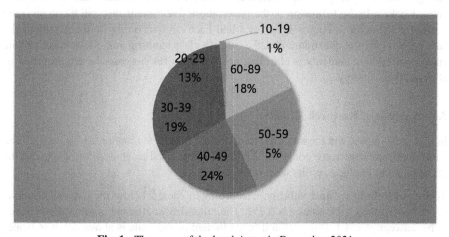

Fig. 1. The score of the level A test in December 2021

3 A Study on English Class Anxiety of Private Vocational College Students

3.1 Research Object

A total of 282 non-English major freshmen from Sanya Aviation and Tourism College were chosen for testing.

3.2 Research Questions

Anxiety level and current situation of non-English majors in English class in Sanya Aviation and Tourism College;

Differences in anxiety levels in English class learning among students of different genders, regions and majors.

3.3 Research Methods and Tools

This study adopts the questionnaire survey method. The questionnaire uses the Chinese version of the Foreign Language Classroom Anxiety Scale (FLCAS) compiled by Horwitz et al. in 1986. Individual measurement questions are modified to make them more in line with the actual situation of students in private higher vocational colleges. FLCAS has a total of 33 questions, which are scored in the form of a five-level Likert scale, "1 point for complete nonconformance, 2 points for less conformance, 3 points for unclear explanation, 4 points for more conformance and 5 points for complete conformance". From five different dimensions, namely, fear of communication, fear of negative evaluation, fear of interaction between teachers and students, lack of confidence in English class and students' negative attitude towards English class, these questions are used to make a quantitative analysis of the English class anxiety level of private vocational college students. The questions in each dimension from the questionnaire are shown in Table 1 below. In the following of the paper, these five dimensions will be represented by Factor 1, Factor 2, Factor 3, Factor 4 and Factor 5. According to the results of the data analysis for the questionnaires, the English class learning anxiety status of private vocational college students is clarified, and some useful suggestions are proposed to alleviate their anxiety.

3.4 Analysis of Research Results

In this study, www.wjx.cn was used to distribute the questionnaire, and a total of 282 valid questionnaires were collected. SPSS 22 0 was used to analyse the data, and the corresponding analysis results were obtained as well.

Testing of Reliability and Validity. According to Table 2 below, the Cronbach's alpha coefficient of the questionnaire is 0.952. The reliability coefficient is generally believed to be between 0 and 1. If the reliability coefficient of the scale is above 0.9, it indicates that the reliability of the scale is very good. According to Table 3, the KMO value of the questionnaire is 0.954, and the Bartlett sphericity test Sig. Value is 0.000. According to Kaiser's KMO metric above 0.9 and the Bartlett test value $0.000 < 0.05$, the questionnaire is suitable for factor analysis with good validity.

The Anxiety Level of Private Vocational College Students in English Classroom
The Overall Anxiety. SPSS 22.0 was used for descriptive statistical analysis of the questions in the questionnaire. Table 4 below shows that the overall mean value of the questionnaire for 33 questions is 87.4043. If the mean value is reached for each question, it is 2.65. According to Oxford and Burry-Stock's interpretation of this questionnaire scale in 1995, those with an average value equal to or higher than 3.5 are considered to have high anxiety, those with an average value equal to or lower than 2.4 are considered to have low anxiety, and those with an average value between 2.5 and 3.4 are considered to have medium anxiety. The students in private vocational colleges generally show moderate anxiety in English class.

Table 1. Questions in each dimension

Dimension	Question
Communicative fear (Factor 1)	21. When the teacher asked me to answer the question, I felt my heart beating fast
	3. I will shiver when the teacher asks me to answer a question in English class
	28. I felt nervous and confused when speaking in English class
	12. I was so nervous in English class that I forgot what I knew
	34. I will feel panic in English class when I make unprepared speeches
	14. I feel embarrassed and shy when answering teacher's questions voluntarily in English class
	17. Even though I am well prepared, I still feel anxious in English class
	1. I'm not confident when speak English in class
	33. I felt nervous when communicating with foreigners in English
	9. I am frightened when answering questions without preparation
Fear of negative evaluation (Factor 2)	24. I always feel that other students speak English better than me
	7. I always thought that other students were better at English than me
	25. I feel embarrassed to speak English in front of other students
	22. The more prepared I am for the English exam, the less sure I am
	19. When I speak in English in English class, I feel confident in myself
	32. I worry that other students will make fun of me when I speak English
	10. I'm worried about the consequences of failing my English class
Fear of Teacher-student interaction (Factor 3)	20. I feel frightened when the teacher corrects every mistake of mine
	16. I get upset when I cannot understand the teacher's instruction and feedback

(continued)

Table 1. (*continued*)

Dimension	Question
	4. In English class, when I do not understand what the teacher says in English, I will feel afraid
	26. The progress of English class is so fast that I'm afraid I cannot keep up
	30. I feel anxious when I can't understand every word of my teacher
Nonconfidence in English class (Factor 4)	2. I am not worried about making mistakes in English class
	5. I feel relaxed in English class
	8. I am usually very relaxed during the English class test
	11. English class increases my interest in English study
	15. I don't feel nervous when speaking English with foreigners
	23. There is no any pressure for me to make good preparations for English class
	29. I feel relaxed before English class
	27. I feel more pleasant in English class than any other classes
Negative attitude towards English class (Factor 5)	18. I often do not want to go to English class
	6. In English class I find myself thinking about things that have nothing to do with the class
	31. I was overwhelmed by learning so many rules to speak English

Table 2. Reliability statistics

Cronbach'Alpha	N of Items
.952	33

English Class Anxiety in Different Dimensions. According to Zhang Jiaqiang and Guo Li (2018), the questions of the questionnaire can be divided into five dimensions for interpretation, namely, communicative fear, fear of negative evaluation, fear of teacher-student interaction, nonconfidence in English class and negative attitude towards English class. In Table 5 below, the total mean value of students' communicative fear is 25.4752. There are 10 questions in this dimension, and the mean value of each question is 2.55. It can be seen that this value is between 2.5 and 3.4, and students' communicative fear

Table 3. KMO and Bartlett test

Kaiser-Meyer-Olkin Measure of Sampling Adequacy		.954
Bartlett's Test of Sphericity	About chi-square	8342.771
	df	561
	Sig.	.000

Table 4. Descriptive statistics

	N	Minimum	Maximum	Mean	Standard deviation
Total Value	282	33.00	165.00	87.4043	26.57735
Valid N (listwise)	282				

in English class presents a medium anxiety state. The mean value of fear of negative evaluation is 19.4326, and there are 7 questions in this dimension, so the mean value of each question is 2.78, which is still between 2.5 and 3.4. Therefore, students' fear of negative evaluation in English class also presents a medium anxiety state. Compared with communicative fear, students' fear of negative evaluation is more than communicative fear. The mean value of teacher-student interaction fear is 12.7943. There are 5 questions in this dimension, and the mean value of each question is 2.56, which is also between 2.5 and 3.4. Students also show moderate anxiety about teacher-student interaction fear in English classes. The mean for the dimension of Nonconfidence in English class is 22.1667, and there are 8 questions in this dimension, so the average value of each question is 2.77. Because the questions in this dimension are reverse scored, the higher the mean, the lower students' English class anxiety level is. According to the value of 2.77, students in the English class show moderate confidence. The mean value of the dimension of negative attitude towards the English class is 7.5355. There are 3 questions in this dimension, so the mean value of each question is 2.51, which is still between 2.5 and 3.4. It can be seen that students' negative attitude towards English class has also reached the medium anxiety state. The average anxiety values of the five dimensions are similar to each other, but the anxiety state of the negative evaluation fear dimension and nonconfidence in English class dimension are slightly higher than that of the other three dimensions. Private vocational college students in many years of English learning, due to poor grades and bad learning habits, are often negatively evaluated by parents, teachers and students, resulting in students' lack of confidence in English class, thus showing a high state of anxiety in these two dimensions.

Correlation Analysis Among Different Dimensions. The results are shown in Table 6. The Sig values of the five dimensions of students' communication fear, negative evaluation fear, teacher-student interaction fear, nonconfidence in English class and negative attitude towards English class are all 0.00 < 0.05, indicating that there is a linear relationship among the five factors. Pearson correlation values between communicative fear and fear of negative evaluation, fear of teacher-student interaction, and negative attitude

Table 5. Descriptive statistics

	N	Minimum	Maximum	Mean	Standard deviation
Factor 1	282	10.00	50.00	25.4752	10.01606
Factor 2	282	7.00	35.00	19.4326	6.49753
Factor 3	282	5.00	25.00	12.7943	5.36433
Factor 4	282	8.00	40.00	22.1667	6.13814
Factor 5	282	3.00	15.00	7.5355	3.15673
Valid N (listwise)	282				

towards English class are 0.889, 0.910 and 0.827, respectively, all of which are greater than 0.8, indicating that students' communicative fear and anxiety are highly correlated with the other three factors and show a positive correlation with the other factors, namely, the more severe communicative fear is, the more serious the fear of negative evaluation is, and the more anxious students are of interacting with teachers in class, which leads to the more serious negative attitude of students towards English class. The Pearson correlation value between fear of negative evaluation and fear of teacher-student interaction is 0.842, which is greater than 0.8, indicating that the more students worry about negative evaluation from teachers or classmates in class, the greater they will hinder the interaction between students and teachers. The Pearson correlation value between fear of negative evaluation and negative attitude towards English class was 0.774, between 0.5–0.8, indicating a moderate correlation between the two. That is, the more students worry about negative evaluation in English class, the more serious their negative attitude towards English class will be. However, the correlation between fear of negative evaluation and fear of teacher-student interaction is greater than the correlation between fear of negative evaluation and negative attitude toward English class. This means that students receive more negative evaluations from teachers, classmates or parents and are more reluctant to interact with others in English classes; however, their influence on students' negative attitudes toward English classes is slightly lighter. The Pearson correlation value between fear of teacher-student interaction and negative attitudes in English class is 0.770, between 0.5–0.8, indicating a moderate correlation between the two. The more students worry about interaction with teachers in English class, the more likely they are to have negative attitudes towards English class. The Pearson correlation values between Nonconfidence in English class and the other four factors are all less than 0.5, indicating that the factor of Nonconfidence in English class shows a negative correlation with the other four factors. If students are confident in English class, they will not worry about communicating with others in English class, receiving negative evaluations, interacting with teachers and bringing negative attitudes to English class.

Differences in Anxiety Levels in English Class Learning Among Students of Different Genders, Regions and Majors. SPSS 22.0 was used to conduct independent sample T test and one-way ANOVA on the research data to determine whether there were significant differences in English class anxiety among students of different genders, regions and majors.

Table 6. Correlation analysis

		Factor1	Factor 2	Factor 3	Factor 4	Factor 5
Factor 1	Pearson Correlation	1	.889**	.910**	.347**	.827**
	Sig.(2-tailed)		.000	.000	.000	.000
	N	282	282	282	282	282
Factor 2	Pearson Correlation	.889**	1	.842**	.411**	.774**
	Sig.(2-tailed)	.000		.000	.000	.000
	N	282	282	282	282	282
Factor 3	Pearson Correlation	.910**	.842**	1	.279**	.770**
	Sig.(2-tailed)	.000	.000		.000	.000
	N	282	282	282	282	282
Factor 4	Pearson Correlation	.347**	.411**	.279**	1	.290**
	Sig.(2-tailed)	.000	.000	.000		.000
	N	282	282	282	282	282
Factor 5	Pearson Correlation	.827**	.774**	.770**	.290**	1
	Sig.(2-tailed)	.000	.000	.000	.000	
	N	282	282	282	282	282

**. Correlation was significant at 0.01 level (two-tailed)

Independent Sample T Test for Students of Different Genders. In Table 7, subjects are divided into two groups by gender for independent sample T test. The Sig. Values for students of different genders whose English class anxiety levels in the five dimensions are 0.145, 0.202, 0.294, 0.419, and 0.158, respectively, which are all greater than 0.05, indicating that the difference between male and female students' English class anxiety in the five dimensions is not significant.

Independent Sample T Test for Students of Different Regions. According to the statistics in Table 8 and 9, the Sig values of the English class anxiety level of students in Hainan province and outside Hainan province from the five dimensions are 0.641, 0.839, 0.958, 0.008 and 0.425 respectively. Among the five values, four are all greater than 0.05, except for the Sig. Value in the dimension of Nonconfidence in English class, which is less than 0.05. It shows that the anxiety degree of students in Hainan province and outside Hainan province is obviously different in the dimension of Nonconfidence in English class. According to the group statistics, the mean value of the 143 Hainan students' nonconfidence in the English class is 21.2672, and the mean value of the 139 out-of-province students in this dimension is 23.2101. Since the questions in this dimension are reverse scored, it can be seen that the out-of-province students are more confident in the English class than the students from Hainan. Due to the backwards basic education resources and the negative transfer effect of local dialects on English learning, the English foundation of students in Hainan province is weaker than that of students outside Hainan

Table 7. Independent-samples T test

	Levene's test forequality of variances		T test for whether the mean is equal					95% Confidence Interval	
	F	Sig	T	df	Sig. (2-tailed)	Mean difference	Standard error	Lower Bound	Upper Bound
Factor 1	1.029	.311	1.462	280	.145	1.84574	1.26271	−.63986	4.33135
			1.526	208.768	.129	1.84574	1.20977	−.53918	4.23067
Factor 2	.007	.935	1.278	280	.202	1.04787	.81986	−.56600	2.66175
			1.312	199.644	.191	1.04787	.79867	−.52704	2.62279
Factor 3	.052	.820	1.052	280	.294	.71277	.67751	−.62089	2.04642
			1.077	198.076	.283	.71277	.66193	−.59258	2.01811
Factor 4	1.358	.245	−.809	280	.419	−.62766	.77586	−2.15493	.89961
			−.863	221.478	.389	−.62766	.72694	−2.06027	.80495
Factor 5	4.151	.043	1.416	280	.158	.56383	.39805	−.21973	1.34739
			1.477	208.352	.141	.56383	.38165	−.18856	1.31622

province, and their English pronunciation is not standard. As a result, students in Hainan province are not confident in English class, and their anxiety level is higher.

Table 8. Independent-samples T test

	Levene's test forequality of variances		T test for whether the mean is equal					95% Confidence Interval	
	F	Sig	T	df	Sig. (2-tailed)	Mean difference	Standard error	Lower Bound	Upper Bound
Factor 1	.054	.816	−.467	278	.641	−.55889	1.19740	−2.91600	1.79823
			−.467	277.741	.641	−.55889	1.19743	−2.91608	1.79830
Factor 2	.117	.733	.203	278	.839	.15779	.77567	−1.36915	1.68473
			.203	276.929	.839	.15779	.77605	−1.36991	1.68549
Factor 3	1.592	.208	.052	278	.958	.03348	.64210	−1.23053	1.29748
			.052	277.538	.958	.03348	.64147	−1.22928	1.29623
Factor 4	1.750	.187	−2.687	278	.008	−1.94254	.72302	−3.36582	−.51925
			−2.690	276.857	.008	−1.94254	.72206	−3.36396	−.52112
Factor 5	.324	.570	−.799	278	.425	−.30169	.37737	−1.04456	.44117
			−.799	275.008	.425	−.30169	.37778	−1.04540	.44202

One-way ANOVA for Students of Different Majors. According to the statistics in Table 10, students from six majors, including Aircraft Electromechanical Maintenance, Airport Operation and Management, Big Data and Accounting, E-commerce, Cruise Service and Art Design, participated in the survey. The Sig. Values of students from different majors in these five dimensions are 0.421, 0.227, 0.648, 0.160, and 0.464 respectively which are greater than 0.05. This indicates that there is no significant anxiety difference among students from different majors in the five dimensions of English class learning.

Table 9. Group statistics

	City	N	Mean	Standard deviation	Standard error mean
Factor 1	Hainan	143	25.2817	10.00806	.83986
	Outside Hainan	139	25.8406	10.02642	.85351
Factor 2	Hainan	143	19.5563	6.38096	.53548
	Outside Hainan	139	19.3986	6.59854	.56171
Factor 3	Hainan	143	12.8451	5.55220	.46593
	Outside Hainan	139	12.8116	5.17934	.44089
Factor 4	Hainan	143	21.2676	6.31885	.53027
	Outside Hainan	139	23.2101	5.75720	.49009
Factor 5	Hainan	143	7.4085	3.03678	.25484
	Outside Hainan	139	7.7101	3.27611	.27888

The overall study motivation of students in private vocational colleges is insufficient. In recent years, due to fierce enrollment competition, many students without complete junior high school backgrounds have become students in many private vocational colleges. They are evenly distributed among majors, so the difference in students' English class learning anxiety in different majors is not significant.

4 Suggestions on Alleviating the Anxiety of Students' English Class Learning in Private Vocational Colleges

Based on the data analysis from the five dimensions in the questionnaire survey, this paper proposes some suggestions to relieve anxiety in the English class of private vocational college students.

4.1 Communicative Fear

English learning includes five basic skills: listening, speaking, reading, writing and translation. However, the English foundation of students in privately-run vocational schools is too weak, especially for some students from minority areas, underdeveloped areas in education, secondary vocational schools or adults without complete school education whose English foundation is almost zero. Most students in private vocational colleges do not know enough words and grammar so that they cannot communicate with teachers or classmates in English. According to the results of the survey in the paper, students of private vocational schools show moderate anxiety in English class when communicating with others. To help students relieve their anxiety, the English teacher should teach students the most basic knowledge, including reading and reciting words, remembering basic sentence patterns, practicing basic grammar knowledge and so on. The complex language points should be simplified, and the simple language points

Table 10. ANOVA

	Quadratic sum	df	Mean square	F	Sig.
Factor 1	499.047	5	99.809	.995	.421
	27691.279	276	100.331		
	28190.326	281			
Factor 2	291.740	5	58.348	1.392	.227
	11571.479	276	41.926		
	11863.220	281			
Factor 3	96.730	5	19.346	.668	.648
	7989.341	276	28.947		
	8086.071	281			
Factor 4	298.193	5	59.639	1.600	.160
	10288.973	276	37.279		
	10587.167	281			
Factor 5	46.217	5	9.243	.926	.464
	2753.928	276	9.978		
	2800.145	281			

should be made alive to help students consolidate the foundation of English learning. Teaching activities in English classes should be designed seriously to attract students' attention. The teaching contents and exercises should be suitable for these students from the aspects of listening, speaking, reading, writing and translation to improve students' English application skills. On the basis of solid basic knowledge, students can slowly overcome their fear of communication in English class and regain their confidence in English learning.

4.2 Fear of Negative Evaluation

Due to poor academic performance, the vast majority of students in private higher vocational colleges have been subject to various negative evaluations from teachers, classmates and parents since primary school, junior middle school or senior high school. These negative evaluations have caused students' nonself-confidence in English learning, and some even have weariness when learning English. Through the data analysis, students in private vocational colleges show greater anxiety in the dimension of negative evaluation fear than in the other four dimensions. As the leaders of students' learning, teachers' evaluations play a vital role for students. Everyone wants to be praised and recognized; therefore, as the English teacher in private vocational colleges, more positive evaluation should be given to students even though they only make a little progress. The more positive evaluation you give to students, the more confident the students will be. In English class, the teacher can give students oral prizes or more grades when they voluntarily answer a question, make a presentation or take part in any activities. Through these methods, students will gradually become confident in class, and then the negative evaluation fear anxiety will be relieved little by little.

4.3 Fear of Teacher-Student Interaction

Due to the weak foundation of English, students in private higher vocational colleges do not have too much interaction with teachers in class. Most of them are passively called by teachers to answer questions. If they cannot answer teachers' questions, they will be more anxious and worried because they are afraid of getting bad evaluations from their teachers and classmates. Therefore, they dare not interact with teachers in class. According to the results of the survey, the anxiety in fear of teacher-student interaction also shows moderate status. To relieve the anxiety in this dimension, English teachers in private higher vocational colleges should be more exposed to new things and avoid using previous experience to educate the new generation of students due to the rapid development of mobile Internet technology. Students born after 2000 and those born in the 1980s grew up in very different environments. Today, Interment is widely available, and mobile phones are widely used on campus. Students are very easily attracted by online games and Tik Tock, and the phenomenon of phubbing in class is serious. Only after a better understanding of the new things that students experience can teachers understand what students are interested in and eliminate the generation gap between teachers and students. At the same time, teachers should often communicate with students, understand students' attitudes and difficulties in English learning, help students solve practical problems from the root, and give students more encouragement and more positive evaluation so that students can be more willing to interact with teachers in English classrooms. Through these methods, the anxiety in this dimension will be gradually reduced.

4.4 Non-confident in English Class

Students in private vocational colleges have a weak English foundation and fail to develop good learning habits. The vast majority of students want to learn English well, but due to inertia or the influence of other students around them and their own limited learning ability, they will eventually give up after a short period of time. They show an indifferent attitude in English class, neither confident nor too anxious, which brings great difficulties to the teaching work of teachers. Sometimes moderate anxiety can encourage students to study actively, but indifferent attitudes will make students lie flat and lack learning enthusiasm. According to the results of the survey, students in the dimension of Nonconfidence in English class also show moderate status. Students from Hainan province are less confident in English class than students from other places. Therefore, in teaching, teachers should first help students establish a correct outlook on life and learning, cultivate students' correct sense of competition, help students find appropriate English learning strategies, cultivate their active learning ability, and then enhance students' confidence in English class learning. For Hainan students, teachers should show more concern to them and understand their actual needs. Extra tasks after class should be designed for them, and feedback should also be given to them in time.

After a period of time, students' anxiety in this dimension will be relieved, and they will become more confident in their English class.

4.5 Negative Attitudes Towards English Class

Due to the moderate anxiety in the above four dimensions, students in private vocational colleges will hold negative attitudes towards English classes. To prevent students from forming negative attitudes toward English classes, teachers should timely adjust the learning content and teaching methods according to the needs of students instead of implementing cramming full teaching so that students can master the basic skills of English learning rather than the memory of language knowledge. Teachers should also enhance personal charm, enrich classroom teaching content, use modern information technology to enrich English class teaching, and let students fall in love with the teacher and English class. Give more encouragement to students, cultivate their active learning ability, pay more attention to the process evaluation in English learning, rather than focusing on academic performance, understand that students' progress is a slow process, and let students' negative attitude towards English class gradually weaken. Through the above methods, students gradually become confident in English class, and they are willing to interact with teachers and other students in class. They will be not too anxious about the negative evaluation from others and like to communicate with others in class. As the alleviation of English class anxiety in the first four dimensions, students' negative attitudes towards English class can also be changed.

5 Conclusion

As one of the important parts of higher education in Hainan, private vocational colleges shoulder the important task of training compound talents who understand foreign languages and technology for the construction of Hainan Free Trade Port. The cultivation of English skills will be affected by many factors, and emotional anxiety is one of the important factors. Teachers should fully understand students' emotional demands in the teaching process and provide timely guidance to help students relieve English class anxiety to improve private higher vocational college students' English learning interest and motivation. In the future, they must be qualified talent for the construction of the Hainan FTP.

References

1. Qin, X.: Quantitative Data Analysis in Foreign Language Teaching Research. Huazhong University of Science and Technology Press, Wuhan (2003)
2. Zhang, J., Guo, L.: A study of classroom anxiety in undergraduate students: a re-examination of the reliability and validity of the Foreign Language Classroom Anxiety Scale (FLCAS). J. North China Univ. Sci. Technol. (Soc. Sci. Ed.) **18**, 105–112 (2018)
3. Guo, Y., Xu, J.: A multidimensional study on English learning anxiety of Non-English majors. J. Foreign Lang. World **4**, 2–11 (2014)
4. Jiang, D., He, J.: On English learning anxiety and English teaching. J. Educ. Res. Teach. Pract. (B) **4**, 45–47 (2014)

5. Li, Y.: A study on the approaches to the relief of the vocabulary-learning anxiety of students who have difficulty in English learning. East China Normal University (2010)
6. Sun, L.: Study on the influential factors of English Classroom learning anxiety of vocational students and its strategies. Hebei Normal University of Science & Technology (2018)
7. Jiang, D., Li, H.: Cooperative learning and relieving oral English anxiety. J. Univ. Shanghai Sci. Technol. (Soc. Sci. Ed.) **35**(1), 36–40 (2013)
8. Yang, Y.: Qualitative Research Methods in Applied Linguistics. Commercial Press, Beijing (2014)
9. Li, X.: SPSS 22.0 Statistical Analysis From Entry to the Master. Electronic Industry Press, Beijing (2015)
10. Horwitz, E., Horwitz, M., Cope, J.: Foreign language class room anxiety. Mod. Lang. J. **70**, 125–132 (1986)
11. Hou, Y.: A review of foreign language anxiety studies. J. Lang. Literat. Stud. **8**, 150–152 (2011)
12. Wu, H.: A study of vocational college students' anxiety in English vocabulary learning. In: Wang, Y., Zhu, G., Han, Q., Zhang, L., Song, X., Lu, Z. (eds.) Data Science. ICPCSEE 2022. Communications in Computer and Information Science, vol 1629, pp 412–426. Springer, Singapore (2022). https://doi.org/10.1007/978-981-19-5209-8_28
13. Ren, D., Xian, D.: Study on vocational college students' communicative competence of intercultural communication. In: Zeng, J., Qin, P., Jing, W., Song, X., Lu, Z. (eds.) ICPCSEE. CCIS, vol. 1452, pp. 443–455. Springer, Singapore (2021). https://doi.org/10.1007/978-981-16-5943-0_36
14. Liu, L., Zhang, Y., Li, M.: An empirical study on teachers' informationized teaching ability in higher vocational colleges. In: Zeng, J., Qin, P., Jing, W., Song, X., Lu, Z. (eds.) ICPCSEE. CCIS, vol. 1452, pp. 419–433. Springer, Singapore (2021). https://doi.org/10.1007/978-981-16-5943-0_34
15. Alpert, R., Haber, R.: Anxiety in academic achievement situations. J. Abnorm. Soc. Psychol. **61**(2), 207–215 (1960)
16. Maclntyre, P., Gardner, R.: The subtle effects of language anxiety on cognitive processing in the second language. Lang. Learn. **44**(2), 283–305 (1994)
17. Chen, C.: A Study on Foreign Language Learning Anxiety of freshmen in higher vocational colleges. Minnan Normal University (2017)
18. Horwitz, E.: Foreign and second language anxiety. Lang. Teach. **43**(2), 154–167 (2010)
19. Guo, Y., Xu, J.: A multidimensional study on English Learning Anxiety of non-English majors. Foreign Lang. World **4**, 2–11 (2014)
20. Chang, H.: A diachronic study of English majors' emotional self-regulation of learning: a dynamic study based on the strategies for eliminating learning anxiety. Foreign Lang. Educ. China **3**(02), 50–57+92 (2020)

Research on the Construction of a Data Warehouse Model for College Student Performance

Juntao Chen[1], Jinmei Zhan[2(✉)], and Fei Tian[1]

[1] Information Technology School, Hainan College of Economics and Business, Haikou 571127, Hainan, China
[2] School of Information Science and Technology, Qiongtai Normal University, Haikou 571127, Hainan, China
jinmeizhan@google.com, jinmeizhan@gmail.com

Abstract. Students' grades not only serve as an effective indicator of their learning achievements but also to some extent reflect the completion of teaching tasks by the instructors. Currently, many universities across the country have collected and recorded various information about students and teachers in the school's information management system, but it is only a simple storage record and has not effectively excavated hidden information, and data have not been fully utilized. Student performance information, enrolment information, course information, teaching plans, and teacher-related information are currently stored in separate databases, which are independent of each other, making it difficult to perform effective data analysis. Data warehousing technology can integrate various information and use data analysis software to excavate more high-value information, which is convenient for teaching evaluation and optimizing teaching strategies. Based on data warehousing technology, the article uses the hierarchical concept of data warehousing to construct the ODS layer, DWD layer, DWS layer and ETL layer. Facing the data warehousing topic, the article designs the data warehousing conceptual model, logical model, and physical model based on student performance, providing a model basis for later data mining.

Keywords: Student Performance · Data Warehouse · Model Construction

1 Introduction

In the information management system of colleges and universities, the management of students' exam performance data is a very important task. Students' grades can not only reflect their learning achievements to a certain extent during a certain period but also serve as an effective basis for evaluating the quality of teachers' teaching. Currently, although the information management system construction of many colleges and universities has achieved preliminary results, various information on students and teachers has been recorded in the database; however, these data have not been fully utilized, and most of the data are idle, gradually becoming cold data [1–4]. Many colleges or universities, when analysing and processing students' grade data, only perform simple arithmetic operations such as calculating the average, maximum, and minimum values and then sort them from high to low. These simple basic operations are extremely insufficient in utilizing this very important data of students' grades and are a very simplistic analysis and processing. Students' grades are influenced not only by personal factors but also by many other factors, such as teacher factors and learning environment factors. Therefore, student grade data can be seen as a type of associative data. A data warehouse (DW) can be established to collect various data generated in teaching tasks, and data analysis techniques can be used to comprehensively and thoroughly analyse student exam grade data, excavating valuable potential information hidden behind student exam grades. This information is used to analyse the execution of tasks at various intermediate stages in teaching work [5]. Comprehensive analysis of student grade data using data warehouse technology facilitates a more comprehensive and realistic understanding of the teaching situation for universities and provides effective data support for future teaching task arrangements.

The essence of student learning and grade analysis technology based on data warehousing is to analyse and evaluate the teaching situation in universities by filtering, cleaning, extracting, categorizing, correlating, analysing, and applying data and to identify potential problems in various stages of teaching tasks. By applying data warehouse technology to teaching tasks in universities, data generated during the teaching process can be selected as the analysis object, and appropriate data analysis software can be used for comprehensive and in-depth analysis of the data. The results of the analysis can provide the basis or reference for task planning in various stages of teaching tasks. Analysing student learning and grades using data warehouse technology can enable universities to have a more comprehensive understanding of the teaching situation, and students can better identify their shortcomings and improve their learning efficiency and grades [6–8].

2 Technical Theories

2.1 Data Warehouse

In 1993, a data warehouse was proposed by William H. Inmonin in the book "Building the Data Warehouse", which is used to store data for information systems. Its main function is to systematically classify and organize various types of data generated in the operation of enterprises or institutions over the years through a special data storage architecture, making it convenient for subsequent data analysis and processing, such as online analytical processing (OLAP) and data mining. It also provides convenience for the creation of subsequent decision support systems (DSSs) and executive information systems (EISs), helping decision-makers quickly and accurately analyse valuable information from a large amount of data and make decisions accordingly [9, 10].

Data warehouses evolved from the development of databases, such as databases, but with many differences. Data warehouses are designed for subjects, while databases are designed for transactions. The construction of the data warehouse model requires a clear subject and, based on actual needs, purposefully collects various departments or different data sources on a higher level of abstraction [11]. The differences between databases and data warehouses are shown in Table 1.

Table 1. Difference between database and data warehouse.

Item	Database	Data Warehouse
Face	Database	Data Analysis
Data Characteristics	Transaction processing	Archive Value, historical Value
Target	Current value, Latest value	Long term information demand and decision support
Data Structure	Daily Operation	Simple, suitable for OLAP
Usage Frequency	Complex	Medium or low
Design Model	Relatively high	Star model, snowflake model, theme oriented
Response Time	Based on ER model, Application oriented	Time-consuming

2.2 ETL Technology

ETL technology is responsible for extracting data from scattered and heterogeneous data sources, cleaning, transforming, and integrating the data in a temporary intermediate layer, and finally loading it into a data warehouse or data mart. The data quality obtained at this stage directly affects the accuracy of subsequent online analysis and decision-making.

2.3 OLAP Technology

OLAP (online analytical processing) is also known as multidimensional data analysis. The OLAP committee defines it as a software technology that enables analysts, managers, or executives to quickly and consistently interact with information that has been transformed from raw databases into data that accurately reflects the characteristics of the enterprise and is easily understood by users from a variety of analytical perspectives to gain deeper insight into the data. Professional data analysts in their field can make scientifically based decisions by utilizing a successfully built data warehouse and combining it with their own experience.

2.4 Layering of Data Warehouse

The main purpose of layering in a data warehouse is to enable developers and users to clearly see the entire lifecycle and relationships of data in the data warehouse. This is because the amount of data in an enterprise is huge and the management of data is complex. Without an effective management method, it will be complex and tedious for subsequent users and developers to manage data [12].

The hierarchy of a data warehouse can be divided into four levels: the data extraction layer (Extract Transform Load, ETL), the data operation layer (Operate Data Store, ODS), the common dimension model layer (Common Data Model, CDM), and the data application layer (Application Data Service, ADS). The common dimension model layer can be further divided into the data detail layer (data warehouse detail, DWD) and the data summary layer (data warehouse service, DWS), as shown in Fig. 1.

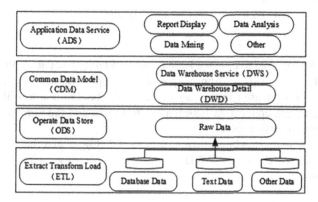

Fig. 1. Data Warehouse Hierarchical Structure.

ODS Layer. The ODS layer is the data operation layer, also known as the source layer. This layer only needs to extract data from various data sources using Flume and Sqoop and then use simple program logic and filtering to remove dirty data. The purpose of doing this is to facilitate the problem of data traceability later, so it is not recommended to perform too much data cleaning work on the data. Therefore, the ODS layer retains the original granularity of the data and is the finest granularity; other layers of the data warehouse are processed or aggregated based on the data in the ODS layer. The main function and role of the ODS layer is to first provide data for the entire data warehouse, which is the source of data; second, by creating this layer, it directly decouples the source data and the data warehouse, which greatly reduces the impact risk of the data warehouse on the business system. Since the ODS layer is basically consistent with the original data and has a huge amount of data, the storage period in the data warehouse is much shorter than that in other layers. The specific situation depends on the company's data scale, data growth rate, and business needs. In the construction of a data warehouse based on student performance (as shown in Table 2), the storage period for log data is 6 months, and the storage period for business data is 12 months.

DWD Layer. The DWD layer mainly performs cleaning and transformation work on the data in the ODS layer and then loads it into the DWD layer. The cleaning and transformation work is described in the ETL design of the data warehouse. The granularity of this layer is basically the same as that of the ODS layer, and there is no aggregation operation in the DWD layer, just pre-processing of log data, business data, historical

Table 2. Design of ODS layer of data warehouse based on student performance.

No.	Example	Font size and style
1	JW_BYSJ_LOG_XS_XTB	Course selection log
2	JW_PJ_JXLLYRZXXB	Liaison officer log information table
3	JW_SXGL_XSZLB	Student internship management data table
4	JW_SXGL_XSZLDLB	Student internship management data category table
5	SP_AUDITING_LOG	Work review log table
6	ZFTAL_XTGL_CZRZB	Operation log table

data, and third-party data to obtain a large dimension table. The data in the DWD layer are processed and divided according to pre-planned topics. Sometimes, the dimension degradation method is used in this layer to degrade the dimension to the fact table, thereby reducing the connection between the fact table and the dimension table and forming a large dimension table.

Based on the understanding of the data in the ODS layer and the grasp of student performance, we will design the DWD layer, and the table names in the DWD layer are shown in Table 3.

Table 3. .

No.	Table Name	Explain
1	JW_CJ_CJBLB	Student supplementary performance table
2	JW_CJ_CJFCSQB	Grade review application form
3	JW_CJ_CJGZKZB	Grade correction control table
4	JW_CJ_CJJGLXSQJLB	Grade result type application record table
5	JW_CJ_XMJDXCJB	Project interim performance table
6	JW_CJ_XMJDXCJB_LS	Project interim performance table (historical)
7	JW_CJ_XMZPCJB	Project final evaluation performance table
8	JW_CJ_XSCJB	Student performance table
9	JW_CJ_XSHKCJB	Student deferred exam performance table
10	JW_CJ_XSQTCJB	Student another performance table
11	JW_GYLDGL_XSCJB	Student Performance Table (Public Service Labour Management)
12	JW_SXGL_XSCJB	Student Performance Table (Graduation Design)
13	JW_SYGL_XSCJB	Student Performance Table (Experiment Management)
14	LSK_JW_CJ_XSCJB	Student historical performance table

DWS Layer. The DWS layer is the data service layer, which aggregates the data in the DWD layer according to analytical topics or stores the results of student performance in the DWS layer for further analysis of the causes of student performance. The specific design of the DWS layer based on the student performance data warehouse is shown in Table 4.

Table 4. DWS layer design of the data warehouse based on student performance.

No.	Table Name	Explain
1	JW_BYGL_BYSFZXXB	Graduate auxiliary information table
2	JW_BYSJ_XTSZ_CJGC	Student performance composition table
3	JW_CJ_BZDMB	Grade remarks code table
4	JW_CJ_CJFCSQB	Grade review application form
5	JW_CJ_CJGZKZB	Grade correction control table
6	JW_CJ_CXMDTJB	Retake list statistics table
7	JW_CJ_DZB	Grade comparison table
8	JW_CJ_JFMXB	Grade bonus detail table
9	JW_CJ_XMCJGZB	Project grade correction table
10	JW_CJ_XSCJGZB	Student performance correction table
11	JW_CJ_XSHKCJB	Student deferred exam performance table
12	JW_CJ_XSQTCJB	Student other performance table
13	JW_CJ_XYYJMXB	Academic warning detail table
14	LSK_JW_BYGL_BYSFZ	Graduate historical auxiliary information table

ETL Layer. The extract-transform-load (ETL) layer is the most expensive and crucial part of data warehouse technology. It is the process of extracting, cleaning, transforming, and loading raw data from various data sources to build a data warehouse. The ETL layer solves the problem of data isolation, lack of integration, and sharing among various business systems in the enterprise. Its main function is to extract, transform, and load data from data sources to the data operational layer. It integrates scattered data from various data sources through cleaning, standardization, and organization. The ETL layer can be seen as the blood of the data warehouse, responsible for data metabolism. The daily management and maintenance of the data warehouse mainly focuses on the ETL layer. The operational data store (ODS) layer is a transitional part from the database to the data warehouse. The structure and classification of data in the ODS layer are basically the same as those in the data source, but data management can be performed by adding fields. The stored data are read-only and cannot be modified. Generally, the data in this layer have a short lifecycle. The common dimensional model (CDM) layer establishes a data analysis model based on the requirements of the theme by obtaining data from the ODS layer. The Analytical Data Store (ADS) layer mainly provides data products or reports and stores them in the data warehouse analysis system, facilitating the use of subsequent data analysis or data mining software.

2.5 Design of ETL in Data Warehouse

There are generally two methods for implementing big data ETLs. The first method is to use ETL tools such as Kettle and Data Stage, which greatly facilitate developers with their visual and graphic interfaces, making the process more user-friendly and easier to operate. The second method is to use programming languages to implement ETL, which has a higher threshold than the first method but is more flexible and reduces enterprise costs. Developers and users can easily implement their own logic using programming languages, which is more convenient for development and cost considerations. ETL is the initial step in the data warehouse, and the success of ETL design and operation directly affects the subsequent development of the data warehouse. The ETL process for student performance management decision analysis data warehouses is divided into three steps: the first step is data extraction. The second step is data cleaning and conversion, in which we use Spark SQL and Spark Core to clean and transform the data, remove invalid data and dirty data, perform data desensitization, and then merge the original data from multiple sources according to business requirements, finally transforming it into structured data. The third step is data loading, which directly loads the structured data into HDFS after the second step and turns it into a Hive table, facilitating subsequent data processing.

3   Design of Student Achievement Data Warehouse Model

Currently, the basic information data of teachers and students in schools and student exam score data are stored in different databases, and these data are important data sources for building a student performance data warehouse. Analysing student performance data is mainly aimed at the theme of teaching management in schools, which has the characteristics of large data volume and complex data structure [9]. In the process of designing a student performance data warehouse, various factors affecting student performance should be fully considered to comprehensively build the student performance data warehouse.

3.1   Data Source

Acquisition The data sources for building a student performance analysis data warehouse are mainly divided into two categories. One category is log data, which mainly comes from data generated on third-party learning platforms, such as student comments, discussions, tests, video playback time and frequency during the learning process on the third-party platform. The second category is business data, which is basically stored in the school's Oracle database, such as student personal information, academic information, exam scores, and various other scores.

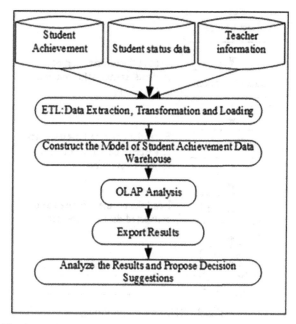

Fig. 2. Design process of student achievement data warehouse.

3.2 Design of Data Warehouse

The design process of the student achievement data warehouse is shown in Fig. 2.

First, the relevant data are subjected to ETL processing. Based on the theme, the selection scope of the data is determined. In this article, student performance data, student academic data, and teacher basic information data are selected as data sources. These data are stored in the database in different data structures. After performing operations such as data cleaning, data integration, data transformation, and data loading on these data, they are updated to the DW. The entire ETL process is shown in Fig. 3.

3.3 Build a Performance Data Warehouse Model

From the definition of a data warehouse, a data warehouse is oriented towards the subject. In the actual development process of the data warehouse, it is necessary to have a clear understanding and control of the company's data and business and divide the data into multiple subjects. The clear and distinct division of subjects is usually a key step in the success or failure of data warehouse design [13–15].

In the student performance data warehouse, a key aspect is the design of the data warehouse model. Its main function is to scientifically and systematically abstract the entities that exist in the real world involved in the data warehouse. A comprehensive and in-depth analysis of student performance should include analysis of different academic years, different majors, different classes, different courses, different teachers, and so on. The student performance data warehouse mainly involves three subjects, namely, students, courses, and teachers, and the data organization table is shown in Table 5.

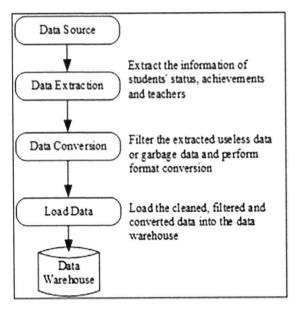

Fig. 3. ETL Processing Process.

Table 5. .

Subject	Information	Data Field
Student	Student status information	StudentID, DepartmentID, MajorID, GradeID, ClassID, Enrollment Time, Student Status, Length of Schooling, Educational Level, ...
Student	Student performance information	Academic Year, Term, CourseID, Student Number, Grade, Grade Point, Course Mark, Course Type, Operation Time, Deferred Examination Result, Deferred Examination Point, ...
Course	Basic Information of the Course	Course Code, Category (Optional and Compulsory), School Year, Semester, MajorID, Course NatureID, Applicant, Course Unit, Examination Form, ...
Teacher	Basic information of teachers	TearchID, Name, Gender, Birthday, Address, Native Place, Nationality, Nationality, ID Card, Education Background, Degree, Category, ...
Teacher	Course information	Teaching Plan Number, Major Code, Major Name, Academic Year, Term, Course Code, Credit, Weekly Class Hours, Assessment Method, Course Nature, Course Unit, TearchID, ClassID, Number of People, ...

4　Model Design of Data Warehouse

4.1　Conceptual Model Design of Data Warehouse

The data model is a method used for data storage and organization. Generally, databases are based on a three-level data model architecture, with the first being the conceptual model, the second being the logical model, and the third being the physical model. The conceptual model is defined from a global consideration of teaching business and data, and it sets the system boundaries and specifies the topics and content of student performance, providing a global view of student performance data to teachers and teaching administrators. The conceptual model is particularly important as the first step in designing the data warehouse model. In building a data warehouse based on student performance, we use the Entity-Relation (ER) diagram to design the conceptual model. The ER modelling method, also known as relational modelling, is a modelling method advocated by the father of data warehousing, Billimmon. It designs a 3NF model from the perspective of school management and can express the relationships between each entity well, paving the way for the design of the logical model. Due to limited space, we will only demonstrate the application of ER model design in the analysis of student performance with this single topic. The conceptual model diagram for student performance is shown in Fig. 4.

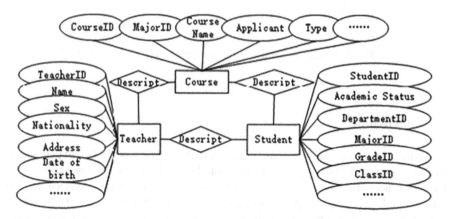

Fig. 4. Student Performance Data Warehouse Conceptual Model.

4.2　Physical Model Design of Data Warehouse

The physical model of a data warehouse is essentially the implementation of the logical model as database tables. When designing the physical model of the data warehouse, we must fully consider the data storage structure and method to ensure that data queries are fast and data storage space is small. Taking student performance data as the centre and radiating outwards, based on the three subject areas previously determined (students, teachers, courses), we have chosen the star schema in the dimensional model for the physical model of the data warehouse based on student performance [16–18]. In the data warehouse for student performance decision analysis, we use the table structure of Hive, and the data are stored in the form of Parquet on HDFS (Fig. 5).

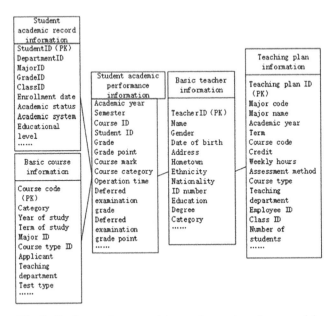

Fig. 5. Student performance data warehouse star schema model.

5   Conclusion

The article designs a data warehouse based on student performance, first explaining the overall system and architecture of the data warehouse. Under this system, we designed the ETL process for the data warehouse. Then, the data warehouse was divided into four layers based on business requirements, and the data warehouse was divided into three main topics and five subtopics. In the logical model of the next three model designs, the star schema in dimensional modelling was selected as the dimensional model, and the fact tables and dimension tables of the three main topics were designed. Next, the key research work in the next step is to realize data mining of the data warehouse based on student grades, fully excavate the potential information hidden under the student grade data, and provide effective technical support for the formulation of talent training programs and the implementation of education and teaching strategies.

Acknowledgement. This work was supported by the Hainan Provincial Natural Science Foundation of China (project number: 622RC723) and the Education Department of Hainan Province (project number: Hnky2023-72).

References

1. Deng, J., Mei, Y.: Clustering analysis of achievement early warning based on behavior data of college students. Mod. Inf. Technol. **7**(6), 35–38 (2023)
2. Li, F., Xu, H., et al.: Research on the integration of university academic information data based on data mining and K-means algorithm. J. Heilongjiang Inst. Technol. **36**(4), 31–36 (2023)
3. Liu, D., Tian, Y.: Application of extension data mining in student achievement analysis. CAAI Trans. Intell. Syst. **17**(4), 707–713 (2022)
4. Liu, X.-Y., Liu, H.-Y., et al.: Research on student achievement prediction based on multiple linear regression. Comput. Technol. Dev. **32**(3), 203–208 (2022)
5. Hu, L.Q., Zhao, G.: Research on influencing factors of machine learning algorithm on student achievement based on data mining. J. Nanchang Hangkong Univ. Nat. Sci. **35**(3), 43–48 (2021)
6. Zhou, S.: The application of data warehouse and data mining in grade analysis of students majoring in computer application in secondary vocational school. China West Normal University (2020)
7. Xu, Q., Li, J., Wang, Y.-M., Zhang, L.-L.: Student achievement analysis and visualization based on apriority algorithm. J. Tonghua Normal Univ. **44**(4), 81–87 (2019)
8. Liu, A.: Application of Data Mining in the Analysis of Colleges Students' Performance, Huaqiao University (2016)
9. Guo, Y.N.: Research on the application of data warehouse technology in the analysis of students' performance. Harbin Normal University (2017)
10. Li, C.: Application analysis of data warehouse and data mining in student achievement analysis. China New Commun. **19**(17), 136–138 (2017)
11. Jia, Y., Yang, G.: A study on Databank's application in students' score analysis. J. Shaanxi Youth Vocat. Coll. (1), 38–42 (2017)
12. Dai, Q., Li, Z.: Research on construction of learning behavior data warehouse. Softw. Guide **17**(10), 187–190 (2018)
13. Zhang, X., Ma, Y., Yu, M.: The research on applying data warehouse to educational administration in institutions of higher learning. J. Anshan Normal Univ. **045**(2), 83–85 (2003)
14. Wang, J.-M., Tang, N., Yang, D.: On the teaching quality according to the analysis of students' scores. J. Chongqing Univ. Soc. Sci. Ed. **12**(1) (2006)
15. Zhuang, Q.: Constructing conceptual model of data warehouse based on E-R schema. Comput. Eng. Appl. **10**(1), 195–200 (2004)
16. Yang, Y., Deng, H., Lai, S.: Application of data warehouse in college student's academic performance management. J. Southwest Univ. Nationalities·Nat. Sci. Ed. **35**(3), 619–621 (2009)
17. Yang, X., Han, X.: Model designing and implementation of data warehouse based on the analysis of students' score. Shanxi Electron. Technol. (1), 11–12+20 (2005)
18. Meng, Y., Huang, Z.: The application of data warehouse technique to educational management of university. J. Xuzhou Normal Univ. **21**(2), 69–78 (2003)

Research and Application of AI-Enabled Education

Zhanquan Wang[1], Yuxin Tian[1], Rui Chen[1], and Linghe Kong[2(✉)]

[1] School of Computer and Information Technology, East China University of Science and Technology, Shanghai 200237, China
zhqwang@ecust.edu.cn, {y80210031,y30201053}@mail.ecust.edu.cn
[2] School of Electronic Information and Electrical Engineering, Shanghai Jiao Tong University, Shanghai 200241, China
linghe.kong@sjtu.edu.cn

Abstract. Artificial intelligence technology has developed rapidly in various fields and has been widely used. Education and teaching are also areas in which artificial intelligence is applied. Research on artificial intelligence-enabled (AI-enabled) education and teaching is emerging, such as educational data mining and intelligent assisted teaching systems. First, research on AI-enabled education is introduced, and then the differences between AI-enabled education and traditional education and cases of educational data mining, learning prediction, learning resource recommendation, and various intelligent-assisted teaching systems are analysed. Our existing research results and future development are proposed, such as research on online learning session dropout prediction and the design and implementation of the zhixin teaching assistance system. Finally, this paper concludes that artificial intelligence has been well integrated into education and teaching activities in various ways and has improved students' learning experience and teachers' teaching quality. AI-enabled education and teaching is efficient and will play an increasingly important role.

Keywords: Education · Artificial Intelligence · Educational Data Mining · Intelligent Teaching System

1 Introduction

The concept of artificial intelligence was first proposed by Professor John McCarthy and others at Dartmouth College in the United States. After more than 60 years of development, artificial intelligence has shown increasing strength. At the beginning of 2022, the Google DeepMind team that developed AlphaGo once again brought a blockbuster product. The team created the programming robot system AlphaCode, and its programming level has reached the average level of human programmers [1]. The rapid development of artificial intelligence has attracted the attention of traditional industries, and the education industry is also looking for a breakthrough.

The application of artificial intelligence in education is relatively late compared to other fields, but it has developed rapidly and has been integrated into teaching activities in different ways. By using machine learning algorithms to mine and analyse educational data, students who have difficulties in learning can be identified in advance, and then they will receive more help. Using resource recommendation systems can give students personalized guidance. Using intelligent-assisted teaching systems can also reduce the heavy teaching tasks of teachers.

This paper aims to study the application of artificial intelligence in education, evaluate the impact of artificial intelligence, and determine whether artificial intelligence has a positive effect on teaching activities. Moreover, studying the application of artificial intelligence in education can help relevant parties in the education field to understand the impact of artificial intelligence on education. This paper can help them better apply artificial intelligence technology to education, thereby improving teaching quality and learning efficiency. Finally, the research will help policymakers develop strategies that promote the deeper application of artificial intelligence in all aspects of education.

2 Research Status of AI-Enabled Education

So-called artificial intelligence in education refers to the deep integration between the traditional education field and emerging artificial intelligence technology. With the maturity of artificial intelligence technology, research related to artificial intelligence in education has gradually increased, and some research results have also begun. Figure 1 shows the number of papers published on the Web of Science using the keywords "artificial intelligence" and "education". Due to the statistical time as of May 1, the number of papers in 2022 is relatively small. It can be seen that the number of papers is increasing yearly.

By reviewing the relevant literature, it is found that the current artificial intelligence has applications in all aspects of education, among which many achievements have been made in educational data mining, learning prediction, personalized guidance, and intelligent assisted teaching. Although the research directions are different, the results all show that the application of artificial intelligence positively impacts the quality of teaching and learning.

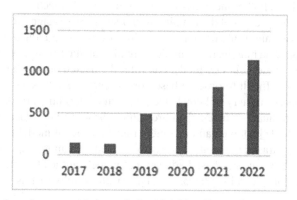

Fig. 1. The number of papers with the topic "artificial intelligence" and "education" in Web of Science from 2017 to 2022

3   The Application of Artificial Intelligence to Education

The current artificial intelligence technology is relatively mature and has applications in many aspects of education. This research will focus on sorting out three areas that have achieved more results: educational data mining, learning personalized guidance, and intelligent assisted teaching. This research then analyses the impact of artificial intelligence on education.

3.1   Educational Data Mining

With the continuous deepening of education informatization, online learning platforms have gradually integrated into teaching activities. Online platforms have collected a large amount of learning behavior data. A large amount of educational data provides better conditions for data-driven artificial intelligence algorithm application and intelligent assisted teaching. Educational data mining technology allows originally sparse data to be used in teaching quality assessment, and the visualization of educational data can also allow teachers to grasp the status of students in real time.

Teaching Quality Performance Evaluation
Teaching quality and learning effect are critical indicators for judging the value of teaching activities. The purpose of educational data mining is to obtain an evaluation of teaching quality based on the analysis of educational data. The appraisal of the validity of educational data involves a large amount of data information, and overall sampling will lead to data confusion. Small sample multivariate data analysis can effectively solve this problem. The multivariate data analysis method of small samples is used to model and analyse the effect evaluation of ideological and political education of college students and the quality evaluation of physical education in colleges and universities [2–6] and realize the processing of massive educational data.

Due to the enormous difference in students' learning ability and course content, it is impossible to timely and accurately assess students' learning situations through tests, homework, and other methods. The use of data mining technology can solve this problem. In [7], the process of data mining is divided into five stages: data cleaning, soft interval, data association rule mining, kernel linear dimension reduction, and teaching effect evaluation. Machine learning models are also applied well in education quality assessment. The hidden Markov model can infer the degree of difficulty of events and variables that are difficult to observe based on observed variables, so it is suitable for evaluating students' mastery of knowledge points in the teaching process.

To monitor and analyse students' progress through the attribute of learning outcomes, K. Mahboob et al. [8] adopted an outcome-based educational model that not only formulates teaching strategies but also indirectly evaluates them to help students achieve desired outcomes. Outcomes-based educational models are analysed to help teachers understand students' learning status. In addition, the results of this analysis provide teachers with a better view of student competencies and help them effectively develop teaching methods [9].

Educational Data Visualization

Educational data visualization is a significant technical means of educational data mining. M. Lu [10] analysed teachers' teaching level, students' learning ability, and curriculum arrangement. He calculated the statistics of students' satisfaction information, graduate distribution data, and the time of students' access to the learning platform. Then, a visual analysis of these data was carried out to obtain the relationship between the quality of educational courses and student development. Finally, solutions are drawn to improve teaching quality based on the analysis.

For scenario-based visualization of educational data, M. P. Macedo et al. [11] presented a suggested guideline for Vis2Learning to support the construction of visualizations of educational data applied in e-learning environments. Vis2Learning provides a set of scenarios for the development of educational data visualization. There is an online questionnaire to evaluate the visualization built according to the guidelines, and the results show that visualization is suitable in an e-learning environment.

3.2 Learning Personalized Guidance

After mining educational data through machine learning algorithms, massive sparse data become easy to count. After preprocessing the data, we can further analyse the data and build models to recommend educational resources or predict academic performance for students.

Personalized Guidance and Resource Recommendation

In traditional teaching activities, learning paths are often consistent. However, there are enormous differences in the learning methods and learning abilities of different students, and adopting a unified teaching method often ignores students' needs for individualized teaching. Using artificial intelligence to provide students with personalized guidance and resource recommendation can solve the problem of neglecting student personalization.

When there are many learning resources, such as learning on MOOCs, learning from start to finish is often not the most appropriate method. At this time, the recommendation of personalized learning paths is particularly important. Constructing the knowledge point map of the course is the key to generating a personalized learning path [12]. Y. Liu et al. [13] first constructed the knowledge point map of the course, then used the RankNet algorithm to generate the preliminary sequence, and finally used the Transformer algorithm to reorder the sequence to generate the recommended sequence. The online learning platform has text and video resources. The text resources are extracted through natural language processing to extract the course concept sequence, and the video can be converted into text through speech recognition. The course knowledge map is constructed according to the course concept sequence and outline [14]. Y. Zhou et al. [15] solved the problem of cold-start recommendation based on clustering and LSTM neural networks, and at the same time, the system can still recommend accurately after the learner changes the learning path.

In an online platform with thousands of learning resources, it is equally vital for students to find suitable resources. Building a student's demand model through artificial intelligence can recommend the most appropriate resources for students. J. Wang et al. [16] used a graph neural network and top-N algorithm to recommend MOOC courses for students and achieved better results than traditional algorithms. Many researchers use knowledge graphs to recommend resources for students. They model student information and knowledge point information when building knowledge graphs and then calculate the similarity between the semantic features of learning resources and users' historical learning resources. Finally, they provide students with personalized resource recommendations [17, 18]. K. Dai et al. [19] completed the diagnosis of students' learning according to the students' learning situation and the structure of the knowledge map and then recommended resources for students through the importance of learning resources. B. Hao et al. [20] proposed an automatic weakly supervised model based on reinforcement learning, which recommended MOOC courses for different jobs. The results showed that the model outperformed traditional recommendation methods. N. Araque et al. [21] used deep learning algorithms to predict the probability of completing their studies under a specific course combination and better help students accomplish their studies. In addition to course recommendation, it is also meaningful to recommend papers related to the research content for researchers. The context-aware citation recommendation model that uses the LSTM neural network makes it easier for researchers to find suitable references [22].

Learning Prediction

Using artificial intelligence to predict students' academic performance is another important topic of educational informatization. In traditional teaching, teachers evaluate students' academic performance through homework and examinations. However, this method lacks real-time performance, making it impossible for students to adjust their learning methods in time, and this phenomenon is particularly evident on online learning platforms. Online learning platforms have the advantages of abundant resources and a low learning threshold, but at the same time, there is a problem of a high dropout rate. Students who do not receive timely guidance when their learning results are not goodwill often choose to interrupt their current learning.

In response to the high dropout rate of MOOCs and the difficulty of timely intervention for at-risk students, many studies have used artificial intelligence to solve them. W. Xing et al. [23] used a generalized boosted regression model to rank students with dropout risk in MOOCs so that students with high dropout risk will receive priority intervention. When predicting student grades in MOOCs, Pigeau et al. [24] used the process mining model, sequence mining model, LSTM neural network, and machine learning algorithms such as AdaBoost. Finally, AdaBoost and logistic regression had the best prediction effect. Many studies use the completion of the course as a yardstick for performance. However, the target of students in the MOOC may not be to complete the course but to learn a method or skill. R. Conijn et al. [25] used the completion of learning objectives to define student performance. The final results showed that although only 2% of students completed the MOOC course, 51% of the students completed the learning objectives. When predicting academic performance, the number of completed assignments and the situation of discussions cannot fully reflect the students' learning status. S. Qu et al. [26] proposed a method to predict grades based on students' time-based learning behavior. The method combines an LSTM neural network and discriminative sequential pattern (DSP) to construct an attention mechanism framework to predict student grades. Aiming at the problem that a large number of college students cannot graduate on time, S. T. Christie et al. [27] established a risk assessment system to automatically identify students who are at risk of dropping out and give early warnings, thus playing the role of early warning of dropouts.

3.3 Artificial Intelligence-Assisted Teaching

There are also many research results of artificial intelligence in intelligent assisted teaching. Regarding the effect of intelligent assisted teaching systems, some studies have noted that in large-scale class teaching, intelligent assisted teaching systems perform better than traditional teaching [28].

In classroom education, many studies use an assisted teaching system based on image recognition, which can help teachers analyse students' learning status in real time, thus giving students better guidance. By analysing students' facial expressions, they can identify students' emotions when they are learning so that teachers can understand the mental state of students in a timely manner to help students adjust their learning state [29, 30]. W. Juan [31] proposed a dynamic gesture recognition method and applied it to classroom teaching. The final research results showed that the method has a good assisted effect in distance education and can help improve the quality of distance education. In distance education, identifying English vocabulary is difficult, and the method based on target visual detection proposed by [32] can assist English vocabulary teaching. Due to the spread of the epidemic, many areas have adopted distance teaching. Compared with traditional offline teaching, distance education has difficulty actively interacting and communicating with students. M. Bulut Özek [33] established a student emotion recognition and learning management system. The results showed that students who have adopted the system have higher learning enthusiasm and better learning effects. For physical education teaching, Q. Zhang and Y. Wang [34] proposed a motion recognition model that combines multiple features and neural networks to give accurate guidance by capturing and analysing the movements of students. This model has achieved good results in football teaching.

In addition to classroom education, artificial intelligence can also play a significant role in other aspects of education. College students in entrepreneurship desperately need to understand the entrepreneurial ecological environment and entrepreneurial performance. Using artificial intelligence can help entrepreneurial college students understand the entrepreneurial environment promptly and better plan future development [35]. For preschool education, AI-assisted education has also achieved good results. X. Chen and G. Jin [36] proposed an interactive system for preschool education based on intelligent sensor image recognition. In addition to recognizing objects that children come into contact with, the system can also interact with children to help children recognize objects. With the improvement of environmental protection awareness, the awareness of cultivating children's garbage classification has gradually received attention. Chen et al. [37] designed a garbage classification education platform based on Raspberry Pi and image recognition technology to educate children about garbage classification through sounds and pictures.

4   Research Results

The author's research team has conducted many types of research on AI-enabled education and teaching, such as learning prediction and intelligent assisted teaching systems. The current research results mainly focus on online learning session dropout prediction and the zhixin recognition system.

4.1   Online Learning Session Dropout Prediction

When students study courses on the intelligent education platform, they often exit the learning platform frequently in the early stage of the learning process. In this case, even if the learning platform provides the optimal learning path and the most appropriate learning resources, it is often difficult to achieve students' learning goals. To help students maintain continuity in online learning, appropriate interventions at the right time can increase their learning time on the platform. It requires accurate prediction of the time when students withdraw from the learning platform and timely detection of students' withdrawal tendency and then reminds students to complete the current learning tasks at the right time or recommends resources for students to help them acquire and consolidate knowledge.

The author's research team proposes a unified online learning session dropout prediction model (Uni-LSDPM). The model diagram is shown in Fig. 2. Using this model, the platform can predict the dropout of learning sessions in an intelligent teaching system. The model is based on the pretraining and fine-tuning paradigm, which uses a multilayer transformer structure. In the pretraining stage, a bidirectional multiattention mechanism is used to learn the sequence of students' continuous behavioral interaction features. In the fine-tuning part, a sequential attention mechanism is adopted to learn the sequential combination of students' continuous behavioral interaction features and dropout states. This pretraining framework is highly scalable and can be adapted to various downstream tasks based on online learned behavioral features.

For the experiments, we directly apply the parameters of the Unified Language Model (UniLM) to the Learning Session Dropout Prediction (LSDP) and then compare with the results using Uni-LSDPM. The results of the comparison are shown in Table 1. The results show that pretraining can greatly improve the performance of LSDP, and Uni-LSDPM outperforms UniLM on ACC and AUC.

In addition, we design experiments to compare Uni-LSDPM with other recent models that have performed well in learning session dropout prediction. These models are Long Short-Term Memory (LSTM), Variable Length Markov Chain (VLMC), Markov Modulated Marked Point Process Model (M3PP), and Deep Attention Session dropout prediction model (DAS). The results are shown in Table 2, which shows that Uni-LSDPM significantly outperforms state-of-the-art baselines on ACC and AUC.

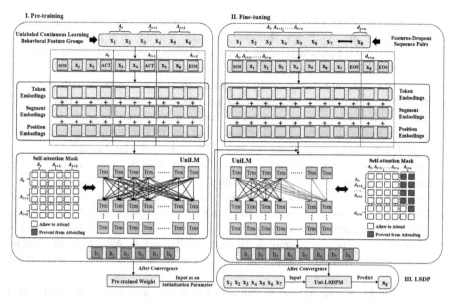

Fig. 2. Overview of the Uni-LSDPM framework. The framework consists of three parts: pretraining, fine-tuning, and downstream task LSDP.

Table 1. Exploring the effect of pretraining.

Pretraining Model	ACC	AUC
No Pretraining	0.7865	0.7547
Language Model (LM)	0.9378	0.9364
Uni-LSDPM	**0.9978**	**0.9979**

Table 2. Comparison experiments with other models

Model	ACC	AUC
LSTM	0.6279	0.5970
VLMC	0.6814	0.6310
M3PP	0.7943	0.7280
DAS	0.8136	0.7661
Uni-LSDPM	**0.9978**	**0.9979**

4.2 Zhixin Recognition System

Intelligent-assisted teaching can not only help children complete common-sense education, such as object recognition but also help parents understand children's thoughts. Therefore, parents can better integrate into children's educational activities. Lego bricks are a great form of teaching for children that can stimulate children's creativity and enthusiasm for exploring things. However, in children's education using Lego blocks, the following two problems will be encountered: (1) Since children's thinking methods are very different from those of adults, children often have some novel ideas when building blocks. The resulting shape may be so abstract that adults cannot visually recognize the image and therefore cannot understand what the child is thinking at this time. (2) Due to the difficulty in recognizing the built image, it is difficult for adults to integrate into the activities of children's building blocks and give guidance to children. Given the above problems, image recognition can be used to identify the built image.

The Zhixin image recognition system is a building block recognition system based on image recognition technology that can automatically identify building block images. After setting up the building blocks, parents take pictures of the shapes of the building blocks and upload them to the system, and then the system will identify the most similar objects in life to match them. As shown in Fig. 3, the child builds a building block in the shape of a "P". The parent uploads the picture to the system. After identification, the system outputs ten results that match the shape and finally outputs P, which best matches the shape.

As shown in Fig. 4, the system can also perform simple image processing, such as moving the image, changing the brightness of the image, and performing grayscale processing on the image. Using this image processing can make image recognition more accurate. The Zhixin image recognition system greatly reduces the difficulty of building block recognition and can make building blocks a more effective form of education.

The neural network model adopted by the Zhixin image recognition system is the Inception_v3 convolutional neural network model. The structure of the neural network model is shown in Fig. 5. Based on inception_v1 and inception_v2, the inception_v3 model splits the two-dimensional convolutional layer into two one-dimensional convolutional layers, which not only reduces the number of parameters but also mitigates overfitting. In addition, based on the Inception_v3 model, fine-tuning is performed, the first 46 layers are extracted by the transfer learning method, and a softmax layer is added to the last layer, for a total of 47 layers.

Ranking	Recognition	Accuracy
1	P	99.589574%
2	T	0.06687195%
3	N	0.045117867%
4	S	0.02757077%
5	D	0.016514742000000002%
6	M	0.013841415%
7	K	0.012390067999999999%
8	0	0.011779974%
9	V	0.01102458%
10	C	0.009085374%

Result   P

Fig. 3. Image recognition function.

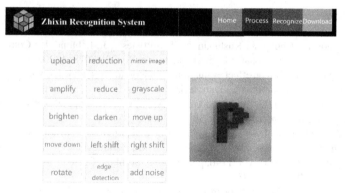

Fig. 4. Image processing function.

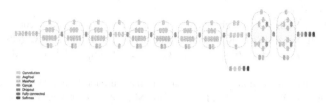

Fig. 5. Structure diagram of the fine-tuned Inception_v3 neural network model.

5   Conclusion

With the deepening of educational informatization, various learning resources have become easier to obtain, increasingly more courses are included in online learning platforms, and the application of artificial intelligence technology in education is gradually maturing. Educational intelligence allows us not only to be limited to traditional learning in schools but also to adopt the most suitable learning method according to personalized guidance. This paper first reviews the literature on AI-enabled education in recent years, studies educational data mining, learning prediction, resource recommendation, AI-assisted teaching, etc., and concludes that AI has a positive impact on education. Then, the research results obtained by this research team are listed in detail, such as online learning session dropout prediction and the zhixin intelligent assistance system. Recently, artificial intelligence-enabled education has achieved initial results. With the gradual maturity of artificial intelligence and the further integration of education and artificial intelligence technology, AI-enabled education will lead to changes in the field of education, thereby further improving teachers' teaching quality and student learning outcomes.

References

1. Li, Y., Choi, D., Chung, J., Kushman, N., Schrittwieser, J., Leblond, R.: Competition-level code generation with AlphaCode. arXiv preprint arXiv:2203.07814 (2022)
2. Wang, S.: Investigation on effect evaluation of undergraduates' education in ideology and politics based on small sample multivariate data analysis. In: Xu, Z., Parizi, R.M., Hammoudeh, M., Loyola-González, O. (eds.) CSIA 2020. AISC, vol. 1147, pp. 397–404. Springer, Cham (2020). https://doi.org/10.1007/978-3-030-43309-3_55
3. Wang, J.: Analysis of physical education quality evaluation model in colleges and universities based on big data analysis. In: Xu, Z., Parizi, R.M., Hammoudeh, M., Loyola-González, O. (eds.) CSIA 2020. AISC, vol. 1146, pp. 588–595. Springer, Cham (2020). https://doi.org/10.1007/978-3-030-43306-2_83
4. Wang, J.: Big data technology in the reform and innovation of ideological and political education in colleges. In: Xu, Z., Parizi, R.M., Hammoudeh, M., Loyola-González, O. (eds.) CSIA 2020. AISC, vol. 1147, pp. 390–396. Springer, Cham (2020). https://doi.org/10.1007/978-3-030-43309-3_54
5. Liu, Y., Luo, Y.: Big-data technology in the reform of ideo-political education in higher education. In: Xu, Z., Parizi, R.M., Hammoudeh, M., Loyola-González, O. (eds.) CSIA 2020. AISC, vol. 1147, pp. 647–652. Springer, Cham (2020). https://doi.org/10.1007/978-3-030-43309-3_94
6. Wang, C.: Analysis method of college student physical education quality based on big data analysis. In: Xu, Z., Parizi, R.M., Hammoudeh, M., Loyola-González, O. (eds.) CSIA 2020. AISC, vol. 1146, pp. 576–581. Springer, Cham (2020). https://doi.org/10.1007/978-3-030-43306-2_81
7. Zeng, Y.: Evaluation of physical education teaching quality in colleges based on the hybrid technology of data mining and Hidden Markov Model. Int. J. Emerg. Technol. Learn. **15**(01), 4 (2020)
8. Mahboob, K., Ali, S.A., Laila, U.E.: Investigating learning outcomes in engineering education with data mining. Comput. Appl. Eng. Educ. **28**(6), 1652–1670 (2020)
9. Ye, J.: Modelling of performance evaluation of educational information based on big data deep learning and cloud platform. IFS **38**(6), 7155–7165 (2020)
10. Lu, M.: Research on data visualization analysis in education curriculum quality management and student development. In: Proceedings of the 2020 International Conference on Computers, Information Processing and Advanced Education, Ottawa, ON, Canada, pp. 490–494 (2020)
11. Macedo, M.P., Paiva, R.O.A., Gasparini, I., Zaina, L.A.M.: Vis2Learning: a scenario-based guide of recommendations for building educational data visualizations. In: Proceedings of the 19th Brazilian Symposium on Human Factors in Computing Systems, Diamantina, Brazil, pp. 1–10 (2020)
12. Chen, H., Yin, C., Fan, X., Qiao, L., Rong, W., Zhang, X.: Learning path recommendation for MOOC platforms based on a knowledge graph. In: Qiu, H., Zhang, C., Fei, Z., Qiu, M., Kung, S.-Y. (eds.) KSEM 2021. LNCS (LNAI), vol. 12816, pp. 600–611. Springer, Cham (2021). https://doi.org/10.1007/978-3-030-82147-0_49
13. Liu, Y., Zhang, Y., Zhang, G.: Learning path recommendation based on Transformer reordering. In: 2020 5th International Conference on Information Science, Computer Technology and Transportation (ISCTT), Shenyang, China, pp. 101–104 (2020)
14. Huang, C., Li, Q., Chen, Y., Zhan, D.: An effective method for constructing knowledge graph of online course. In: 2021 4th International Conference on Big Data and Education, London, United Kingdom, pp. 12–18 (2021)
15. Zhou, Y., Huang, C., Hu, Q., Zhu, J., Tang, Y.: Personalized learning full-path recommendation model based on LSTM neural networks. Inf. Sci. **444**, 135–152 (2018)

16. Wang, J., Xie, H., Wang, F.L., Lee, L.K., Au, O.T.S.: Top-N personalized recommendation with graph neural networks in MOOCs. Comput. Educ. Artif. Intell. **2**, 100010 (2021)

17. Fang, C., Lu, Q.: Personalized recommendation model of high-quality education resources for college students based on data mining. Complexity **2021**, 1–11 (2021)

18. Wei, Q., Yao, X.: Personalized recommendation of learning resources based on knowledge graph. In: 2022 11th International Conference on Educational and Information Technology (ICEIT), Chengdu, China, pp. 46–50 (2022)

19. Dai, K., Qiu, Y., Zhang, R.: The construction of learning diagnosis and resources recommendation system based on knowledge graph. In: 2021 IEEE International Conference on Progress in Informatics and Computing (PIC), Shanghai, China, pp. 253–259 (2021)

20. Hao, B., Zhang, J., Li, C., Chen, H., Yin, H.: Recommending courses in MOOCs for jobs: an auto weak supervision approach. arXiv preprint arXiv:2203.07814 (2022)

21. Araque, N., Rojas, G., Vitali, M.: UniNet: next term course recommendation using deep learning. In: 2020 International Conference on Advanced Computer Science and Information Systems (ICACSIS), pp. 377–380 (2020)

22. Yang, L., et al.: A LSTM based model for personalized context-aware citation recommendation. IEEE Access **6**, 59618–59627 (2018)

23. Xing, W., Du, D.: Dropout prediction in MOOCs: using deep learning for personalized intervention. J. Educ. Comput. Res. **57**(3), 547–570 (2019)

24. Pigeau, A., Aubert, O., Prié, Y.: Success prediction in MOOCs: a case study. In: 12th International Conference on Educational Data Mining, pp. 390–395 (2019)

25. Conijn, R., Van den Beemt, A., Cuijpers, P.: Predicting student performance in a blended MOOC. J. Comput. Assist. Learn. **34**(5), 615–628 (2018)

26. Qu, S., Li, K., Wu, B., Zhang, S., Wang, Y.: Predicting student achievement based on temporal learning behavior in MOOCs. Appl. Sci. **9**(24), 5539 (2019)

27. Christie, S.T., Jarratt, D.C., Olson, L.A., Taijala, T.T.: Machine-learned school dropout early warning at scale. In: 12th International Conference on Educational Data Mining, pp. 726–731 (2019)

28. Du Boulay, B.: Artificial Intelligence as an effective classroom assistant. IEEE Intell. Syst. **31**(6), 76–81 (2016)

29. Li, Q., Liu, X., Gong, X., Jing, S.: INDReview on facial expression analysis and its application in education. In: 2019 Chinese Automation Congress (CAC), Hangzhou, China, pp. 4526–4530 (2019)

30. Sun, A., Li, Y., Huang, Y.M., Li, Q.: The exploration of facial expression recognition in distance education learning system. In: Wu, T.-T., Huang, Y.-M., Shadiev, R., Lin, L., Starčič, A.I. (eds.) Innovative Technologies and Learning, vol. 11003, pp. 111–121. Springer, Cham (2018). https://doi.org/10.1007/978-3-319-99737-7_11

31. Juan, W.: Gesture recognition and information recommendation based on machine learning and virtual reality in distance education. IFS **40**(4), 7509–7519 (2021)

32. Wu, J., Chen, B.: English vocabulary online teaching based on machine learning recognition and target visual detection. IFS **39**(2), 1745–1756 (2020)

33. Bulut Özek, M.: The effects of merging student emotion recognition with learning management systems on learners' motivation and academic achievements. Comput. Appl. Eng. Educ. **26**(5), 1862–1872 (2018)

34. Zhang, Q., Wang, Y.: Construction of composite mode of sports education professional football teaching based on sports video recognition technology. In: 2020 5th International Conference on Mechanical, Control and Computer Engineering (ICMCCE), Harbin, China, pp. 1889–1893 (2020)

35. Xia, Y., Lin, Z.: Application of image recognition technology in the field of ecological environment entrepreneurship education for college students. In: 2021 4th International Conference on Information Systems and Computer Aided Education, Dalian, China, pp. 1224–1228 (2021)
36. Chen, X., Jin, G.: Preschool education interactive system based on smart sensor image recognition. Wirel. Commun. Mob. Comput. **2022**, 1–11 (2022)
37. Chen, G., Wang, H., Zheng, J.: Application of image recognition technology in garbage classification education platform. In: 2019 5th International Conference on Control, Automation and Robotics (ICCAR), Beijing, China, pp. 290–294 (2019)

A Study on the Online Teaching Input of Higher Education Teachers Based on K-Means Analysis

Zhi-peng Ou[✉], Han Zhang, and Xia Liu

Sanya Aviation and Tourism College, Sanya 572000, Hainan, China
hainanozp@126.com

Abstract. Online teaching has become an important form of regular teaching in higher education institutions, and teachers' input in online teaching is directly related to the quality of online teaching. A K-means cluster analysis of teachers' teaching behaviours in the online teaching platform of Sanya Aviation and Tourism College and a comparison of the mean values of each semester's behaviours reveal that teachers' overall online teaching input is insufficient and their online teaching behaviours need to be optimized; their teaching energy has improved and their online teaching resources are more completed; their teaching emotional input is seriously lacking and their online interaction needs to be strengthened. To address these problems, the following measures can be taken: improve teachers' incentives to increase the overall investment in online teaching; change teachers' roles to strengthen the emotional investment in online teaching; and strengthen teachers' training to enhance online teaching design skills.

Keywords: higher education teachers · online teaching · teaching input · data analysis

1 The Problem

With the deep integration of online teaching and offline teaching, online teaching has become an important form of regular teaching in higher education institutions, and is an important part of the overall quality of education and teaching in higher education institutions. However, the overall situation of online teaching in higher education institutions is not optimistic: for teachers, the course teaching is online only for the online teaching, and there is a lack of special teaching design for the online learning environment, and the online teaching behavior is reduced to the "three axes" of resource release, forum topic and online test, with a single learning path; for students, due to the lack of independent learning habits and motivation, online self-learning is often overwhelming and confusing, and the quality of online resources varies, making the overall online learning experience more difficult. These problems are closely related to teachers' commitment to online teaching and learning, in terms of building online teaching resources, designing teaching activities and providing feedback on learning assessment. A number of studies have shown that teachers' commitment to teaching has a significant impact on the effectiveness of teaching [1–3], and that the greater the commitment to teaching, the

Z. Yu et al. (Eds.): ICPCSEE 2023, CCIS 1880, pp. 433–446, 2023.
https://doi.org/10.1007/978-981-99-5971-6_31

higher the quality of teaching. Therefore, the first task to improve the quality of online teaching is to improve teachers' online teaching input. Since 2018, Sanya Aviation and Tourism College has been using the comprehensive online teaching platforms such as Cloud Classroom, Smart Tree and Smart VET for online teaching (hereinafter referred to as the platforms). Over the years, what has been the overall input of teachers in online teaching and learning? Is the focus on building and publishing resources, or is it more on interacting with students without boundaries? How has teachers' commitment to online teaching changed after the epidemic? Through a comprehensive analysis of teachers' teaching behaviour data from platforms such as Cloud Classroom, Smart Tree and Smart VET, the author will present an objective picture of teachers' investment in online teaching and provide a reference basis for improving the effectiveness of online teaching.

2 The Meaning of Online Teaching Inputs

Online teaching input is derived from the concept of teaching input in the context of the increasing popularity of online teaching, so it is necessary to first elaborate on the connotation and structure of teaching input, and then define the connotation in the light of the characteristics of online teaching. According to Liu Zhentian and Liu Yourong, teachers' teaching input is the sum of time, energy and emotion they invest in educational activities, and the more teachers invest in teaching, the better their teaching quality will be [4, 5]. Mou Zhijia et al. studied the measurement of teachers' teaching input in offline face-to-face teaching and classified teaching input into three dimensions: preparation input, cognitive input and affective input from the perspective of process attributes [6]. Li Shuang et al. argued that teaching engagement is the sum of energy, emotion and cognition invested in teaching, reflecting the voluntary distribution of teachers' energy, emotion and cognition within the scope of tasks corresponding to their teaching functions, which not only has individual variability but also changes according to teachers' internal and external environments, and further investigated the relationship between teachers' teaching engagement and learning effectiveness in online teaching [7]. Yuan Shuai et al. were consistent with Li Shuang et al. in defining the connotation of teaching input while attempting to construct an evaluation index of teachers' online teaching input [8]. This shows that scholars have not reached a broad consensus on the concept of teaching engagement, and research on teachers' teaching engagement in online teaching is still in the exploratory period.

In online teaching, the main tasks of teachers are to build online teaching resources, design online teaching sessions, organize and implement teaching activities, and complete online interactive feedback. This study tends to classify the teaching input into three levels: time, energy and emotion. Among them, time input refers to the total time spent by teachers on the course platform for a series of teaching-related activities, such as building resources, organizing activities, answering questions and discussing, etc. It is a basic physical quantity to measure the extent to which teachers attach importance to online teaching. Energy input reflects the efficiency, effectiveness and quality of teachers' online teaching input, i.e. the degree of effort, dedication and exertion, but unlike time input, it is difficult to be expressed explicitly through direct physical quantities, and can only be reflected indirectly through other forms, such as the relevance of online teaching activities, the richness of teaching content, and the quality of teaching resources such as videos and courseware etc. Affective input refers to teachers' attitudes towards and commitment to online teaching, which is reflected in whether teachers pay attention to students' online learning performance and give them positive feedback. In terms of the hierarchical relationship, time commitment is the foundation of online teaching, energy commitment ensures that teachers can perform the basic online teaching tasks externally, and emotional commitment represents teachers' internal psychological recognition of online teaching. The three factors are not entirely independent of each other, but rather influence each other, with time and energy inputs being the external manifestations of emotional inputs and emotional inputs being the internal drivers of time and energy inputs.

3 Data Analysis of Online Teaching Inputs

(i) Selection of indicators

The basis of online teaching is a series of online teaching resources and activities carefully designed by teachers. The data generated by each resource and activity is a mapping of teachers' online teaching input. Combining the statistics of Cloud Classroom, Smart Tree and Smart VET, the indicators of this study are selected based on as follows.

"Online time (minutes)" and "Number of course accesses" are the basis for teachers' online teaching behaviours, and these two indicators are firstly included in the statistics. Secondly, the platform can provide two types of functions: teaching resources management and teaching activities organization. In terms of teaching resources, the most common teaching resources used in courses include electronic courseware, videos, texts and other basic forms, among which video resources are the most difficult to produce and are currently the most important medium for students to acquire knowledge when conducting online independent learning, and their quantity and quality best reflect the level of teachers' commitment to online course construction. For this reason, two indicators were included in this study: "number of videos added to the module" and "size of resources (MB) added to the module". In terms of the organization of teaching activities, notifications, assignments, tests and forums are all regular online teaching activities and are the main forms of interaction between teachers and students, therefore, "the number of course notifications", "the number of course assignments" and "the number of course assignments corrected" are included in this study. The number of course assignments corrected" "the number of online tests posted" "the number of topics posted in the course discussion forum" "the number of replies to the course discussion forum" "the number of course questionnaires posted" are also included in the statistics (online tests are all objective questions and are automatically marked by the system, so the act of test marking is not involved here). Finally, the "total number of course selections", although not a teacher's teaching behaviour, represents the range of influence of teachers' online teaching input. Therefore, 12 research indicators were finally identified in this study.

Based on the previous theoretical analysis of online teaching engagement, the 11 indicators of teaching behaviour other than 'total number of lesson attendees' were classified into categories and the results are shown in Table 1.

Table 1. Types of online teaching inputs represented by each teaching behaviour indicator

category		Indicators
Time input		Length of time online (minutes)
Energy input	Course Resource Building	Number of adding video clips in the broadcast module, the size of adding the resources in the broadcast module (MB)
	Organization of teaching activities	Number of course accesses, number of course announcements, number of course assignments, number of online tests posted, number of topics posted on course discussion boards
Emotional input		Number of course questionnaires posted, number of course discussion forum responses, number of course assignments corrected

Time input: the time teachers spend on online teaching, as reflected in the "minutes spent online".

Energy Input: This includes both the construction of course resources and the organization of teaching activities, and includes "Number of videos added to podcasts", "Size of resources added to podcasts (MB)", "Number of course announcements issued", "Number of assignments issued" and "Number of online tests issued". ", "Number of course assignments", "Number of online tests" and "Number of topics posted in the course discussion forum" are included in the category of online teaching energy. The six indicators are included in the category of online teaching effort. In addition, "Number of course accesses" is a prerequisite for teachers to carry out online teaching activities and is therefore also included in this category. Overall, the higher the value of these indicators, the better the online teaching resources, the richer the teaching activities, and the more effort teachers put into online teaching. It is important to note that these indicators focus more on teachers' one-way delivery of resources and activities, and do not involve feedback to students, so they do not correlate with the number of students taking the course.

Emotional Input: In contrast to the one-way focus on the dissemination of information by teachers, the indicators of emotional engagement focus on the two-way subjective interaction between teachers and students online, i.e. the expectation of receiving feedback from students on the course and the willingness to expend energy on giving feedback to students on their teaching behaviours. Specific indicators include three indicators: "Number of course questionnaires posted", "Number of course discussion board responses" and "Number of course assignments corrected". The indicators "Number of responses to course discussion boards" and "Number of course assignments corrected" are closely related to the number of students taking the course. Usually the higher the number of students taking a course, the higher the number of student discussion board comments and assignments submitted, and naturally the higher the two indicators for teachers.

(ii) **Overall cluster analysis**

In this part, we focus on the overall situation of teachers' online teaching input and the differences between various input types since Sanya Aviation and Tourism College launched online teaching five years ago. The remaining 280 items were valid after removing 246 invalid items (including those with zero access to courses, teaching supervisors, student assistants, course make-up exams and other unconventional teaching behaviours). It is customary to classify behavioural data into four categories: excellent, good, moderate and poor. Therefore, in this study, the K-means clustering analysis was carried out with a K value of 4, and the final clustering result had a contour coefficient of 0.5, with good clustering quality. See Table 2 for details.

Clustering 1: 199 people (71.1%), a superficial type, with an overall lack of commitment to online teaching. The average time spent online was 11,053. 14 min, less than 1/6 of the other clusters, and the number of accesses to courses was seriously insufficient, at 190.04. Other indicators were also very unsatisfactory. Teachers in this group do not use Cloud Classroom, Smart Tree and Smart VET to implement effective online teaching and learning, and they are seriously under-invested in online teaching. Some teachers are more traditional in their teaching methods and are not yet fully aware of the advantages of online teaching, and are unwilling to spend too much time and energy on online teaching.

Clustering 2: 49 teachers (17.5%) are task-submitting teachers, with an overall average investment in online teaching. The average time spent online was 68, 104.73 min, second only to cluster 3, which was the best performer, and the teachers' time commitment to online teaching was good. The indicators for "Size of resources added to podcast units" (14,696.30) and "Number of videos added to guest units" (214.33) are also very close to Cluster 3, indicating that video resources are reasonably well constructed. It is worth noting that this cluster has the best performance among all clusters for the indicators of "Number of course assignments assigned" (161. 12) and "Number of course announcements posted" (44.31), indicating that teachers are putting good effort into online teaching activities. In terms of emotional engagement, although the "Number of course announcements" (44.31) was the best across the cohorts, it is clear that teachers are well engaged in online teaching activities. In terms of emotional engagement, while the "Number of course assignments corrected" (6096.02) was the highest among the clusters, the "Number of course discussion forum replies" (21.43) and "Number of course questionnaires posted" (3.22) were the highest among the clusters. However, two important indicators, "Number of course discussion forum responses" (21.43) and "Number of course questionnaires posted" (3.22), are not as good as they could be and need to be improved. Such teachers tend to focus their platform on the submission of course assignments, and there is insufficient in-depth communication between teachers and students.

Table 2. Results of cluster analysis of teachers' online teaching behaviour

Category	Clustering 1	Clustering 2	Clustering 3	Clustering 4
Number of cases, %	199, 71.1%	49, 17.5%	14, 5.0%	18, 6.4%
Add Podcast Module Resource Size (MB) (Importance = 1.00)	6151.45	14 696.30	15 939.15	87 372.52
Number of course discussion forum replies (Importance = 0 97)	6.60	21.43	201.50	36.83
Length of time online (minutes) (importance = 0 .95)	11 053.14	68 104.73	73 445.71	57 218.22
Number of accesses to courses (Importance = 0 .92)	190.04	1269.47	1542.50	1089.72
Number of videos added to the podcast module (Importance = 0 81)	32.96	214.33	219.57	744. 17
Number of assignments corrected (Importance = 0.64)	469. 11	6096.02	2360.50	4227.39
Number of course assignments assigned (Importance = 0.55)	18.50	161.12	83.29	107.28
Number of course announcements (Importance = 0 46)	6.84	44.31	42.57	19.11

(*continued*)

Table 2. (*continued*)

Category	Clustering 1	Clustering 2	Clustering 3	Clustering 4
Number of course discussion forum responses (Importance = 0.40)	6.06	48.98	69.71	26.28
Number of online tests posted (Importance = 0.29)	7.27	44.08	34.43	57.72
Total Course Attendance (Importance = 0.25)	2918.16	4226.90	11 827.79	5511.83
Number of course questionnaires posted (Importance = 0. 18)	0.38	3.22	4.64	3.83

Clustering 3: 14 people (5%), experienced type, with the best time and emotional commitment to teaching online and average energy commitment. With 73,445.71 min of online time, the most teaching time was spent. In terms of teaching effort, the two indicators "Number of times accessing the course" (1542.50) and "Number of times posting topics in the course discussion forum" (69.71) are the best among the clusters, but the construction of video resources is average (adding resources for podcast units size = 15939. 15, number of videos added to podcast units = 219.57), and the indicators of "Number of course assignments assigned" (83.29) and "Number of online tests posted" (34.43) were both unsatisfactory. This shows that teachers in this cohort are generally devoted to teaching and learning, and there is still more capacity for optimizing the construction of teaching resources and the organization of online teaching activities. In terms of emotional engagement in teaching, the two indicators of "Number of course questionnaires posted" (4.64) and "Number of course discussion forum responses" (201.50) performed best, especially "Number of course discussion forum responses". This indicates that teachers expect feedback from students and are able to give it to them positively, with more in-depth interaction between teachers and students. Further analysis of the list of teachers in this group and the courses they offer shows that the teachers in this group are mostly teachers of public courses such as computing and Civics or teachers of professional courses with a large number of majors, so the total number of people taking the courses is the highest, and teachers can effectively improve their teaching efficiency by adopting online teaching, which is also an important reason why teachers can continue to invest in online teaching. Take the example of Civics courses. These courses are not suitable for objective online tests, but require in-depth

discussions with students and positive feedback, so they are more emotionally involved. The large number of students taking the course makes it difficult for teachers to assign and correct assignments frequently, so less effort is devoted to teaching activities such as tests and assignments. In addition, most teachers have many years of experience in using platforms such as Cloud Classroom, Smart Tree and Smart VET, and have gradually developed a good ability to migrate to information-based teaching, so when they build and use other course platforms such as China University MOOC and Classroom School, they are still able to make good use of the various functions of the platforms and interact with students online.

Clustering 4: 18 people (6.4%), quality-engineering-driven type, with optimal energy investment in resource building and insufficient emotional investment. With 57,218.22 min of online time, teachers' online teaching time commitment is average. In terms of teaching energy investment, "Size of resources for adding podcast units" (87,372.52) "Number of videos for adding podcast units" (744. 17) was much higher than other clusters with very complete course video resources The "Number of online tests posted" (57.72) also performed very well, but in terms of "Number of course accesses" (1089.72) "Number of course assignments assigned" (107.28), "Number of course announcements" (19.11), "Number of topics posted in the course discussion forum" (26.28), the teaching effort was good. In terms of emotional engagement in teaching, "Number of replies to course discussion boards" (36.83) "Number of assignments corrected" (4227.39) "Number of course questionnaires posted" (3.83) three indicators of average performance, the overall teaching emotional investment is not enough, and deep interaction with students needs to be strengthened. By comparing the lists and looking at the course pages, we found that most of the courses taught by teachers in this group have participated in the construction review of various high-quality courses, so the online teaching resources of the courses are completed and of high quality. However, because of the late start of online teaching, the number of course logins and the length of time online are not high. As for the low number of forum replies, it may be due to the fact that the discussion forum of the platform is not as convenient and timely as social networking software such as WeChat, so this function of the course platform has not been fully used. This section of teachers has not yet developed a proactive online teaching input based on improving learning outcomes.

(iii) **Comparative analysis of teaching inputs by sessions**

In order to further compare the changes in teachers' online teaching input in each semester before and after the epidemic, this study took semesters as the unit again, and selected 1431 data from two semesters during the epidemic and two semesters before and after the epidemic, i.e. five semesters from spring of 2019 to spring of 2021, for comparative analysis of teaching behaviour data. After data cleaning, pieces of data were removed, such as the number of course logins, lesson preparation, examinations and other non-teaching data, leaving 931 pieces of valid data. Since the courses offered in the spring semester and the autumn semester are usually not the same in higher education, and the lead teacher usually only generates effective online teaching input in the semester in which the course is offered, the mean test was conducted with the course instructor as an independent sample. The final results show that out of the 12 indicators of online teaching input, only five indicators, namely "Length of time spent online (minutes)", "Number of times course assignments were corrected", "Number of videos added to podcasts", "Number of times courses were accessed" and "Number of times course assignments were assigned", showed significant differences across academic periods (see Table 3), while the other seven indicators did not show significant differences.

Table 3. Comparison of Means for 5 Indicators with Significant Differences in Teaching Behavior by Semester Line, Spring of 2019 to Spring of 2021

Indicators	2019 Spring	2019 Autumn	2020 Spring	2020 Autumn	2021 Spring	F Check	df	Importance
Length of time online (minutes)	1853 3057 246 155	2065 2749 210 172	7578 11240 909 153	6387 7996 530 228	4402 6025 403 223	23	4, 926	1 (important)
Number of assignments corrected	286 861 69 155	239 633 48 172	544 1225 99 153	213 618 41 228	203 443 30 223	6	4, 926	1 (important)
Number of adding video clips in the broadcast module	9 34 3 155	29 77 6 172	18 78 6 153	40 108 7 228	15 57 4 223	5	4, 926	1 (important)
Number of course accesses	64 105 8 155	69 97 7 172	98 168 14 153	87 127 8 228	53 72 5 223	5	4, 926	1 (important)
number of course assignments	9 20 2 155	8 16 1 172	13 27 2 153	7 17 1 228	6 14 1 223	4	4, 926	1 (important)

Note: The cell contents are in the following order: mean, standard deviation, standard error, count

The indicator "Minutes spent online" increased significantly in the spring of the 2020 semester (M = 7578) and maintained good continuity in the two subsequent semesters, with faculty spending more than twice as many hours online overall as in the previous two semesters, indicating a significant increase in faculty time devoted to online teaching during the epidemic and post-epidemic periods. The three indicators of "Number of course assignments corrected", "Number of courses accessed" and "Number of course assignments assigned" were outstanding only in the spring of the 2020 semester, but in the post-epidemic period, these indicators dropped significantly, even below the two semesters before the epidemic. Once offline teaching resumed, teachers preferred offline interaction in the classroom, with online feedback taking a back seat.

The indicator "Number of videos added to podcast units" increased significantly in the fall of 2020 semester, with an average value of 40. This is mainly due to the introduction of a pilot policy on blended learning at Sanya Aviation and Tourism College in this semester, which has given teachers an incentive to carry out blended learning as their online workload is officially recognised. Given the clear direction of the resource building assessment in the acceptance indicators, teachers' efforts in building course resources increased significantly, effectively contributing to the increase in the quantity and quality of online teaching resources.

4 Conclusions of the Study

(i) **The overall input of teachers is insufficient and online teaching is yet to be optimized**

More than 70% of teachers (Cluster 1) did not rely on platforms such as Cloud Classroom, Smart Tree and Smart VET for effective online teaching, and their online teaching either started late or was superficial. Random interviews with some teachers revealed that the main reasons for their lack of commitment to online teaching were:

Firstly, teaching inertia formed by years of traditional teaching experience. According to Wu Daguang, teachers will slowly enter a collective unconscious state of "teaching inertia" or "teaching memory" [9] during their long-term interaction with the surrounding teaching environment, forming a solidified teaching pattern, and this teaching inertia will then become the biggest obstacle to innovation. To reform online teaching means changing the inertia that has developed over the years, stepping out of the comfort zone of teaching, re-adapting to a new environment, learning new technologies and exploring new experiences. When faced with an unfamiliar online environment with many uncertainties, teachers are often more intimidated and therefore more reluctant. In particular, the older they are, the more inertial they are to teaching and the more resistant and fearful they are to new technologies, methods and ideas.

Secondly, there is a lack of training and guidance on effective online teaching methods. Some teachers wish to engage in useful online teaching practices, but they do not know how to design and organise their teaching online, and they cannot find proven reference models to draw on. At the same time, because teachers do not place enough emphasis on online teaching and learning, there is a lack of deep interaction between teachers and students on the platform, and students' experience of online learning is not good.

Thirdly, teachers' expectations of online teaching are not optimistic. Under the premise of rational thinking, teachers will anticipate the output of each teaching input before they make it, and if the output of online teaching is not expected, teachers will naturally reduce the corresponding teaching input or not invest at all. At present, teachers are most concerned about promotion and semester appraisals, as long as they are required to complete a basic teaching load, and whether or not they teach online does not affect the quality of their teaching, nor does it have a material impact on their promotion appraisal. This has led to a serious lack of investment in online teaching by teachers as a whole.

(ii) **Energy investment has improved and resources are becoming increasingly available**

By comparing the session data for the five semesters before and after the epidemic, it was found that the number of teachers starting classes online increased year by year and teachers' online teaching behaviours such as "Adding the number of videos of podcast units" "Assigning the number of course assignments" "Correcting the number of course work"has increased significantly indicating that teachers' energy investment in online teaching is improving and more and more teachers are gradually adapting to and recognising the role of online teaching. This is influenced by two main factors Firstly, in terms of the external environment, the new crown epidemic prevention and control has become a strong driver for online teaching. The objective urgency of "stopping classes without learning" did not allow teachers to hesitate. Although most teachers were "forced to go online", they gradually adapted to the online teaching format, changed their teaching concepts and maintained a good habit of devoting time to online teaching in the post-epidemic period. Secondly, in the first semester after the epidemic, a large-scale pilot of blended learning was launched at the college level, and a policy was introduced to formally recognize teachers' workload in online teaching, and the acceptance criteria led to a significant increase in teachers' commitment to online teaching resources ("Number of videos added to podcasts" $M = 40$). The level of effort has increased significantly. It is worth noting, however, that the "Number of course assignments assigned" did not differ from the semesters before the epidemic, suggesting a rebound in the energy devoted to the organization of online teaching activities. While teaching resources are often built to last for several semesters, the organization of teaching activities requires more sustained input from teachers, as they are designed to be tailored to different learning contexts at each session and require effective feedback from teachers. It is also the part of teaching that is most likely to 'bounce back', making it difficult to maintain a good level of inertia.

(iii) **Emotional engagement is seriously lacking, and deep interaction needs to be strengthened**

While the overall investment in online teaching continues to improve, most teachers still publish only one-way teaching resources and activities such as videos, tests and assignments, lacking in-depth interaction and feedback with students such as question and answer discussions. The teachers' emotional commitment to online teaching is seriously lacking, and there is even a situation of "online just only for online". Because of the different percentages of online courses required in the blended teaching pilot courses, such as 25%, 50%, 75%, etc., teachers have forced the segmentation of the teaching content and simply "moved" some of the offline content to online, lacking effective teaching design and organization of teaching activities. The teachers simply summarize the students' online learning data in offline teaching and complete the online learning assessment by adding up the weights, and rarely discuss and communicate with students in depth about the problems that arise in online learning. In fact, the lack of time and space constraints in online learning makes it even more important for students to interact emotionally with their peers and teachers in order to increase their drive for independent learning. A number of studies have shown that teachers' emotional engagement in online teaching is effective in enhancing students' online learning. Therefore, interactive feedback from teachers in the form of question and answer sessions and discussions is an important way to enhance students' motivation for self-directed learning. The overall cluster analysis in the previous section found that only 5% (cluster 3) of teachers were able to interact well with students online and were emotionally engaged, with the majority of teachers performing poorly in this area, which is an important constraint to the overall effectiveness of online teaching.

5   Suggestions for Measures

(i) **Improve incentives and increase overall investment in online teaching**

Firstly, the assessment indicators for the construction of various courses were optimised. The analysis of Clustering 4 shows that external factors such as the accreditation of various high-quality courses have effectively driven teachers' investment in online teaching, especially in resource building, thus improving the quality of online course resources and ensuring a rich resource base for online teaching. After the first stage of resource building, the evaluation indicators for the later stages of course building should be tilted towards teacher-student interaction and student-student interaction, guiding teachers to increase their investment in the organization of teaching activities and their emotional investment in online interactive feedback. Secondly, the output of online teaching should be clarified, and the output should be used to motivate teachers to invest in online teaching. By comparing the list of teachers in the previous cluster analysis, we found that teachers with higher online teaching input are also those who have won more teaching awards and achieved more in course construction, including various course awards and teaching competitions at the provincial level and above, and even teaching-related research projects. This is because online resources provide the resources for course evaluation, information technology provides the experience base for teaching competitions and the behavioural records of teaching processes provide the data for research. Therefore, the output of online teaching is not limited to the assessment of a particular course, but also includes the enhancement of all aspects of teaching, teaching reform and research, which can drive the professional growth of individual teachers in the long term.

(ii) **Change the role of teachers and strengthen the emotional input of online teaching**

Under the "new normal" of online teaching, learning channels and learning methods have become more diverse and varied, with various learning resources readily available, and learning can take place anywhere and anytime. The teacher's role in transmitting knowledge is gradually being replaced by various information technologies, and his or her role and status needs to change from that of a monopolist of knowledge learning to that of a manager of learning resources and a guide of learning activities. As a consequence, the investment of energy in building online learning resources becomes less necessary and less important than it once was, and is replaced by an investment in the organization of teaching activities and an emotional investment in guiding students' online learning. The teacher's role has gradually receded from the front of the stage to the backstage, where the teacher's creative design, the creation of teaching scenarios, and the organization of teaching activities stimulate and sustain the intrinsic motivation of the student as a learning agent. Therefore, teachers need to strengthen the organisation and implementation of online learning activities based on the availability of teaching resources, highlight the diversity of learning pathways, focus on immediate diagnosis and feedback, and pay attention to the iterative value-added of learning outcomes, which is currently the focus of teachers' online teaching work nowadays.

(iii) **Strengthen teacher training and enhance online instructional design skills**

Professional teachers in higher education institutions often have never been exposed to systematic learning of educational theories, and lack professional theoretical guidance in daily teaching design, and do not even understand the basic ideas and methods of teaching design. This depends on continuous and planned training and seminars to help teachers gradually reduce the inertia formed by years of traditional teaching experience and fundamentally change their teaching concepts. With the idea of blended teaching, the training of teachers' teaching design skills can also be completed through a combination of online theoretical learning and offline case demonstrations: using high-quality courses on teaching methods, teaching research and educational information technology in various catechism platforms, teachers can receive online theoretical training and be assessed and certified in the teachers' semester assessment; organizing offline experience sharing and teaching demonstrations with outstanding teachers to provide face-to-face guidance and practical exercises for teachers to improve their teaching skills.

Acknowledgments. Project supported by the Education Department of Hainan Province, project number: Hnky2022ZD-25.

Project supported by Sanya Aviation and Tourism CollegeThe ideological and political specialproject number: SATC2023SZ-04.

References

1. Xuming, H.: A case study of teachers' teaching engagement affecting students' learning engagement. J. Educ. Scholarsh. **7**, 93–99 (2014)
2. Yunping, C., Ting, L., Xuebin, L.: Research trends and future prospects of teachers' work engagement. Mod. Educ. Manage. **4**, 61–66 (2018)
3. Shuang, L., Yao, Z.: The impact of online teachers' teaching engagement on students' learning performance - based on the perspectives of teachers and students. Open Educ. Res. **3**, 99–110 (2020)
4. Liu, Z.: The theory, current situation and strategies of teachers' teaching commitment in higher education. China High. Educ. Res. **47**(8), 14–19 (2013)
5. Yourong, L.: Teachers' "insufficient teaching input": conceptual connotation, realistic problems and practical orientation. Jiangsu High. Educ. **11**, 66–74 (2020)
6. Mou Zhijia, S., Xiuling, Y.D.: Modeling research on teacher teaching engagement assessment based on teaching behavior in classroom environment. Mod. Distance Educ. **3**, 61–69 (2020)

7. Shuai, Y., Huan, H., Tingting, H., et al.: Evaluation of online teaching input of university teachers based on teaching behavior analysis: an example of Z University in Hubei Province. Mod. Educ. Technol. **3**, 57–63 (2021)
8. Daguang, W., Wen, L.: An empirical study on the characteristics of large-scale online teaching and learning in Chinese universities based on questionnaires from students, teachers and teaching staff. J. East China Normal Univ. (Educ. Sci. Ed.) **7**, 1–30 (2020)
9. Hao, M.C., Flat, G.I., Zhou, X.: The network direct seeding in line for a sustained use and subjective well-being: based on the phase at acknowledge model and standard social relations theory perspective. J. Journalism Commun. Rev. (2), 29–46

Research and Exploration on Innovation and Entrepreneurship Practice Education System for Railway Transportation Specialists

Jiu Yong[1,2,3,4(✉)], Jianwu Dang[1,3], Yangping Wang[1,3], Jianguo Wei[2], and Yan Lu[3]

[1] The National Virtual Simulation Experiment Teaching Center of Railway Traffic Information and Control, Lanzhou Jiaotong University, Lanzhou 730070, China
yongjiu@mail.lzjtu.cn
[2] College of Intelligence and Computing, Tianjin University, Tianjin 300350, China
[3] The School of Electronic and Information Engineering, Lanzhou Jiaotong University, Lanzhou 730070, China
[4] Key Laboratory of Four Electric BIM Engineering and Intelligent Application Railway Industry, Lanzhou 730070, China

Abstract. With the background of concept of rail transit major engineering, guided by industry demand, and based on the cultivation of practical abilities for outstanding engineering and technology talents in rail transit characteristic majors, we are committed to creating a national strategic innovation and entrepreneurship practical education system for rail transit characteristic professionals, continuously improving the practical teaching mode. We will carry out the construction of a "matrix style" practical teaching system that integrates the four modernizations, the construction of a virtual simulation practical teaching platform that integrates expertise and innovation, and open sharing, as well as the reform and exploration of the entire chain of integrated practical education. The scheme focuses on improving students' ability to solve complex engineering problems and their ability to apply composite knowledge and innovate and start businesses, cultivating a large number of outstanding rail transit engineering talents with solid professional foundations, outstanding practical abilities, and innovative and entrepreneurial qualities.

Keywords: Rail transportation · Innovation and entrepreneurship · Practical teaching · Teaching platform · Teaching team · Integration of specialization and innovation

1 Introduction

With the rapid development of high-speed rail and urban rail transit worldwide, it has triggered a large shortage of high-end engineering talent for rail transit [1]. On the one hand, the rapid development of high-speed, heavy-haul and urban rail transportation has increased the requirements for system safety, reliability, efficiency and environmental protection. The lifecycle of rail transit system concerns interdisciplinary integration including design, construction, and operation management, which are facing rapid industrial upgrading with the development of artificial intelligence technology and the arrival of Industry 4.0, which puts forward higher requirements for the practical and innovative comprehensive ability of engineering talents [2–5]. On the other hand, the process of rail transit operation has characteristics of "huge complexity, multi-link, strong coupling, high risk and irreversibility", which makes it difficult for scientific research and professional education to carry out many practical aspects at the operation site. The use of artificial intelligence technologies such as virtual reality and human-computer interaction, combined with the Internet, to timely transform innovation and creation in engineering practice into new educational technology products, and to construct a highly simulated and diverse virtual simulation experimental system have become an inevitable path for the cultivation of rail transit professionals [6, 7].

This article is aimed at cultivating talents with distinctive characteristics in rail transit. With the background of concept of rail transit major engineering and the guidance of industry demand, it is based on the cultivation of practical abilities for outstanding engineering and technology talents with distinctive characteristics in rail transit. It is committed to creating a national strategic innovation and entrepreneurship practical education system for rail transit specialty talents, continuously improving the practical teaching mode, We will carry out the construction of a "matrix style" practical teaching system that integrates the four modernizations, construction of a virtual simulation practical teaching platform that integrates expertise and innovation, and open sharing, as well as the reform and exploration of the entire chain of integrated practical education. We will focus on improving students' ability to solve complex engineering problems and their ability to apply composite knowledge to innovate and start an undertaking.According to the requirements for cultivating certified talents in engineering majors, we will combine virtuality and reality, integrate learning and practice with creativity, building a practical teaching platform of "experiment $\rightarrow$ innovation $\rightarrow$ entrepreneurship", connecting the entire chain of practical education of "cognition $\rightarrow$ skill $\rightarrow$ application $\rightarrow$ innovation", integrating artificial intelligence technology into professional practical education, developing new engineering disciplines, and improving students' practical innovation ability [9–11]. Furthermore, we aim to cultivate a large number of outstanding engineering talents in rail transit with solid professional foundations, outstanding practical abilities, and innovative and entrepreneurial qualities.

2 Key Problems in Innovation and Entrepreneurship Practice Education for Talents with Rail Transportation Characteristics

Innovative and entrepreneurial practice education for talents with rail transportation characteristics is needed to cultivate excellent engineering talents who can adapt and lesd the new round of scientific and technological revolution and industrial changes. However, there are currently problems in the cultivation of rail transit talents.

2.1 Curriculum Knowledge System is Solidified, and It is Difficult to Adapt to the Interdisciplinary Needs of Rail Transportation Industry Development

With the widespread application of new generation information technology such as virtual reality, artificial intelligence and big data in rail transportation industry, it puts forward higher requirements for talents' disciplinary vision and frontier professional knowledge acquisition. The existing knowledge system is solidified and slow to update, which is difficult to adapt to the demand of rapid development of new rail transportation technology.

2.2 The Single Practical Teaching Mode is Difficult to Meet the Diversified Needs of Talents in Modern Rail Transportation Industry

The solidification of the cultivation goal of professional setting for talents with rail transportation characteristics has hindered the compound and diversified development of students. The lack of carriers for cultivating students' interests, inspiring inspiration, and realizing creativity has weakened students' autonomy, cooperation, and exploratory learning, making it difficult to support the demand for engineering and composite talents in the rail transit industry.

2.3 Insufficient Practical Ability of Students to Meet the Demand of Railway Transportation Construction and Operation

The engineering problems in the context of national rail transportation large engineering construction and operation have put forward higher requirements for the engineering practice innovation ability of talents in rail transportation industry. Students are short of systematic understanding of job responsibilities, industry chain, etc., and lack training of students' ability to solve practical engineering problems in the training process.

3 The Main Contents of the Innovation and Entrepreneurship Practice Education System for Talents with Rail Transportation Characteristics

Adhering to the fundamental approach, facing the needs of cultivating talents in rail transportation, integrating learning and doing, combining reality to build a three-level practical teaching platform of "experiment → innovation → entrepreneurship", penetrating the four-level practical education chain of "cognition → skill → application → innovation", improving students' comprehensive practical innovation ability, and cultivating excellent engineering talents with a big engineering view of rail transportation.

3.1 Build a "Matrix" Practical Teaching System of Innovation and Entrepreneurship for Talents with Rail Transportation Characteristics by Taking Industry Demand as the Guide and Integrating Four Aspects

The teaching system is insisted on the background of big engineering view of rail transportation, oriented by industry demand, based on the requirement of cultivating practical ability of compound excellent engineering science and technology talents in rail transportation specialties, guided by the construction of new engineering discipline, to aim at cultivating college students' innovation spirit, improve service ability and strengthen social responsibility [12]. Based on the teaching concept of combined science and technology, intersected engineering, and integrated engineering and literature, we construct a matrix practical teaching system for talents with rail transportation specialties. Furthermore, we build a practical teaching platform of "experimentation → innovation → entrepreneurship", and the teaching content is designed according to the modularization of "case-research-practice- innovation". Through modularization of experimental teaching, case-based practical teaching, diversification of industry training, and project-based practical innovation, the "four modernizations" are achieved, allowing students to fully cultivate their professional qualities and exercise their professional skills. And their innovation and entrepreneurial ability can be promoted through multi-dimensional innovation training, so that their employment and entrepreneurial ability can be effectively improved.

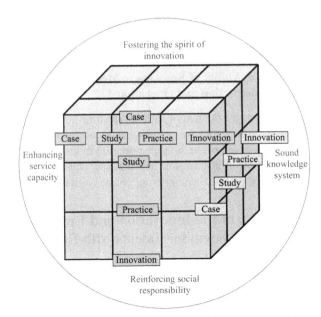

Fig. 1. Four-in-one "matrix" practical teaching system of innovation and entrepreneurship for professionals with rail transportation characteristics

3.2 Strengthen Top-Level Design, Combine Virtuality with Reality, and Build a Practice Teaching Resource Platform of "Four-Wheel Drive, Specialization and Innovation Integration, Open Sharing"

Based on the integration of the four modernizations, we will construct a "matrix" practical teaching system for innovation and entrepreneurship of characteristic professionals in rail transit. We adhere to the top-level design principle of "platform comes from engineering, function is close to the site, and technology keeps up with the forefront". In response to the characteristics of "huge complexity, multiple links, strong coupling, high risk, and irreversibility" in the rail transit operation process, we will adopt independent development, school enterprise co-construction, research achievement transformation, enterprise investment donations by introducing high-quality resources and building cross disciplinary projects [13]. We aim to build a virtual simulation teaching resource and network sharing platform that includes four modular levels of rail transit locomotive control, traction power supply, communication signal, and operation management, which can create national virtual simulation experimental teaching first-class undergraduate courses, enrich practical teaching contents, break through the limitation of time and space of practical education, and reduce the cost and safety risk of practical teaching. Combine the discipline characteristics and advantages to promote the mechanism innovation of industry-education integration and school-enterprise cooperation, cultivate the ability of teachers to combine artificial intelligence with education methods to develop information-based education, and adopt new generation information technology such as artificial intelligence and virtual reality with professional education to carry out the construction of new engineering to implement intelligent education. Deepen the integration of professional practical education and innovation and entrepreneurship education, and promote the construction maintenance and development sharing of the practical teaching resource platform through four-wheel drive of credit-driven, activity-driven, project-driven and competition-motivated.

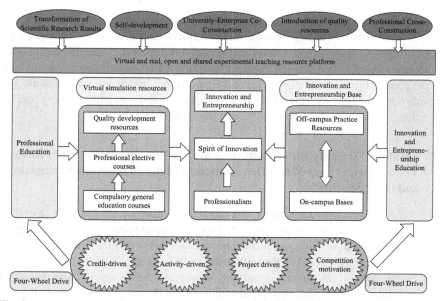

Fig. 2. Four-wheel drive, specialization and innovation integration, open sharing" practice teaching resource platform.

3.3 Strengthen the Process Management and Build the Whole Chain of "Cognitive-Skill-Application-Innovation" Through Practice Education Step by Step

In accordance with the goal of cultivating talents with rail transportation characteristics, the students' professional knowledge, practical education and innovation ability are cultivated step by step from general education to professional education, unity to individuality. The chain of practical education is divided into four stages of "cognitive-skill-application-innovation", and the personalized training and independent learning of students are strengthened.

4 Implementation Plan of Innovation and Entrepreneurship Practice Education System for Talents with Rail Transportation Characteristics

The training of talents with rail transportation characteristics should integrate "teaching" and "learning" into "doing". Through the implementation plan of graded progression, combination of virtuality and reality, modular design, build the innovative and entrepreneurial education system of rail transportation professionals with characteristics of "learning, doing and creating, unity of knowledge and action".

4.1 Tiered Progression, Constructing Three-Level Practical Teaching System

General education aims at students majoring in rail transit, offers practical courses in general education, such as cognitive internships in rail transit engineering, and education on the history of computer and network technology development; Professional education combines with training plans and curriculum syllabi, offers practical elective modules for different subject directions to achieve differentiation; Innovation education aims to provide innovative and entrepreneurial quality expansion and simulation training courses for willing students, with mentors providing personalized guidance.

4.2 Combine Virtuality with Reality, and Build a Network Platform of Practice Teaching with "Four-Wheel Drive, Specialization and Innovation Integration, and Open Sharing"

Take the construction of national virtual simulation experimental teaching first-class undergraduate courses as an opportunity to build a complete network of shared experimental openness and service system, and continuously develop virtual simulation experimental teaching resources [14]. Strengthen the multi-level and multi-forms cooperation with enterprises and institutions in industry-university-research-application, construct the innovation and entrepreneurship practice based on the integration of specialization and innovation, and realize the convergence of education and teaching elements.

4.3 Modular Design to Realize the Integration of Four Modules of "Case, Research, Practice and Innovation" in Practical Teaching Content

Relying on the practice bases inside and outside the university, students choose to carry out industry training in a menu style around the direction of choosing a career, and implement projects for practice and innovation with the guidance of tutors throughout the whole process, highlighting the cultivation of practice and innovation ability with the carrier of projects. At the level of teaching methods, driven by the needs of enterprises and industries, with resource construction as the core, a combination of independent development and high-quality resource introduction is adopted to strengthen the dual line integration of online and offline teaching through online and offline integration.

5 Features of Innovation and Entrepreneurship Practice Education System for Talents with Rail Transportation Characteristics

Guided by the demand of high-level talents in rail transportation specialties, the system builds a practice teaching resource platform, strengthens the process management, and builds a whole chain of through practice education step by step, thus enhancing the comprehensive practice and innovation ability of students.

5.1 Build a "Matrix" Practical Teaching System of Innovation and Entrepreneurship for Talents with Rail Transportation Characteristics by Taking Industry Demand as the Guide and Integrating Four Aspects

Against the background of concept of rail transit major engineering and guided by industry demand, we cultivate the practical ability of outstanding engineering and technology talents with complex characteristics in rail transit. Guided by the construction of new engineering disciplines, we aim to cultivate the innovative spirit, improve service ability, and strengthen social responsibility of college students.Based on the teaching philosophy of combining science and engineering, intersecting engineering and literature, we construct a matrix based practical teaching system for rail transit characteristic professional talents.

5.2 Strengthen the Top-Level Design, Combine Virtuality with Reality, and Build a Practice Teaching Resource Platform of "Four-Wheel Drive, Specialization and Innovation Integration, and Open Sharing"

We build a network platform of practice teaching with "four-wheel drive, professional and creative integration, and open sharing" by combining virtuality with reality, artificial intelligence with professional education. Taking the construction of the national virtual simulation experiment teaching project as an opportunity, we will establish a complete network sharing experiment opening and service system, and continue to develop virtual simulation experiment teaching resources.

5.3 Strengthen the Process Management and Build the Whole Chain of "Cognitive-Skill-Application-Innovation" Through Practice Education Step by Step

According to the goal of cultivating talents for rail transportation specialties, we create a whole chain of practical education of "cognition-skill-application-innovation" as the main line, "special activities, special forums and special programs" as the core and "experiment internship, design research, training competition, innovation entrepreneurship" as the integration.

6   Conclusion

To meet the needs of cultivating talents in the field of rail transit, we aim to integrate learning, practice, and innovation, and combine virtual and real experiences. We aim to build a practical teaching platform that combines experimentation, innovation, and entrepreneurship, connecting the entire chain of practical education from "cognition, skills, application, and innovation", and enhancing students' comprehensive practical and innovative abilities.Facing the needs of cultivating talents in rail transportation, we aim to build a practical teaching platform of "experiment → innovation → entrepreneurship" by integrating learning and practice, combining virtuality and reality. The whole chain of practical education with "cognition → skills → application → innovation" will enhance the comprehensive practical innovation ability of students.By building a first-class undergraduate course in national virtual simulation experimental teaching, we will apply artificial intelligence technology for rail transit operations, continue to develop virtual simulation resources with real data and scene representation, leverage the advantages of intelligent education, cultivate students' perspective on big engineering, and achieve the goal of "strengthening basic principles, training design abilities, and enhancing innovation capabilities".

Acknowledgment. This work was financially supported by the National Natural Science Foundation of China (62067006), The Ministry of education of Humanities and Social Science project(21YJC880085), Gansu Natural Science Foundation (23JRRA845), Gansu University Innovation Fund Project (2021B-092), Open fund project of Key Laboratory of four power BIM engineering and intelligent application railway industry(BIMKF-2021-05), and China University industry university research innovation fund(2021BCB02001).

References

1. Zhang, Z., Chen, Y., Wu, X., Lin, J., Zhang, Y.: Reform and exploration of practical teaching mode of rail traffic signal and control under the background of new engineering. Laboratory Res. Explorat. **38**(09), 190–193 (2019)
2. Peng, L., Hu, D., Wang, S.: Research on the reform of innovation and entrepreneurship practice teaching in mechanical majors of applied universities. Educ. Teach. Forum **36**, 195–196 (2020)
3. Xiao, J.-d, Wang, Z.: The construction of innovation and entrepreneurship practice platform under the guidance of the concept of "specialization and innovation integration. Laboratory Res. Explorat. **39**(08), 223–227 (2020)

4. Lu, C., Zhao, M.: Analysis of the characteristics of the practical programs of entrepreneurship education in the world's top universities - taking MIT and Stanford University as examples. Res. Higher Eng. Educ. (04):174–179 (2020)
5. Wang. Y., Yu, Z.: Research on the practice model of innovation and entrepreneurship education in universities in the context of Guangdong-Hong Kong-Macao Greater Bay Area. Laboratory Res. Explorat. **39**(07):240–243+299 (2020)
6. Xin, Z., Cao, H.: The difficulties and countermeasures of "5 + 3" innovation and entrepreneurship practice teaching system. J. Adv. Educ. Res. **5**(3) (2020)
7. Wu, X., Wang, F., Hu, Y.: Strategies for cultivating innovation and entrepreneurial practice in computer science majors in local undergraduate institutions. Educ. Mod. **7**(53), 48–51 (2020)
8. Yang, Q.-L., Tang, X.-J., Huang, G.-Y.: Research on teaching reform of innovation and entrepreneurship practice courses in applied colleges and universities. J. Higher Educ. **18**, 46–50 (2020)
9. Zu, Q., Wei ,Y.: Discussion on the logic of generation and construction significance of provincial innovation and entrepreneurship practice education center for college students. Experim. Technol. Manag. **37**(06), 15–18+23 (2020)
10. Liang, G., Guang, L.: The Model construction of college students' innovation and entrepreneurship practice under task driven method. J. Residuals Sci. Technol. **13**(3) (2016)
11. Alum, R.A.: Book-review innovation and entrepreneurship: practice and principles. Public Productivity Rev. **10**(1) (1986)
12. Luo, J.-g., Zhao, Y.-q.: The cooperative development of college students' innovation and entrepreneurship practice and teachers' ability training. DEStech Trans. Soc. Sci. Educ. Hum. Sci. (2018)
13. Xu, Z., Wu, J.: Analysis of influential factors of teaching effect of innovation and entrepreneurship practice in comprehensive university based on ISMP. In: Proceedings of the 2019 5th International Conference on Social Science and Higher Education (ICSSHE 2019) (2019)
14. Wang, Y., Yong, J., Wang, W., Yue, B.: Construction and exploration of a model virtual simulation experiment project. Comput. Educ. **09**, 38–41 (2019)

Author Index

Printed in the United States
by Baker & Taylor Publisher Services

Printed in the United States
by Baker & Taylor Publisher Services